Geometry at Work

A Collection of Papers
Showing Applications of Geometry

Catherine A. Gorini

Editor

Published and Distributed by The Mathematical Association of America

The MAA Notes Series, started in 1982, addresses a broad range of topics and themes of interest to all who are involved with undergraduate mathematics. The volumes in this series are readable, informative, and useful, and help the mathematical community keep up with developments of importance to mathematics.

MAA Notes

11. Keys to Improved Instruction by Teaching Assistants and Part-Time Instructors, *Committee on Teaching Assistants and Part-Time Instructors, Bettye Anne Case*, Editor.
13. Reshaping College Mathematics, *Committee on the Undergraduate Program in Mathematics, Lynn A. Steen*, Editor.
14. Mathematical Writing, by *Donald E. Knuth, Tracy Larrabee, and Paul M. Roberts*.
16. Using Writing to Teach Mathematics, *Andrew Sterrett*, Editor.
17. Priming the Calculus Pump: Innovations and Resources, *Committee on Calculus Reform and the First Two Years,* a subcomittee of the Committee on the Undergraduate Program in Mathematics, *Thomas W. Tucker*, Editor.
18. Models for Undergraduate Research in Mathematics, *Lester Senechal*, Editor.
19. Visualization in Teaching and Learning Mathematics, *Committee on Computers in Mathematics Education, Steve Cunningham and Walter S. Zimmermann*, Editors.
20. The Laboratory Approach to Teaching Calculus, *L. Carl Leinbach et al.*, Editors.
21. Perspectives on Contemporary Statistics, *David C. Hoaglin and David S. Moore*, Editors.
22. Heeding the Call for Change: Suggestions for Curricular Action, *Lynn A. Steen*, Editor.
24. Symbolic Computation in Undergraduate Mathematics Education, *Zaven A. Karian*, Editor.
25. The Concept of Function: Aspects of Epistemology and Pedagogy, *Guershon Harel and Ed Dubinsky,* Editors.
26. Statistics for the Twenty-First Century, *Florence and Sheldon Gordon*, Editors.
27. Resources for Calculus Collection, Volume 1: Learning by Discovery: A Lab Manual for Calculus, *Anita E. Solow,* Editor.
28. Resources for Calculus Collection, Volume 2: Calculus Problems for a New Century, *Robert Fraga*, Editor.
29. Resources for Calculus Collection, Volume 3: Applications of Calculus, *Philip Straffin*, Editor.
30. Resources for Calculus Collection, Volume 4: Problems for Student Investigation, *Michael B. Jackson and John R. Ramsay,* Editors.
31. Resources for Calculus Collection, Volume 5: Readings for Calculus, *Underwood Dudley*, Editor.
32. Essays in Humanistic Mathematics, *Alvin White,* Editor.
33. Research Issues in Undergraduate Mathematics Learning: Preliminary Analyses and Results, *James J. Kaput and Ed Dubinsky,* Editors.
34. In Eves' Circles, *Joby Milo Anthony*, Editor.
35. You're the Professor, What Next? Ideas and Resources for Preparing College Teachers, *The Committee on Preparation for College Teaching, Bettye Anne Case*, Editor.
36. Preparing for a New Calculus: Conference Proceedings, *Anita E. Solow*, Editor.
37. A Practical Guide to Cooperative Learning in Collegiate Mathematics, *Nancy L. Hagelgans, Barbara E. Reynolds, SDS, Keith Schwingendorf, Draga Vidakovic, Ed Dubinsky, Mazen Shahin, G. Joseph Wimbish, Jr.*
38. Models That Work: Case Studies in Effective Undergraduate Mathematics Programs, *Alan C. Tucker*, Editor.
39. Calculus: The Dynamics of Change, *CUPM Subcommittee on Calculus Reform and the First Two Years, A. Wayne Roberts*, Editor.
40. Vita Mathematica: Historical Research and Integration with Teaching, *Ronald Calinger*, Editor.
41. Geometry Turned On: Dynamic Software in Learning, Teaching, and Research, *James R. King and Doris Schattschneider,* Editors.

42. Resources for Teaching Linear Algebra, *David Carlson, Charles R. Johnson, David C. Lay, A. Duane Porter, Ann E. Watkins, William Watkins,* Editors.

43. Student Assessment in Calculus: A Report of the NSF Working Group on Assessment in Calculus, *Alan Schoenfeld,* Editor.

44. Readings in Cooperative Learning for Undergraduate Mathematics, *Ed Dubinsky, David Mathews, and Barbara E. Reynolds,* Editors.

45. Confronting the Core Curriculum: Considering Change in the Undergraduate Mathematics Major, *John A. Dossey,* Editor.

46. Women in Mathematics: Scaling the Heights, *Deborah Nolan,* Editor.

47. Exemplary Programs in Introductory College Mathematics: Innovative Programs Using Technology, *Susan Lenker,* Editor.

48. Writing in the Teaching and Learning of Mathematics, *John Meier and Thomas Rishel.*

49. Assessment Practices in Undergraduate Mathematics, *Bonnie Gold,* Editor.

50. Revolutions in Differential Equations: Exploring ODEs with Modern Technology, *Michael J. Kallaher,* Editor.

51. Using History to Teach Mathematics: An International Perspective, *Victor J. Katz,* Editor.

52. Teaching Statistics: Resources for Undergraduate Instructors, *Thomas L. Moore,* Editor

53. Geometry at Work: A Collection of Papers Showing Applications of Geometry, *Catherine A. Gorini,* Editor.

MAA Service Center
P. O. Box 91112
Washington, DC 20090-1112
800-331-1622   fax: 301-206-9789

Preface

This collection of papers is a resource for students and teachers of geometry who are interested in the practical applications of their subject. The wide range of topics presented here by leading researchers and teachers will appeal to students of every background and interest.

The papers are organized according to the area of application, beginning with art and architecture and culminating with science and even mathematics itself. Within each section, the papers are organized roughly according to the level of background required. The papers can serve many purposes: supplementary materials for teachers of geometry, resources for student projects, ideas for special lectures, inspiration for further research, or simply to broaden one's awareness of geometry and its applications.

Although the papers are organized by the areas of application, the reader will find common threads or themes that appear in many different parts of the book. Uses of graph theory, polytopes, convex hulls, and lattices can be found in many of the papers. The more familiar and elementary themes of similar triangles, trigonometry, and squaring the circle have a significant role to play in at least half of the papers.

The reach of the papers is from ancient to modern, concrete to abstract, familiar to cutting edge. All of them, however, show geometry to be the lively and exciting discipline that fascinates us all.

Contents

Preface vii

Introduction xi

Part 1: Art and Architecture

Spirals and the Rosette in Architectural Ornament, *Kim Williams* 3

Sun Disk, Moon Disk, *Paul Calter* 12

Façade Measurement by Trigonometry, *Paul Calter* 20

A Secret of Ancient Geometry, *Jay Kappraff* 26

Part 2: Vedic Civilization

Square Roots in the Śulba Sūtras, *David W. Henderson* 39

Applied Geometry of the Śulba Sūtras, *John F. Price* 46

Part 3: The Classroom

Ethnomathematics for the Geometry Curriculum, *Marcia Ascher* 59

Education with Fascination: Teaching Descriptive Geometry with Applications,

    Marina V. Pokrovskaya 64

Part 4: Engineering

Making Measurements on Curved Surfaces, *James Casey* 71

Mathematics to the Aid of Surgeons, *Ramin Shahidi* 76

The Geometry of Frameworks: Rigidity, Mechanisms and CAD, *Brigitte Servatius* 81

Geometry and Geographical Information Systems, *George Nagy* 88

On the Other Hands: Geometric Ideas in Robotics, *Bud Mishra* 105

Part 5: Decision-Making Processes

Decisions through Triangles, *Donald G. Saari* 121

Geometry in Learning, *Kristin P. Bennett and Erin J. Bredensteiner* 132

Part 6: Mathematics and Science

The Geometry of Numbers, *Antonie Boerkoel* 149

Statistical Symmetry, *Charles Radin* 157

Three-Dimensional Topology and Quantum Physics, *Louis H. Kauffman* 162

Bridges between Geometry and Graph Theory, *Tomaž Pisanski and Milan Randić* 174

Polytopes in Combinatorial Optimization, *Thomas Burger and Peter Gritzmann* 195

Introduction

There is no branch of mathematics, however abstract, that will not eventually be applied to the phenomena of the real world.

—N. I. Lobachevsky[1]

Geometry is the study of shape and form, with origins in surveying land, designing buildings, and measuring volumes. More importantly, geometry is a point of view that sees the shapes and forms that are intrinsic to any mathematical concept or relationship. Indeed, almost every area of mathematics incorporates geometric concepts and the geometric viewpoint in a fundamental way. For example, Descartes used analytic geometry to display the lines and curves associated with algebraic equations. A differential equation has inherent within its formulation a geometric phase space. When confronted with a new group, an algebraist might view it in terms of the symmetries of a higher-dimensional polyhedron to better understand its structure. Binary relations can be understood in terms of trees, and even probability has its foundation in the properties of measure spaces.

Today, geometry encompasses many different approaches, techniques, and theories, including Euclidean and non-Euclidean geometries, projective geometry, finite geometries, transformational geometry, computational geometry, differential geometry, discrete geometry, tilings, and knot theory. The papers in this collection show that the geometric point of view, in all its many different varieties, has many diverse applications going far beyond its origins in measuring distances, areas, and volumes.

Euclidean geometry, which has dominated the development of Western geometry, uses compass and straightedge to study the flat plane or three-dimensional space. We see the influence of Euclidean geometry in many of the papers in this collection. It is used in the field of art in papers by Kim Williams and Jay Kappraff and we see it used by the engineer as a vital tool for design and communication in the paper by Marina Pokrovskaya. Both Ramin Shahidi and Paul Calter show that measuring distances with Euclidean geometry continues to find new applications.

Non-Euclidean geometry takes the view that the intrinsic shape of space need not be flat but may be curved. Many of the shapes in our environment are curved, and Jim Casey shows how the engineer can make use of Riemannian geometry to understand these curved shapes.

Analytic geometry integrates algebra and geometry, and it has allowed the geometer to make use of the powerful tools of algebra. Papers by Thomas Burger and Peter Gritzmann, Kristin Bennett and Erin Bredensteiner, Ton Boerkoel, Bud Mishra, and Don Saari demonstrate the effectiveness of integrating algebraic formulations with the study of geometric information.

---

[1]N. I. Lobachevsky, quoted in the *American Mathematical Monthly*, February 1984, p. 151.

Recent research in geometry has focused on invariants and other properties subtler than those of measurement. This work includes tilings, knot theory, discrete geometry, computational geometry, and graph theory. Applications of these approaches can be found in the papers by Lou Kauffman, Charles Radin, George Nagy, Brigitte Servatius, and Tomaž Pisanski and Milan Randić.

Finally, cultures different from our own have seen geometry quite differently. In papers by Marcia Ascher, Jay Kappraff, John Price, and David Henderson we see that our own understanding of geometry can be used to analyze the work of other cultures and to bridge what gaps there may be in time or space between these cultures and ourselves.

The papers in this collection, written by pioneers and leading experts in their fields, are a valuable resource for geometry teacher and student alike. All of the papers are accessible to anyone having a college-level course in geometry and it is hoped that they will provide a broad vision of applied geometry—geometry at work.

1   The Nature of Applications of Knowledge

Some have the attitude that knowledge exists for its own sake, that applications need not be of concern to the mathematician, and in fact may not even be possible for some areas of mathematics. G. H. Hardy espoused the view that number theory, which he considered to be the most beautiful and profound area of mathematics, had not the "slightest 'practical' importance."[2] The comment by N. I. Lobachevsky, one of the founders of non-Euclidean geometry, given at the beginning of this introduction, takes quite a different stance. History has proved Hardy wrong—today the very mathematics he cited as an example of the unproductive is the heart of the widely used RSA cryptosystem. Lobachevsky, on the other hand, displayed remarkable prescience, since the revolutionary geometry that he discovered, opposed to the contemporary view of space and originally viewed as unnatural and invalid, turned out to be precisely the viewpoint needed by Albert Einstein in his general theory of relativity. Similarly, many new ideas and discoveries in mathematics and science have eventually proven to have many useful applications despite unpromising beginnings. For this reason, it is worthwhile to understand how the theoretical knowledge of mathematics can have real-world applications.

Although in all disciplines the general processes of gaining and applying knowledge are similar, there are significant differences between mathematics and the sciences which can serve to highlight the unique role of mathematics in applying knowledge generally and in the types of applications we see here. In the sciences, physical phenomena are the objects of study, while in mathematics, the objects of study are purely intellectual concepts and constructs. The real number line, for example, is purely conceptual, having no physical existence itself. In the sciences, observations of physical phenomena are made by conducting experiments and recording measurements. For the mathematician, computations, examples, counterexamples, special cases, and diagrams replace microscopes and telescopes as a way of observing and measuring the structure and behavior of mathematical objects. From observations, principles of knowledge are derived by the scientist or mathematician which describe patterns of behavior common to a significant class of examples. In science, these

---

[2]Godfrey H. Hardy, *A Mathematician's Apology*, Cambridge, Cambridge University Press, 1976.

principles are verified by experimentation and further measurement; in mathematics, principles are verified intellectually by mathematical proof.

Applying knowledge involves extending these general principles to guide progress in some specific area of life. By its very nature, knowledge will be relevant to that area of life from which it is derived and can serve as the basis for applications in that area. However, it is our experience that mathematics has applications far beyond the boundaries of the area in which it was first developed. Indeed, mathematics is striking in that the concepts, principles, and techniques developed for the understanding of purely non-physical mathematical constructs provide the essential tools that scientists in all areas use to understand their observations and measurements of the physical world. Eugene Wigner, in discussing the success of mathematics in physics, says,

> It is important to point out that the mathematical formulation of the physicist's often crude experience leads in an uncanny number of cases to an amazingly accurate description of a large class of phenomena.[3]

Moreover, once the scientist's observations have been given a mathematical formulation, the methodology and computations of mathematics can extend these formulations to predictions about behavior that has not yet been observed. This is of enormous importance to the scientist because, as Richard Hamming points out,

> Constantly what we predict from the manipulations of mathematical symbols is realized in the real world. ... For glamour, I can cite transistor research, space flight, and computer design, but almost all of science and engineering has used extensive mathematical manipulations with remarkable success.[4]

The fact that abstract mathematical concepts, which are often only suggested by observations of the physical world and depend mostly on the imagination of the mathematician for their creation, have such universal success in the world of applications is indicative of a common source for both mathematics and the physical world.

As a subjective discipline, mathematics depends on the creativity and aesthetic sensibilities, as well as the intelligence, of the mathematician. Mathematical progress is always in the direction of locating deeper and more abstract concepts, structures, and relationships, ever further removed from the physical world. Yet when we look at mathematics in general, or at geometry in particular, we see that the deeper and more abstract the concepts, the greater is the range of application in the real world. In other words, the greater the subjective component of a mathematical theory, the more effective is that theory in its objective role of applications. William Thurston sees this as a natural phenomenon:

> My experience as a mathematician has convinced me that the aesthetic goals and the utilitarian goals for mathematics turn out, in the end, to be

---

[3]Eugene P. Wigner, "The Unreasonable Effectiveness of Mathematics in the Natural Sciences," in E. P. Wigner, *Symmetries and Reflections: Scientific Essays,* The M.I.T. Press, Cambridge, MA, 1967, p. 230.

[4]R. W. Hamming, "The Unreasonable Effectiveness of Mathematics," *American Mathematical Monthly,* 87, 1980, p. 82.

quite close. Our aesthetic instincts draw us to mathematics of a certain
depth and connectivity. The very depth and beauty of the patterns
make them likely to be manifested, in unexpected ways, in other parts
of mathematics, science, and the world.[5]

Mathematical formulations of abstract patterns and relationships appear to be in
many cases our deepest understanding of principles that exist throughout nature.
Indeed, we can make the case that the wide applicability of mathematics suggests
the interconnectedness of all spheres of life, from the abstract to the concrete. The
beauty, orderliness, and universality that we see in all areas of mathematics are
reflected in the beauty and orderliness that scientists find in the physical world.

2   The Role of Applications in the Study of Geometry

The full range of geometry is from the most theoretical and abstract theorems based
on axioms and undefined terms to varied applications in science and technology, as
well as in other areas of mathematics. In recent years, applications of geometry
have taken on a more prominent and exciting role, both within and outside of
mathematics. For example, computers have spurred the development of many new
areas, including computational geometry, image processing, visualization, robotics,
and dynamic geometry.

Today more than ever, the study of applications is an important component in
the study of geometry, a subject traditionally valued for its practicality. We gain
a broader and richer understanding of geometrical concepts when we see the unexpected applied contexts in which they appear. In applied settings, geometrical
concepts can take on new and quite different interpretations; for example, a finite
geometry can become a graph or a knot can become a description of a quantum
mechanical operator. Finally, there is charm in seeing familiar theorems and principles showing their value in many different roles. Without studying applications, a
student will never see the complete character of geometry.

This collection is an abundant resource for those wishing to include applications
in their study of geometry. We see here geometry used to describe and understand
the shapes that we see in the world around us and geometry used to design the
shapes that we construct to enrich our environment. We also see how geometry is
used in other branches of mathematics and in science to give shape and form to
mathematical data or concepts that are not inherently or intrinsically geometric.

3   Geometry Used to Understand the Environment

Many papers in this collection show ways of applying the tools of geometry to
the analysis of shapes that exist in our environment. An important such use is to
transform measurements that one is capable of obtaining into the kind of information
that one can really use. An old example of this is, of course, surveying, but there
are very modern examples as well.

Ramin Shahidi in "Geometry to the Aid of Surgeons" and Paul Calter in "Façade
Measurement by Trigonometry" both use similar geometric techniques. In the first

---

[5]William P. Thurston, "Mathematical Education," *Notices Amer. Math. Soc.,* 37, 1990, p. 848.

case, measurements of a patient's anatomy made by medical imaging machinery are converted into a potential surgical trajectory. In the second case, measurements made of the façade of a building by surveying instruments are converted into distances between specific points on the façade.

George Nagy has an analogous problem, that of converting the geographical measurements in a Geographical Information System into a useful format. In a GIS, however, there is so much data that the role of geometry is to synthesize the data into a format that can readily be interpreted by the researcher. For example, elevation data can be transformed into visibility graphs that can then be used for the optimal placement of fire towers or radio transmitters.

Of a subtler nature is the question of the arrangement of atoms in a quasicrystal, a kind of material having a new and surprising symmetrical structure as measured by X-ray diffraction. Studying the possible arrangements of atoms in a quasicrystal, Charles Radin has developed the concept of statistical symmetry, a way of measuring regularities in an arrangement of shapes that is not, strictly speaking, symmetrical.

The traces left behind by other cultures are sometimes difficult to understand and they can easily be misinterpreted according to the learned fashions of the day. When cultural legacies have shape and form, a geometrical analysis can give us a quite reliable basis from which to make an interpretation of what has been left to us. Marcia Ascher and Jay Kappraff demonstrate how to undertake a geometrical analysis of our cultural legacies. They lead us to a broad-minded appreciation of the possible interpretations that can legitimately be given to what we see in other cultures.

In her paper on spirals, Kim Williams explains techniques for constructing spirals, volutes, and rosettes so that we can understand the challenges facing architects who have incorporated these beautiful geometric shapes into their work. With this, we gain insight into the intentions of the architects and deeper appreciation for their work.

David Henderson and John Price both look at writings left by the Vedic civilization and, through close geometric analyses, give insight into the possible reasonings, computations, and motivations behind the geometric constructions given in the Śulba Sūtras. As with the examples given by Ascher, Kappraff, and Williams, we see that this expands our view of the achievements of the past.

4   Geometry Used to Build the Environment

Geometry is essential for designing the shapes that we build to mold our environment. Descriptive, or projective, geometry has been the main tool used by the engineer to develop, record, and communicate plans and designs. Marina Pokrovskaya connects for us the theoretical aspects of descriptive geometry to simple but realistic examples of the practical applications of descriptive geometry. She shows how these examples can be used in the classroom to train those who will be using descriptive geometry in their work.

In her paper on rigidity of frameworks, Brigitte Servatius examines the structure of rectangular grids to determine whether or not they are rigid. She shows how the determination of the rigidity of a structure begins with the side-side-side theorem of Euclidean geometry, and then goes on to show that many other geometric ideas can be applied to the analysis of frameworks. In particular, graph theory is very effective in this area since a framework can be viewed as a collections of rods (edges) connected at joints (vertices).

In addition to flat surfaces, the designer or engineer needs curves and curved surfaces. Jim Casey develops our understanding of curvature and Riemannian geometry through measurement and experimentation and then introduces us to the analysis of the structural properties of curved surfaces.

A recent engineering creation is the robot. How should a robot hand be designed so that it can securely grasp any shape? Using the properties of convex sets, Bud Mishra is able to determine the exact number of fingers a robot hand must have, and further, he shows how to determine the placement of those fingers to grasp any given object.

Kim Williams and Jay Kappraff have used geometrical analysis to deduce the intentions of artists from geometrical evidence in their work. In Paul Calter's paper "Sun Disk, Moon Disk," the artist himself explains his experience of integrating his aesthetic purposes with his mathematical thinking in the design and construction of a massive sculpture. This paper shows firsthand the nature of the mathematical thinking that can go into creating a work of art.

5   Geometry Used to Create Form from the Formless

A significant way that geometry is used in more theoretical applications is to first create form and structure out of abstract relationships and then to use geometrical techniques in the analysis of those shapes. A good example of this is the way that Don Saari converts all possible outcomes of an election with n candidates into an $(n-1)$-simplex. Each voting procedure breaks the simplex into different ranking regions, where the winners are different. Geometrical arguments can then be used to compare different voting procedures, with rather surprising and disquieting conclusions.

Thomas Burger and Peter Gritzmann look at combinatorial optimization problems, such as the matching problem or the traveling salesman problem, which have a finite, discrete set of feasible solutions. The authors form a polytope whose vertices correspond to these feasible solutions, recasting the discrete problem as a geometric problem in which the approach and techniques of linear programming can be applied.

Experts in an area routinely classify objects that have been measured into classes; for example, tumors can be classified as malignant or benign. Kristin Bennett and Erin Bredensteiner set out to train computers to classify new data points as effectively as expert technicians by training them on sets of data points that have already been classified by experts. Their method is to plot the given set of data points in Euclidean n-space and then to find a hyperplane that separates the two types of points in an optimal way. New data points can then be classified by the computer according to which side of this hyperplane they lie on.

From another perspective on graph theory, Tomaž Pisanski and Milan Randić are interested in the way that geometric shapes such as polytopes can give rise to specific types of graphs. The geometry of the generating polytope provides a basis for classifying and understanding the resulting graphs.

Number theory is often regarded as the purest and most abstract branch of mathematics. In his paper, Ton Boerkoel shows that even here, geometry has a special role to play. Minkowski's Convex Body Theorem, based on a geometric version of the pigeonhole principle, gives conditions that ensure that a convex set contains a point of a lattice. Allowing lattice points to correspond variously to rationals, pairs of integers, or quadruples of integers, Boerkoel proves three standard theorems from number theory.

Another beautiful example of converting abstract relationships into geometric structure is given by Lou Kauffman. The creation and annihilation operators of quantum mechanics can be represented graphically by a cup (lower semicircle) and cap (upper semicircle), and the quantum mechanical amplitudes for a composition of operators turn out to correspond precisely to knot invariants. This connection has already proven to be quite fruitful, both for physicists and mathematicians.

6   This Book

The purpose of this book is to provide resources for the student and teacher of geometry. Available here are additional topics for standard courses, material for advanced work by individual students, and the stimulus for further research by students and faculty. The papers assume a minimal amount of standard geometry and give ample references for those who wish to pursue a topic further. This collection is organized into six sections according to the main areas of application in each paper: Art and Architecture, Vedic Civilization, the Classroom, Engineering, Decision Making Processes, and Mathematics and Science.

7   Acknowledgements

This volume would not have been possible without the invaluable help of Ken Hince, who assisted in all aspects of the design and typesetting. Help with the typesetting and graphics was also provided by H. Güneş Yücel, Christine Ports, Arthur Bichler, Luv Passi, Elizabeth Bergun, and Beverly Ruedi. Elizabeth Bergun designed the cover based on artwork by Kim Williams.

Part 1

Art and Architecture

Spirals and the Rosette
in Architectural Ornament[*]

Kim Williams
Via Mazzini 7
50054 Fucecchio Italy
k.williams@leonet.it

The occurrence of spirals in our natural environment has been by now so well examined and catalogued as to be familiar to most of us. Spiral patterns of growth in the animal world are found in the nautilus shell, the snail's shell, and in the ram's horn; in the plant world we may find them in the sunflower, the pine cone, and the pineapple. Natural spirals may be as large as spiral nebulae or as tiny as the spiral structure of the cochlea in the human ear.[1] For centuries architects have incorporated spiral patterns from nature, as well as patterns inspired by spirals if not constructed from them, into the built environment. This paper will discuss how the spiral has been used to express concepts of structure and space in architecture.

Greek architects devised three orders of architecture, Doric, Ionic, and Corinthian; Roman architects adopted the Greek orders and produced two of their own, the Tuscan and Composite. An "order" of architecture is defined by the style of the columns used and the details which accompany the columns. The archetype of the logarithmic spiral in architectural decoration is perhaps found in the volute of the capital belonging to the Ionic order of Greek architecture.[2] See Figure 1. One school of thought about the origins of the orders of architecture holds that the individual elements of the order were designed not only to perform a structural function but also in some way to express that function. Consider for a moment the column as a load-bearing element. The shape of the column is expressive of its being placed in compression by the weight which it is to transfer to the base: instead of

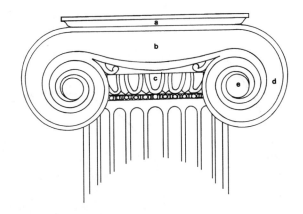

Figure 1: The Ionic capital and its components: a) abacus; b) cushion; c) echinus; d) volute; e) eye

a uniform cylinder, the diameter increases towards the lower half of the column, producing a slight deformation known as *entasis*. The volute of the Ionic capital may express compressive force in a similar way: in this case, the element below the *abacus* is forced into a downward curve. The reader may be familiar with the lotus capitals of Egyptian architecture. To visualize the genesis of the volute, it may help to imagine the result of compression on the lotus bud, that is, that its petals are forced to curl downward under the weight of the load it must bear.

It is to a Roman architect and theoretician that we owe the first description of the Ionic volute and its construction, found in the *Ten Books on Architecture* by Vitruvius. Vitruvius' description is difficult to interpret, and after the rediscovery of the classical orders in the Renaissance, various formulae for constructing the volute, all purported to be based upon the of technique their Roman predecessor, were published. Without citing an ex-

---

[*]All figures are by the author except Figures 5, 6, and 7.

[1]For more about spirals, cf. Philip J. Davis, *Spirals: From Theodorus to Chaos*, Wellesley, Massachusetts: A. K. Peters, 1993, and Theodore Cook, *The Curves of Life*, 1914, reprinted, New York: Dover Publications, 1986.

[2]The spiral volute is found also in the Greek Corinthian order and in the Roman Composite order. For descriptions of all the orders, see [3].

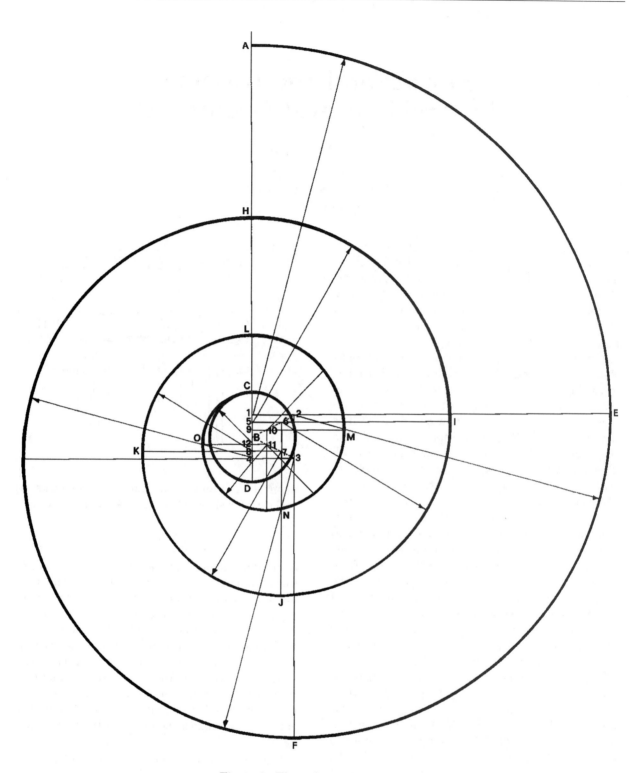

Figure 2: The volute construction.

haustive list all who championed various methods, some of the most important include Leon Battista Alberti in the 15th century, Andrea Palladio, Sebastiano Serlio and Giacomo Vignola in the 16th century, and William Chambers and James Gibbs in the 18th century. The method which I have chosen to present for the purposes of this study is that presented in Bannister Fletcher's authoritative

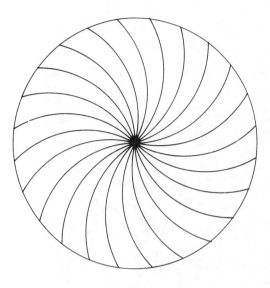

Figure 3: The fan pattern.

Figure 4: The rosette pattern.

survey of historic architecture.[3] This method, attributed to Goldman by Fletcher, concentrates on an approximate construction of the logarithmic spiral which forms the volute. In a second step, to the first spiral of the volute is added a second which forms the *fillet*, or thickness, of the volute.

The procedure involves establishing a given module for the construction, which is related to the diameter of the column. The dimension of the eye of the volute is determined in proportion to the module. The method discussed here establishes the eye as $1/9$ of the module, or $3\ 1/3$ parts of a module which has 30 parts. Vertical line A–B equal to $1/2$ module is drawn. The construction proceeds by drawing the eye concentric with the center of the module (point B in Figure 2). The diameter of the eye is divided into 4 equal lengths (C–1, 1–B, B–4, and 4–D); 1–B and B–4 are divided into thirds. Three squares are constructed with sides 1–4, 5–8, and 9–12 respectively. Beginning with the compass point at 1 and a radius equal to 1–A, an arc is drawn to point E. At E the compass point is moved to point 2, and an arc of radius 2–E is drawn to point F. The construction continues in this manner, changing compass location and radius at each $90°$ junction, until at last, with compass at point 12 and radius O–12, the spiral closes upon the eye of the volute.

This rather lengthy introduction to the logarith-

mic spiral found in the volute brings us around now to my particular area of study, which is the changing nature of paving design in the history of Italian architecture. Usually seen as a minor decorative art, pavements often shed new light on concepts of mathematics and space design in any given epoch. Consider that the floor surface is the largest unbroken plane surface in a work of architecture: it can become, practically speaking, a kind of canvas upon which ideas relating to the architecture may be expressed. Designs based upon spirals form a subgroup of pavement designs which I have found especially interesting. It is speculation on my part, though not, I believe, untoward speculation, that the pavement designs based upon spirals were inspired by the spiral's having become a part of the architectural vocabulary through its use in the volute.[4]

The mosaic pavements of Pompeii demonstrate a range of patterns which seem to suggest a progression of uses of the spiral motif for creating new paving patterns. One pattern is based upon similar curves rotated about the center of a circle, forming a fan pattern.[5] See Figure 3. A next step be-

[3]Cf. Bannister Fletcher, *A History of Architecture on the Comparative Method*, 12th ed., New York: Charles Scribner's Sons, 1945, p. 100, Figure Q.

[4]Spirals, of course, have been used to decorate objects of many specialties, such as pottery, metal work, fabrics and paintings, and in many ancient cultures, including Egyptian, Greek, Roman, Etruscan, and Minoan. I have chosen to emphasize the volute, however, as an object specifically related to architecture.

[5]Cf. Asher Ovadiah, *Geometric and Floral Patterns in Ancient Mosaics*, Rome: L'Erma di Bretschneider, 1980, p. 153.

Figure 5: The Pompeii rosette pavement panel.

yond the fan pattern involves reversing the direction of the curves and then superimposing the new set of curves upon the first set. This produces a set of curvilinear regions which, when rendered in two colors, forms a design of concentric bands of curvilinear triangles.[6] Because Albrecht Dürer refers to a circular design of this sort as a "rosette," I have come to call this whole class of related designs by this name. See Figure 4. A Roman example of such a design is found in a pavement discovered during the excavations of Pompeii and dated to the first century B.C.[7] See Figure 5. An even earlier Greek pavement with the rosette pattern may have existed: one plan of the Tholos, or round temple, at Epidaurus in Greece, dated 350 B.C., shows a pavement design based upon the rosette, but it is unclear whether it is original or reconstructed.[8] A medieval example of the rosette may be found in an emblema or decorative insert in the pavement of the Baptistery of Florence.[9] See Figure 6. In the

Renaissance, rosettes seems to have enjoyed special favor. In Florence, a rosette pattern is found in a panel of the pavement in the Cathedral, while the Sacristy of the church of S. Spirito is entirely covered by a rosette pattern. In the sixteenth century, Michelangelo is credited with the design of two pavements featuring rosettes, that of the Laurentian Library in S. Lorenzo in Florence and that of the famous Campidoglio or Capitol in Rome.[10] Architectural theory of the Renaissance placed particular importance on centrally-planned churches, identifying the center point of the space with the Creator of the cosmos. This emphasis on center may explain the frequent application of the rosette in pavement designs. As a motif with strong visual references to its center, it communicates in a clear way the importance of this point in the architecture to the observer who moves through the space.

A cleverly constructed optical design created by British psychologist J. Frazer and reproduced in Ralph Evans' *Introduction to Color* illustrates in a

[6]Ibid., p. 144.

[7]This pavement panel is now on display in the National Museum, Rome.

[8]Cf. Cyril Harris, ed., *Historic Architecture Sourcebook*, 1977, reprinted as *An Illustrated Dictionary of Historic Architecture*, New York: Dover Publications, 1983, illustration under heading "Tholos," p. 531.

[9]The pavement of the Baptistery of Florence is interesting

as a geometric study in itself. Cf. Kim Williams, "The Sacred Cut Revisited: The Pavement of the Baptistery of San Giovanni, Florence" in *Mathematical Intelligencer*, Vol. 16, no. 2 (Spring 1994), pp. 18–24.

[10]The rosette of this last is framed by an ellipse rather than a circle.

Figure 6: The rosette panel in the Baptistery of Florence.

Figure 7: Frazer's rosette and circle optical illusion.

striking way the relationship between the rosette pattern and the spiral.[11] See Figure 7. In this figure, when concentric circles represented in two colors are superimposed upon a rosette design, a strong optical illusion of a logarithmic spiral is created. But, reader beware, it must be pointed out that although the basic rosette pattern is easily identified, not all rosettes are constructed the same way, that is, only some of them are constructed using a genuine logarithmic spiral. The others are constructed using a "shortcut" method which looks like it is formed of spiralling curves, but has the advantage of being much more quickly constructed. Sebastiano Serlio, architect and theoretician of the sixteenth century, was candid in admitting that, for its aesthetic effect, architecture is dependent upon workmen whose good fortune exceeds their skill, lamenting the difficulty of explaining complicated constructions to unlearned laborers. Perhaps it is for this reason that a shortcut for constructing the ancient and oft-repeated rosette design gained popularity. The shortcut method is based upon a simply constructed set of circles, which we shall examine in some detail presently. But first, how can one

tell the difference between a logarithmic rosette and a circular rosette? The answer lies in the proportions of the interstices created by the overlapping curves. If the interstices grow larger as they move farther away from the center of the design but remain of the same proportions, then the rosette is based upon a true logarithmic spiral. A circular rosette creates interstices which change in both size and shape according to their location in the design. Compare now the Pompeii rosette (Figure 5) with that of the Baptistery of Florence (Figure 6). Though they are colored in the same way, the difference in the configuration of the curvilinear triangles is clear.

Now let us look at the circular rosette. The construction is based on overlapping "radial circles" whose centers lie on a "centrum ring" which is concentric with an outer "reference circle." Proportions of the interstices of the rosette depend upon the relationships between the reference circle and the radial circles. Albrecht Dürer constructed the rosette for a wooden pavement in 1525. In his design, the diameter of each of the twelve circles, as well as that of the centrum ring, is equal to the radius of the reference circle. See Figure 8. Using the same proportions while increasing the number of radial circles produces a denser rosette. See Figure 9.

Three further examples illustrate the relationships among the elements of the construction.

---

[11]This figure is reproduced in Harold Jacobs, *Mathematics: A Human Endeavor*, 2nd ed., New York: W.H. Freeman, 1982, p. 351.

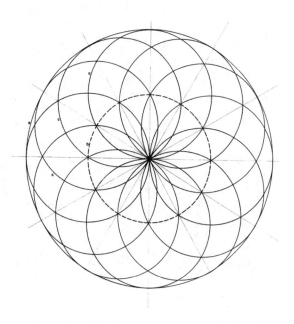

Figure 8: The rosette proposed by Dürer and the elements of the rosette: a) reference circle; b) centrum ring; c) radial circles.

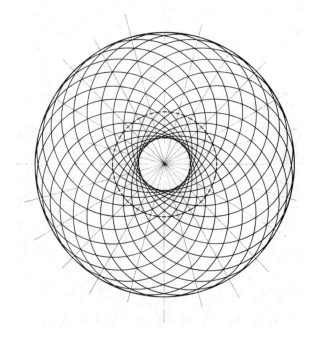

Figure 10: Rosette with radial circles of radius 6/10.

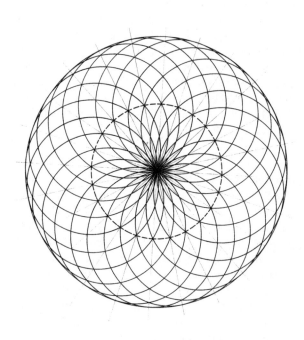

Figure 9: A double Dürer rosette.

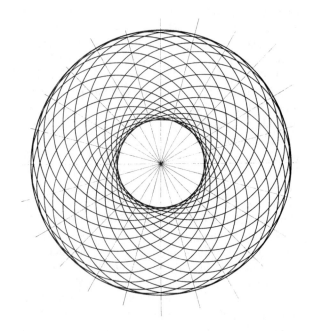

Figure 11: Rosette with radial circles of radius 2/3.

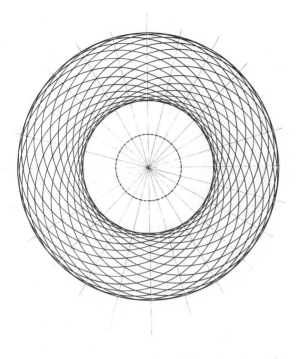

Figure 12: Rosette with radial circles of radius 3/4.

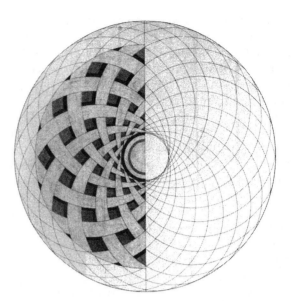

Figure 13: Basketweave rosette, Cathedral of Florence.

Where the perimeters of the radial circles pass through the center of the reference circle, a flower-like shape is created in the center. See Figures 8 and 9. Where the diameter of the radial circles is greater than the radius of the reference circle, an empty inner circle is formed, as for instance in a rosette where the diameter of the radial circles is equal to 6/10 of that of the reference circle, while the diameter of the centrum ring is equal to 4/10 of the diameter of the reference circle. See Figure 10. By increasing the diameter of the radial circles to 2/3 of the diameter of the reference circle (and decreasing the diameter of the centrum ring to 1/3 of the diameter of the reference circle) the centrum ring can be made to coincide with the circumference of the empty inner ring. See Figure 11. The greater the ratio of the diameter of the centers of the radial circles to the diameter of the reference circle, the greater the diameter of the empty inner circle. In a last example, the diameter of the radial circles is equal to 3/4 of the diameter of the reference circle (while the diameter of the centrum ring is equal to 1/4 of the diameter of the reference circle). Note that the diameter of the centrum ring is now smaller than that of the empty inner ring. See Figure 12. It may be noted in all cases that the diameter of the centrum ring is always equal to the diameter of the reference circle minus the diameter of the radial circles. This property of the circular

rosette construction is helpful when analyzing an existing rosette to determine its structure.

The empty inner ring created within the rosette design is one key to its appearing to be formed out of overlapping logarithmic spirals, for the inner ring then reads as the eye upon which the curve appears to close. But in order to accomplish fully the illusion that the design is constructed of logarithmic spirals, the rosette must be cut off before the interstices may be seen to decrease. This is indeed how the rosette is treated in almost all pavement designs I have studied. Figure 6 is a good example of this treatment. The reader will notice that in the circular rosette, the interstices created by the overlapping radial circles increase in size as they move further from the center up to a certain limit (changing all the while in proportion, as we noted earlier), at which point they begin to decrease in size as they near the circumference of the reference circle. In the true logarithmic rosette, the interstices do not decrease in size, but constantly increase.

One property of the rosette which makes it valuable as a graphic device is that the basic rosette may be articulated through the use of color in such a way as to produce designs of varying character. The simplest use of color in the pattern is found when two colors are applied to alternating interstices, resulting in a sort of spiralling checkerboard. This is how the pavement of the Tholos in Epidaurus is depicted. Perhaps the most common treatment is to use two colors, one for the bottom half of each interstice and one for the top half. This is found in the Roman mosaic pavement as well as in the emblema in the Baptistery of Florence. A more complex color treatment is used in the paving panel in the Cathedral of Florence, where contiguous groups of three interstices form strips which are interwoven to produce a basketweave design. Depth is then added to the design by adding a shadow line. See Figure 13. Michelangelo's Laurentian Library pavement panel is similar to the basketweave found in the Florence Cathedral, but the strips which separate the interstices of the design are created not from a single rosette pattern, but rather by the addition of an identical second rosette, the radials of which are rotated slightly away from those of the basic rosette.

In addition to its ease of construction and its capacity for variation through the use of color, a third characteristic of the circular rosette recommends its use as a paving design, and that is that it may be efficiently built, as is any pattern that is repetitive. The actual building procedure goes something like this: an artist in the marbleworking shop draws the pattern and indicates the templates to be cut out of wood; the templates are used to make lines on the slabs of marble; the marble is cut and roughly finished in the shop, then transferred to the site; the marble pieces are assembled on the site like a jigsaw puzzle, set in mortar, and finally, finely polished. Looking at the circular rosette of the inset panel in the Baptistery of Florence (Figure 6) we can determine by counting that there are 409 pieces in all which compose the panel.[12] However, there are only 18 individual shapes requiring separate templates. An examination of the basketweave panel in the Cathedral of Florence (Figure 13) reveals that it is less efficient to build, having a slightly smaller number of pieces, 396, but a greater number of shapes, 25.

Another characteristic of the rosette design is that in addition to producing the interstices which increase and decrease in size, it also produces sets of points which may be joined to form a series of concentric circles with ever-greater diameters. In the circular rosette, the diameters of the circles increase only up to a certain limit, before decreasing to the point where the outermost circle finally coincides with the reference circle. This is analogous to how the interstices of the circular rosette increase in size, then decrease. In the logarithmic rosette, the diameters of the circles continually increase. These

---

[12]I count the star-shaped piece in the center as one piece and one shape.

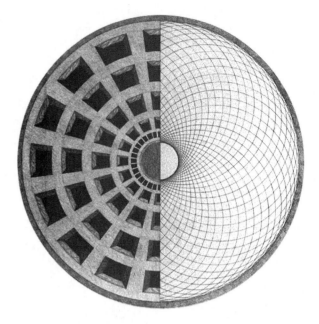

Figure 14: "Coffered" design derived from the rosette.

concentric circles may serve as a proportional device in the creation of new designs. One such design projects onto the pavement an arrangement of distorted squares such as one might see looking up into a dome articulated by coffers: three concentric rings are separated from one another by ever-greater distances as they move away from the center. The spacing of these rings coincides with that created by a circular-based rosette. See Figure 14.

We have now seen that the spiral has been used as a stylized expression of compressive force on architecture elements, as well as the basis for an important group of pavement designs. Both of these have come about as the result of architects' observations of the natural world, a world which has equally fascinated mathematicians. In the construction of the volute and the rosette, the two disciplines are united, producing works of art which can still communicate their truths to us today.

References

[1] Leon Battista Alberti. *The Ten Books of Architecture*. Dover, New York, 1986. Reprint of 1755 edition.

[2] Paul Calter. *Technical Mathematics*. Prentice Hall, Englewood Cliffs, New Jersey, 1990. 2nd ed.

[3] Robert Chitham. *The Classical Orders of Architecture*. Rizzoli, New York, 1985.

[4] Bannister Fletcher. *A History of Architecture on the Comparative Method*. Charles Scribner's Sons, New York, 1945. 12th ed.

[5] Matila Ghyka. *The Geometry of Art and Life*. Dover, New York, 1977.

[6] Jay Hambidge. *The Elements of Dynamic Symmetry*. Dover, New York, 1967. Reprint of 1926 edition.

[7] Cyril Harris, editor. *An Illustrated Dictionary of Historic Architecture*. Dover, New York, 1983. Reprint of *Historic Architecture Sourcebook*, 1977.

[8] Harold Jacobs. *Mathematics, A Human Endeavor*. W.H. Freeman, New York, 1982. 2nd ed.

[9] Jay Kappraff. *Connections: The Geometric Bridge Between Art and Science*. McGraw-Hill, New York, 1991.

[10] Jay Kappraff. The spiral in nature, myth, and mathematics. In C. A. Pickover and I. Hargittai, editors, *Spiral Symmetry*. World Scientific, 1992. Special issue of *Symmetry*.

[11] Asher Ovadiah. *Geometric and Floral Patterns in Ancient Mosaics*. L'Erma di Bretschneider, Rome, 1980.

[12] Andrea Palladio. *The Four Books of Architecture*. Dover, New York, 1965. Reprint of 1570 edition.

[13] Sebastiano Serlio. *The Five Books of Architecture*. Dover, New York, 1982. Reprint of 1611 edition.

[14] John Summerson. *The Classical Language of Architecture*. Thames and Hudson, London, 1980.

[15] Vitruvius. *The Ten Books on Architecture*. Dover, New York, 1960.

[16] Kim Williams. *Patterns and Space, An Architect's Study of Pavements in Italian Architecture*. Unpublished.

Sun Disk, Moon Disk

Paul Calter

Professor of Mathematics, Emeritus, Vermont Technical College
Visiting Professor of Mathematics, Dartmouth College

1 Introduction

We probably visualize a sculptor at work in a studio, chipping at a block of marble or shaping a lump of clay, guided only by the muses, but for many modern artists a more accurate image might be of the sculptor seated at a computer. In this paper I will show how geometry, trigonometry, astronomy, and the computer have all contributed to the design of an abstract sculpture. Here, geometry provided interesting forms and contributed to the meaning of the piece; trigonometry was needed to calculate the relationships of the parts to each other, which required a surprising amount of mathematics; astronomy gave the sculpture a function, that of a calendar; and the computer was used both for numerical computation and for design layout.

2 Earlier Geometric Sculpture

Geometry has been inseparable from art and architecture since the beginnings of recorded history. There are many books and articles detailing the overlap of these fields, some of which are given in the bibliography. They outline the interdependence of art and mathematics from the geometry of the pyramids to the influence of the computer and of fractal geometry on contemporary art. My own interest in geometric art started, quite naturally, when I began teaching mathematics, and my first geometric work was *Focus I* (Figure 1) which contains parabolas and an ellipse.

This was followed by a great number of smaller works, both paintings and sculpture. Larger public pieces include *Armillary VII* (Figure 2) which has

Figure 1: *Focus I.*

Figure 2: *Armillary VII.*

Figure 3: *Mandala II.*

Figure 4: *Sun Disk, Moon Disk.*

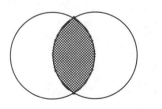

Figure 5: *Vesica Pisces.*

an ellipse and a crescent and is a functioning armillary sundial, and *Mandala II* (Figure 3) which has a circular marble slab whose surface is carved in a fractal pattern, which rotates inside an elliptical chamber.

3 Overall Description

My latest geometric sculpture, *Sun Disk, Moon Disk,* (Figure 4) consists of four elements: a vertical *moon disk*, the tilted stone *sun disk*, the *gnomon* which pierces the sun disk at right angles, and the *polar rod*. The moon disk is vertical and provides the main structural support for the sculpture. It is aligned to be in a north-south plane. The sun disk is tilted from the horizontal at an angle of 43°42′, the local latitude. Thus the polar rod points to the north star while the gnomon points to the intersection of the local meridian and the celestial equator. The sun disk squarely faces the equator, so that at noon on the equinoxes the sun's rays strike the sun disk at right angles.

The twenty-four stones of the sun disk are carved from white marble, verde antique (serpentine), and pink granite. The remainder of the structure is of hot-rolled steel, painted to prevent rusting. The sculpture is twelve feet high and weighs about 3200 pounds. *Sun Disk, Moon Disk* is located at the entrance to the Montshire Museum of Science in Norwich, Vermont, not far from Dartmouth College in Hanover, New Hampshire.

4 The Crescent

The moon disk is named for its crescent shape. I have used this shape in many earlier works, such as in *Armillary VII.* I like the crescent's appearance, but I am also intrigued by its lunar associations. It is also mathematically interesting because certain lunes were the first curved figures whose area was exactly computed by Hippocrates, without the use of π [1].

5 The Vesica

One of the main forms on the upper surface of the sun disk is the figure called the *vesica pisces*, or fish bladder, (Figure 5) formed by the intersection of two circles. It is also called a *mandorla*, after its almond shape. The vesica is an ancient symbol, often used as an aureole around depictions of Christ. It also represents the *yoni* in Hindu cosmology, and

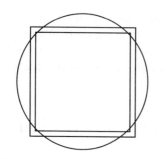

Figure 6: Approximate squaring of the circle.

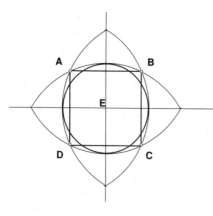

Figure 7: Orthogonal vesicae approximately square the circle.

Figure 8: Rear surface of the sun disk.

cell division. Mathematicians will recognize it as a Venn diagram of symbolic logic, representing the intersection of two sets.

Four such vesicae produce the eight-pointed star seen on the sun disk in Figure 4. Keith Critchlow links this construction to the method for laying out Indian temples, to the layout of stone circles in Great Britain, and to the floor plan of Glastonbury Cathedral in Britain [6]. As the pattern is aligned to the north, it functions as a *compass rose*, pointing out the cardinal directions. The pattern is also intended to suggest a botanical shape, to relate to the natural science component of the Montshire's activities.

6 Squaring the Circle

The ancient problem of *squaring the circle* means to construct, with just compass and straightedge, a square having the same perimeter or area as that of a given circle. Thus in Figure 6 the smaller square has the same perimeter as that of the circle, while the larger square has the same area as the circle.

Figure 9: Carving the analemma.

Zodiac	Equation of Time (Deg.)	Declination of Sun (Deg.)	Zodiac Sign
January 1	-0.8	-23.1	Capricorn
February 1	-3.4	-17.3	Aquarius
March 1	-3.1	-7.8	Pisces
March 21	-1.9	0	Aries
April 1	-1.0	4.3	Aries
April 16	0	9.9	Aries
May 1	0.7	14.9	Taurus
June 1	0.6	22.0	Gemini
June 14	0	23.2	Gemini
July 1	-0.9	23.2	Cancer
August 1	-1.6	18.2	Leo
September 1	-0.1	8.5	Virgo
September 2	0	8.2	Virgo
September 23	1.8	0	Libra
October 1	2.5	-3.0	Libra
November 1	4.1	-14.2	Scorpio
December 1	2.8	-21.7	Sagittarius
December 25	0	-23.4	Capricorn

Table 1: Equation of Time and Declination of Sun at Different Times of the Year. Each zodiac sign extends from approximately the 21st of one month to the 21st of the following month. Thus *Capricorn* applies from December 21 to January 21, approximately.

Leonardo da Vinci seems to have been fascinated by this problem[31]. According to Ladislao Reti [24], "The problem in geometry that engrossed Leonardo interminably was the squaring of the circle. From 1504 on, he devoted hundreds of pages in his notebooks to this question of quadrature ... "

It has been proved that an exact squaring of the circle is impossible but there are many approximate solutions. For example, a square drawn from the points of intersection of two orthogonal vesicae will have approximately the same perimeter as the inner circle. Thus in Figure 7, square $ABCD$ has approximately the same perimeter as circle E. Using trigonometry we can determine that the perimeter of the square is approximately $6.583r$, where r is the radius of the inner circle, compared to a circumference of $2\pi r$ for that circle, a difference of nearly 5%. This is much less accurate than the squaring obtained in the Great Pyramid of Ghiza, where, either by design or by accident, the perimeter of the square base agrees with the circumference of a circle with a radius equal to the pyramid's height to within 0.1% [29].

Squares connecting the intersections of the vesicae can be seen on the underside of the sculpture, Figure 8. Two such squares were included to create a pleasing star pattern. The circle has often been taken to represent the cosmos or heaven, perfect and without end, while the square represented earth and the works of man. Thus squaring the circle came to symbolize the harmonizing of the heavenly and the earthly, and other pairs of dualities, and the rationalizing of the irrational. In the context of the Montshire sculpture, I interpret the circle to represent the universe and the square to represent mankind, so that squaring the circle represents mankind's attempts to understand the universe—a fitting symbol for a Museum of Science.

7 The Analemma

The gnomon is tilted at the local latitude angle and aligned to north, so that it points to the intersection of the celestial equator and the local meridian. Four times during the year, on approximately April 16, June 14, September 2, December 25, the noontime shadow of the gnomon, called the *noon mark*, falls on a vertical line through the sun disk center. But because of the difference between clock time and solar time, the noon mark will be either left or right of this line at other times of the year. This difference between clock time and solar time is called the *equation of time*. It is sometimes engraved on a sun dial as a table of corrections to be applied to the dial reading.

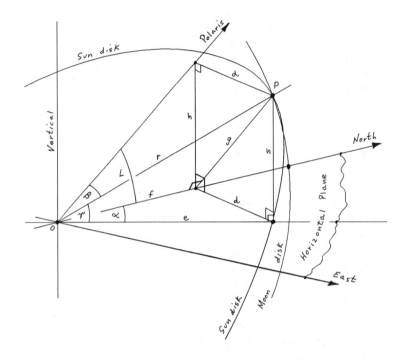

Figure 10: Calculation of intersection of sun disk and moon disk.

Twice each year, at the equinoxes, the noon mark falls on a horizontal line through the center of the sun disk. During the summer the noon mark is below this line, and in winter, above it. The angle of the sun above or below the equator is called its *declination*. Both the equation of time and the declination of the sun are shown in Table 1 for different times of the year, along with the sign of the zodiac [30].

Plotting the equation of time versus the declination results in a figure-eight shaped figure called an *analemma*. This figure, often found on older globes, was carved into the surface of the sun disk (Figure 9), along with the months of the year and the corresponding signs of the zodiac. The sculpture thus functions as a crude calendar. The same analemma figure is also repeated in rotated positions where it has no calendar function but only serves the design.

8   Trigonometry

In addition to geometry, trigonometry was sometimes needed, for instance in locating the point P where the sun disk attaches to the moon disk, Fig-

ure 10. In Figure 10:

O = center of sun disk and center of the circular opening in the moon disk

L = angle of tilt of sun disk from the horizontal (the latitude angle of the Montshire Museum)

α = horizontal angle between sun disk and moon disk

β = angle between north and the attachment point P in the plane of the sun disk.

Noting that $\sin \beta = d/r$ gives

$$\sin^2 \beta = \frac{d^2}{r^2}.$$

But $r^2 = g^2 + f^2 = d^2 + h^2 + f^2$, so

$$\sin^2 \beta = \frac{d^2}{d^2 + h^2 + f^2}.$$

But $h = f \tan L$ and $d = f \tan \alpha$, so

$$\sin^2 \beta = \frac{f^2 \tan^2 \alpha}{f^2 \tan^2 \alpha + f^2 \tan^2 L + f^2}$$

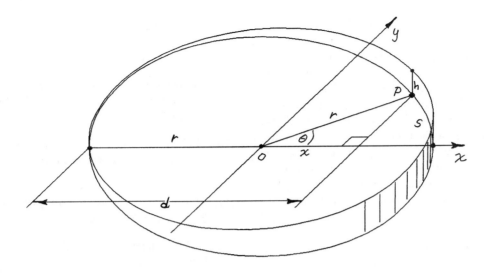

Figure 11: Calculation of widths on the moon disk rim.

or

$$\sin^2 \beta = \frac{\tan^2 \beta}{\tan^2 \alpha + \tan^2 L + 1}.$$

Substituting 43.7° for the latitude angle L and 45° for the angle α between the sun disk and the moon disk gives $\sin \beta \approx 0.586$, or

$$\beta \approx 35.9°.$$

Turning now to the moon disk, we let $\gamma =$ vertical angle from the horizontal to P, measured in the plane of the moon disk. Then

$$\cos \gamma = \frac{e}{r}.$$

But $e = d/\sin \alpha$ and $d = r \sin \beta$, so

$$\cos \gamma = \frac{\sin \beta}{\sin \alpha}.$$

Substituting $\alpha = 45°$ and $\beta \approx 35.9°$ gives $\cos \gamma \approx 0.829$, or

$$\gamma \approx 34.0°.$$

Thus we have the location of one attachment point P on each disk. The other attachment point is simply the reflection of P across the center of each disk. Since an error in this calculation would seriously affect the sculpture, I checked these angles with cardboard models.

9 The Rim

Another place where trigonometry was indispensable was in making the pattern for the rim that is wrapped around the moon disk. If the edge of this rim is to be in a plane, after being wrapped into a cylinder, what is the shape of the steel strip before wrapping? My reflexive answer was, of course,

$$\text{cylinder} + \text{plane} = \text{ellipse},$$

which was wrong, as the following calculation will show. Figure 11 shows the height h of the rim at point P, a distance d from the narrow edge of the moon disk. If the slope of the rim is m, then

$$h = md.$$

But $d = r + x = r + r \cos \theta$, where r is the radius of the moon disk, x and θ are the distance and angle to P, respectively, from the center of the moon disk. Substituting,

$$h = mr(1 + \cos \theta).$$

To get h in terms of the linear distance s along the rim, we note that $\theta = s/r$, where θ is in radians, so

$$h = mr[1 + \cos(s/r)].$$

Choosing $m = 2/15$ and $r = 57$ in., a table of widths h versus distance s is then easily generated. To my surprise the resulting curve was a sinusoid. That this shape is correct is demonstrated simply by cutting a roll of paper at an angle and unrolling it.

10 The Computer as a Sculpture Design Tool

Finally, I want to mention the use of the computer in the design of this sculpture. I used *Derive* to check the trigonometric equations for the attachment points and for the Moon Disk rim widths, and a spreadsheet to perform the numerical calculations. I also used AutoCAD to create the drawings needed to proportion the parts, such as the Sun Disk design, and for making the dimensioned drawings needed for cutting the steel. But AutoCAD also served as a powerful tool for finding missing dimensions. For example, instead of calculating the location of the concrete mounting pads, I simply drew them in and had AutoCAD compute their locations and place the dimensions on the drawing.

11 Summary

I have tried to show how geometry, trigonometry, astronomy, and the computer have all contributed to the design of an abstract sculpture: to provide interesting forms, to enhance the meaning of the piece, to calculate the relationships of the parts to each other, and for design drawings. For me, this sculpture was particularly satisfying because it enabled me to combine many of my interests in a single project. If art is one circle in a Venn diagram and mathematics is another, then this project lies at their intersection.

References

[1] Asger Aaboe. *Episodes from the Early History of Mathematics*. Random House, New York, 1964.

[2] Martha Boles and Rochelle Newman. *The Golden Relationship: Art, Math and Nature*. Pythagorean Press, Bradford MA. 4 Vols.

[3] Charles Bouleau. *The Painter's Secret Geometry*. Harcourt, New York, 1963.

[4] John Briggs. *Turbulent Mirror*. Harper, New York, 1989.

[5] Tons Brunés. *The Secrets of Ancient Geometry and its Use*. International, Copenhagen, 1967. 2 vols.

[6] Keith Critchlow. *Time Stands Still*. Fraser, London, 1979.

[7] Magdalena Dabrowski. *Contrasts of Form: Geometric Abstract Art 1910–1980*. Museum of Modern Art, New York, 1985.

[8] William V. Dunning. *Changing Images of Pictorial Space: A History of Spatial Illusion in Painting*. Syracuse U. Press, Syracuse, 1991.

[9] Samuel Edgerton. *The Renaissance Rediscovery of Linear Perspective*. Basic Books, New York, 1975.

[10] Michelle Emmer, editor. *The Visual Mind: Art and Mathematics*. MIT Press, Cambridge, 1993.

[11] Matila Ghyka. *The Geometry of Art and Life*. Dover, New York, 1977.

[12] E. H Gombrich. *Art and Illusion*. Pantheon, New York, 1960.

[13] Jay Hambidge. *The Elements of Dynamic Symmetry*. Dover, New York, 1967.

[14] Linda Dalrymple Henderson. *The Fourth Dimension and Non-Euclidean Geometry in Modern Art*. Princeton U. Press, Princeton, 1983.

[15] William Ivins. *Art and Geometry: A Study in Space Intuitions*. Harvard U. Press, Cambridge, 1946.

[16] Lesley Jones, editor. *Teaching Mathematics and Art*. Stanley Thornes, Cheltenham, 1994.

[17] Jay Kappraff. *Connections: The Geometric Bridge Between Art and Science*. McGraw-Hill, New York, 1990.

[18] Martin Kemp. *The Science of Art*. Yale U. Press, New Haven, 1990.

[19] Gyorgy Kepes. *Structure in Art and Science*. Brazille, New York, 1965.

[20] Michael Kubovy. *The Psychology of Perspective and Renaissance Art*. Cambridge U. Press, Cambridge, 1986.

[21] Robert Lawlor. *Sacred Geometry*. Thames and Hudson, London, 1982.

[22] George Markowsky. Misconceptions about the golden ratio. *College Mathematics Journal*, page 2, Jan 1992.

[23] Dan Pedoe. *Geometry and the Visual Arts*. Dover, New York, 1976.

[24] Ladislao Reti. *The Unknown Leonardo*. McGraw-Hill, New York, 1974.

[25] Irma Richter. *Rhythmic Forms in Art.* John Lane, London, 1932.

[26] Colin Rowe. *The Mathematics of the Ideal Villa and Other Essays.* MIT Press, Boston, 1976.

[27] Doris Schattschneider. *Visions of Symmetry: Notebooks, Periodic Drawings, and Related Works of M.C. Escher.* W. H. Freeman, New York, 1990.

[28] Leonard Shlain. *Art and Physics: Parallel Visions in Space, Time, and Light.* Morrow, 1991.

[29] Peter Tompkins. *Secrets of the Great Pyramid.* Harper, New York, 1971.

[30] Albert E. Waugh. *Sundials: Their Theory and Construction.* Dover, New York, 1973.

[31] Herbert Wills. *Leonardo's Dessert—No Pi.* NCTM, Reston, VA, 1985.

Façade Measurement by Trigonometry

Paul Calter

Professor of Mathematics, Emeritus, Vermont Technical College
Visiting Professor of Mathematics, Dartmouth College

1 Introduction

We are all familiar with the trigonometry text-book problem, *"The angle of elevation to the top of a building from a point 200 feet from ... Find the height of the building,"* and such methods are hardly new (Figure 1). Here we describe a trigonometric method that not only measures heights of points on a building, but widths and depths of those points; it gives the measurements as x, y, and z coordinates with respect to a specified origin.

This method was developed for researchers in the history of architecture to measure Medieval and Renaissance structures in Italy. To measure a building, a historian would be most likely to use a tape measure from scaffolding set up for that purpose, a direct but costly and laborious method.

For this method, the equipment needed is a surveyor's tape, a plumb bob, and a theodolite. The theodolite is an instrument used for measuring horizontal and vertical angles; it consists of a bubble level to establish the horizontal and vertical and a telescope that can rotate vertically in a mount that can also turn horizontally, with precise scales for reading angles.

To get the depth dimension of a selected point on a façade, this procedure requires two theodolite setup positions, with a set of readings taken from each location giving *two values* for each dimension (six figures for each point). The pairs of x coordinates and of z coordinates are not independent, and can serve as a check on the calculation. The two y coordinates are independent and can be averaged to give a final value.

This procedure does *not* require the theodolite to be at the same height at each position, so is suitable for sighting from sloping ground. Further, it is not required that the two theodolite positions be at the same distance from the wall. Thus, this method will work with walls that are leaning out of plumb, have offsets, are curved, or have projecting elements like sills or cornices.

A literature search revealed few references to

Figure 1: Surveying exercise, from Fillipo Callandri, *De arithmetica* (Florence, 1491).

trigonometric methods for the indirect measurement of buildings. Martin Kemp, when talking about Fillipo Brunelleschi, the architect of the cupola of the Cathedral in Florence, says "On his first visit to Rome, as described in his biography, he made measured drawings of Roman buildings, using his understanding of standard surveying techniques 'to plot the elevations', using measurements 'from base to base' and simple calculations based on triangulation. The basis for such procedures would have been the 'abacus mathematics' he learnt as a boy."[1] His source for this information is Antonio Manetti's *Life of Brunelleschi.*[2] A search of Manetti's biogra-

[1] Martin Kemp, *The Science of Art*. New Haven: Yale University Press, 1990, p. 11.

[2] Antonio Manetti, *The Life of Brunelleschi.* Ed. H. Saal-

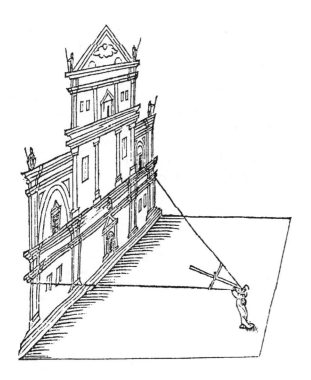

Figure 2: The *radio astronomico* used to measure the width of a façade.

Figure 3: An astrolabe used for surveying a building.

phy found reference to a visit to Rome, but no mention of his use of trigonometry to measure façades. In fact, there is some doubt expressed by the editor, Howard Saalman, that Brunelleschi ever went to Rome; this passage may have been added to enhance the stature of Manetti's subject.

Also according to Kemp,[3] Leonardo recorded in the Codice Atlantico a cross-shaped measuring staff which he called the *bacolo* of Euclid, which was used to establish similar triangles. This instrument was perfected in the sixteenth century as the *radio astronomico* by the geographer and astronomer Gemma Frisius, who commends it for terrestrial as well as astronomical measurements[4] (Figure 2). Kemp speaks of medieval instruments of considerable elaboration and precision, most notably quadrants and astrolabes, which could be used for terrestrial mensuration, although this was not their prime function. Cosimo Bartoli shows an astrolabe being used for the measurement of a building[5] (Figure 3). Raphael described a circular instrument used for

his survey of ancient Rome, at the center of which is a compass, and a peripheral scale with arms that carry sight-vanes.[6] A similar device called a *bussola* is described in Bartoli's book (Figure 4). The theodolite used for our present method is nothing but a more precise version of the bussola.

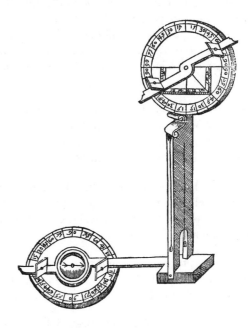

Figure 4: A compound *bussola* for taking horizontal and vertical bearings.

More recently, dimensioned drawings of façades have also been made by stereophotogrammetry, such as those for Independence Hall in Philadel-

---

man, Trans. C. Engass. London and Pennsylvania: Penn. State Press, 1970, pp. 152–3.

[3]Kemp, pages 168–170.

[4]Gemma Frisius, *De radio astronomico et geometrico liber*, Antwerp, 1545.

[5]Cosimo Bartoli, *Del Modo di misurare*, Venice, 1589.

[6]V. Golzio, *Raffaelo nei documenti e nelle testimonianze del suo secolo*, Vatican, 1936.

phia.[7] The Pantheon in Rome was surveyed by a method that used two electronic theodolites to simultaneously sight a point on the structure, their output being fed to a computer to give an instantaneous readout of coordinates.[8] This paper describes a low-tech alternative to these methods, suggested to me by architect Kim Williams.

2   The Method

1. Study the façade. Take photos, measure whatever can be easily reached by manual taping, and make a preliminary drawing. Choose and number the target points. Place adhesive targets on the wall, wherever possible.

2. Select or lay out a base line. The intersection of the façade and pavement makes a good base line if it is straight and horizontal. A stretched cord can be used as shown in Figure 5 if no suitable physical base line is available. This figure shows possibly the most difficult measuring situation, a curved building on sloping ground. Mark two theodolite setup points A and B on the ground or pavement, which can be at different heights and at different distances from the base line. Record their horizontal distance c apart and their horizontal distances d_A and d_B to the base line.

3. Set up and level the theodolite at location A. With the telescope horizontal, sight and mark a point T at any place on the wall that will also be visible from location B.

4. Set a plumb line over the other theodolite location. Sight the plumb line with the theodolite and adjust the horizontal scale of the theodolite to read zero.

5. Sight each target point and record the horizontal angle α and the vertical angle θ for each, as shown in Figure 6.

6. After all target points have been sighted, move the theodolite to the second location. With the telescope horizontal, sight a point R on the wall vertically in line with point T, found in step 3. Measure the vertical distance from that point to T.

[7]HABS/HAER Annual Report, 1992, p. 33.

[8]Marco Pelletti, *Note al Rlievo del Pantheon (Notes on the Survey of the Pantheon)*. Quaderni d'Instituto di Storia dell'Architettura, V. 13, 1989, pp. 10–18. Trans by Kim Williams.

7. Repeat steps 4 and 5 for all target points, recording the horizontal angle β and the vertical angle ϕ for each of them.

8. Enter all measurements into the computer spreadsheet and print out the x, y, and z coordinate of each target point.

9. Make a final dimensioned drawing by hand or by use of a CAD program.

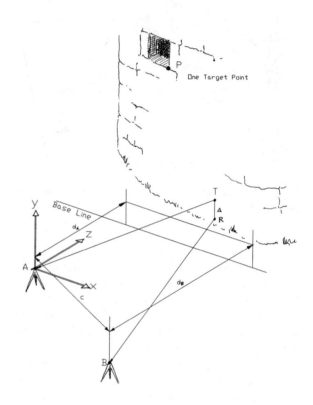

Figure 5: Base line and target points.

3   Derivation of Façade Equations

The equations that the spreadsheet uses to reduce the data are easily derived. The three original taped measurements are c, d_A, and d_B, shown in Figure 5, where c is the horizontal distance between theodolite locations and d_A and d_B are the horizontal perpendicular distances from the baseline to the theodolite locations. Let Δ be the vertical offset between theodolites that was measured in step 6. From these we get the following distances, which

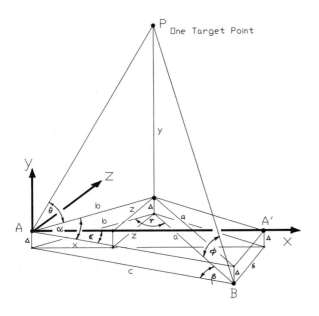

Figure 6: Locating a target point.

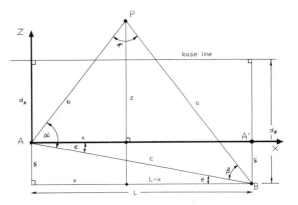

Figure 7: Plan view.

Referring to Figure 6,

$$\begin{aligned}
\gamma &= 180 - \alpha - \beta \\
a &= c\frac{\sin\alpha}{\sin\gamma} \\
b &= c\frac{\sin\beta}{\sin\gamma}.
\end{aligned}$$

From the plan view (Figure 7) we see that

$$\begin{aligned}
\cos(\alpha - \epsilon) &= \frac{x}{b} \\
x &= b\cos(\alpha - \epsilon).
\end{aligned}$$

From position B:

$$\begin{aligned}
\cos(\beta + \epsilon) &= \frac{L - x}{a} \\
x &= L - a\cos(\beta + \epsilon).
\end{aligned}$$

Note that the x-coordinate obtained from position B is not independent of that obtained from A, but is useful for checking the computation.

Next we find the y-coordinate of point P from the measurements at position A,

$$\begin{aligned}
\tan\theta &= y/b \\
y &= b\tan\theta
\end{aligned}$$

and from the measurements at position B:

$$\begin{aligned}
\tan\phi &= \frac{y + \Delta}{a} \\
y &= a\tan\phi - \Delta.
\end{aligned}$$

Here, the values of y found from each setup position *are* independent.

are shown in Figure 7:

$$\begin{aligned}
\delta &= \text{horizontal offset} = d_B - d_A \\
\epsilon &= \text{angular offset} = \arcsin\left(\frac{\delta}{c}\right) \\
L &= \text{distance } AA' \text{ between } A \text{ and } B \\
&\quad \text{parallel to the base line} \\
&= \sqrt{c^2 - \delta^2}.
\end{aligned}$$

For each target P (Figure 6) we have,

$$\begin{aligned}
\alpha &= \text{horizontal angle at } A \text{ from } B \text{ to target} \\
\theta &= \text{vertical angle at } A \text{ from horizontal} \\
&\quad \text{to target} \\
\beta &= \text{horizontal angle at } B \text{ from } A \text{ to target} \\
\phi &= \text{vertical angle at } B \text{ from horizontal} \\
&\quad \text{to target.}
\end{aligned}$$

Our coordinate axes are as shown in Figures 5, 6, and 7, with the origin at A, the x-axis parallel to the baseline and directed to the right, the y-axis vertical and directed upwards, and the z-axis perpendicular to the x- and y-axes and directed towards the building. A simple translation of axes will later place the origin at any selected point such as a corner of the building.

We now calculate the x-coordinate of point P.

Next we find the z-coordinate of point P, first from the measurements taken at position A,

$$\sin(\alpha - \epsilon) = \frac{z}{b}$$
$$z = b\sin(\alpha - \epsilon)$$

and then from the measurements taken at position B,

$$\sin(\beta + \epsilon) = \frac{z + \delta}{a}$$
$$z = a\sin(\beta + \epsilon) - \delta.$$

Just as with the calculation for x, these two values of z are not independent.

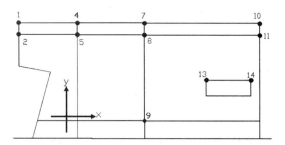

Figure 8: Green Academic Center.

4   Field Test at VTC

This method was tested by taking measurements of the front of Green Academic Center (Figure 8) at Vermont Technical College.[9] This figure shows the eleven target points, all visible from both theodolite locations, that were sighted. A Wild T2 theodolite, capable of a precision of about 0.2 seconds of arc, was used. The baselines were taped three times using a standard surveyor's tape graduated in millimeters, and the readings were averaged. Some of the distances measured by theodolite were also taped for comparison. The data were reduced using a Lotus 123 spreadsheet, giving the three coordinates of each target point in meters; see Table 1.

Note that a pair of values is given for each y-coordinate, corresponding to the two equations used for their calculation. These, of course, should be identical for each target point, and their difference gives us some measure of the precision of the method. For point 1, for example, the two values

---

[9]The measurements were made with the able assistance of Douglas Pennington of Vermont Technical College

Point	x	y		z
1	−3.231	7.310	7.337	15.091
2	−3.230	6.451	6.443	15.091
4	1.381	7.363	7.351	13.271
5	1.381	*6.423*	*6.153*	13.212
7	6.779	7.348	7.342	13.254
8	6.776	6.445	6.444	13.263
9	6.770	−0.275	−0.274	13.280
10	16.208	7.335	7.350	15.024
11	16.208	6.450	6.457	15.024
13	11.319	2.945	2.946	14.579
14	14.913	*3.042*	*2.942*	14.508

Table 1

of 7.310 m and 7.337 m have an average of 7.324 m. Each differs from this by 0.0135 m, or 0.18%. For the other points, the deviation from the average is also less than 0.2%. The figures shown in italic type are considered measurement errors because of the large difference between the y values and were discarded.

Next let us compare points that are expected to be at the same height on the building, at the same depth, or on the same vertical. For example, points 1, 4, 7, and 10 are at the top of the building, and should all have the same y-coordinate. The average y value for these four points is 7.324 m, with a maximum deviation from this value of 0.032 m. These deviations may represent inaccuracies in the measurements or actual differences in the heights of these points. Note that points 1 and 10 are at opposite corners of the building, nearly 20 meters apart, and that points 4 and 7 are on the offset portion of the façade. Other comparisons are given in Table 2.

Target Points	Average distance (meters)	Maximum deviation from average (meters)
Horizontal 1, 2	3.2305	0.0005 (0.015%)
Horizontal 4, 5	1.381	0
Horizontal 7, 8, 9	6.775	0.005 (0.07%)
Horizontal 10, 11	16.208	0
Vertical 1, 4, 7, 10	7.342	0.0032 (0.44%)
Vertical 2, 5, 8, 11	6.448	0.009 (0.14%)
Depth 1, 2, 10, 11	15.058	0.0034 (0.23%)
Depth 4, 7, 8, 9	13.269	0.0013 (0.10%)

Table 2

On the basis of this one test, it would appear that, with moderate care, accuracies within a few centimeters, or within 0.5%, are easily obtained.

There is no theoretical limit to the accuracy of the method.

With two theodolite setup positions, only one independent value is obtained for x and for z. In order to check these values a third setup C is recommended. The readings from C can be combined with both A and B to give three sets of coordinates, AB, AC, and BC. It is also advisable to tape as many dimensions as can be reached, as a good check on the computed values.

5 Measurements in Italy

The first real use of the method was on the western façade of the *Torre Bernarda*, a Medieval tower (Figure 9) in the town of Fucecchio, near Florence. An aerial view of the tower, located on the Rocca fiorentina (Figure 10) shows that the only available place to set up the instruments was on a patio which was lower than the base of the tower and offset to the south. The approximate positions of the setup points A and B are shown.

Figure 10: Aerial view of the *Rocca fiorentina* (Photo G. Pierozzi).

Since the terrain did not enable us to establish a base line parallel to the façade, a modified procedure was used. The line AB between the two setup points was taken as the baseline, so that the resulting origin of axes was at A, the x-axis was horizontal in the direction AB, the y-axis was vertical through A, and the z-axis was horizontal and perpendicular to the x-axis. The same equations used before apply here, but with the horizontal offset and the angular offset both equal to zero. Then by a simple translation of axes, the origin was placed at a convenient point on the façade, and a rotation of axes put the x-axis in the plane of the façade. Accuracies obtained were of the same order as for the VTC measurements.

In conclusion, we have here a fast, inexpensive, low-tech tool for measuring façades that is capable of giving accuracies with less than 1% error. For good measure, this method has its roots firmly planted in the history of architecture.

Figure 9: Torre Bernarda, Fucecchio.

A Secret of Ancient Geometry

Jay Kappraff
New Jersey Institute of Technology
Newark, New Jersey

1 Introduction

The architecture of antiquity has a sense of harmony and proportion rarely equaled even in the greatest works of other eras. The quality of the work of an architect or designer is determined by how he or she comes to grips with the mathematical constraints on space inherent in all designs—"what is possible," in contrast with the designer's intention, "what ought to be." The history of architecture reflects the history of ideas in that "what ought to be" has changed from metaphysical "natural" world views to explorations of the individual "artist." Additionally, the history of technology is reflected in changes of "what is possible."

There are two kinds of constraints on space that the architect or designer must confront:

> constraints imposed on a design because of the geometrical properties of space;

> constraints imposed on a design by the designer who creates a geometrical foundation or scaffolding as an overlay to the design. The designer's choice is based on the context of the design and on the effect that he or she wishes to achieve.

Without constraints, a design is chaotic, irrelevant and lacking in focus. Where do the designer's constraints come from? In ancient times they were derived either from spiritual contexts or handed down from generation to generation by tradition. The results were cathedrals such as Chartres and Hagia Sophia or structures such as the Egyptian Pyramids and the Great Temple of Jerusalem or the temples of ancient Greece such as the Parthenon, the Theseum, and the temples of Poseidon and Ceres.

Modern architecture has replaced spiritual—and tradition-bound—contexts with the private vision of the designer or architect and substituted diversity for tradition. However, the designer is left with few tools to deal with such a lack of constraint. After all, what should the designer do when each de-

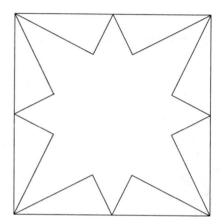

Figure 1: The Brunes star.

sign breaks new ground? In an effort to recover the principles of ancient architecture, many researchers have studied the geometric and spiritual bases of ancient structures [7, 16, 17, 19].

This paper will discuss the work of Tons Brunes, a Danish engineer, who hypothesized a system of ancient geometry that he believed lay at the basis of many of the temples of antiquity [4]. It was Brunes's belief that there existed until about 1400, a network of temples and a brotherhood of priests originating in ancient Egypt which had a secret system of geometry. At the basis of Brunes's theory is the eight-pointed star illustrated in Figure 1. Brunes claimed to have seen this star in ancient temples (but gave no references) where he stated that it was often mistaken for ornamentation. From the geometry of this star he was able to reconstruct reasonably close facsimiles to the plans and elevations of the ruins of ancient temples such as the Pantheon, the Theseum, and the Temples of Ceres and Poseidon, noting that certain intersections coincide with features of these temples [10]. Unfortunately, although the examples he uses to illustrate his theories are cleverly rendered, there is no historical record to support his claims. As a result his research has met considerable skepticism. Nevertheless, as we shall see, the Brunes star reveals a geometry consistent with ancient architecture, folk art, and the musical scale. Even though it is unlikely to have played the all-pervasive role for temple construction that Brunes conjectured, it may well have been one of the organizing tools. At any rate, the beauty of its geometry is reason enough to study it.

2 The Concept of Measure in Ancient Architecture

One thing that we can say for sure about the thought processes of antiquity is that they differed markedly from our own. Until Aristotle introduced observation and measurement as the only way to arrive at truth, it appears that reality was best described by numbers, music, and poetry.

R. A. Schwaller di Lubicz [5] felt that the combination of myth and symbol conveyed by ancient writings was the only way information about the workings of the universe could be conveyed without reducing its true meaning. According to Di Lubicz [5] the ancient Egyptians felt that:

> Measure was an expression of Knowledge that is to say that measure has for them a universal meaning linking the things of here below with things Above and not solely an immediate practical meaning—quantity is unstable: only function has a value durable enough to serve as a basis [for description]. Thus the Egyptians' unit of measurement was always variable—measure and proportions were adapted to the purpose and the symbolic meaning of the idea to be expressed. [For example] the cubit will not necessarily be the same from one temple to another, since these temples are in different places and their purposes are different.

Before a standard unit of measure can be introduced, there must be a sophisticated means of transportation in order for its users to be able to travel to a central location to retrieve the standard measure for their own purposes.

In societies without access to standard measures, other methods were developed to enable the craftsman to build or the architect to create structures without need of standard measure. Even when measuring rods were available, they may have been used only as an adjunct to the use of pure geometry in the design of sacred structures. In place of numbers to describe a measurement, a kind of applied geometry was developed in which lengths were constructed without the need to measure them. All that was needed was a length of rope and a straightedge (the equivalent of our compass and straightedge). Methods were then devised to subdivide any length into sublengths, always by construction. Evidence of construction lines have been discovered on the base of the unfinished Temple of Sardis in Turkey and also in the courtyard of the Temple of Zeus of Jerash in Jordan [18]. Artmann [2] has shown how such methods were used to construct the Gothic cathedrals. The geometry needed to build these cathedrals was learned from boiled-down versions of the first books of Euclid, known as pseudo-Boethius [3] which highlighted the constructive methods while eliminating the proofs of the theorems. The knowledge to implement this geometry was taught to the guilds of masons, other artisans, and builders and then passed on from generation to generation by oral tradition. One can imagine easily learned constructive techniques based on the Brunes star being transmitted by this tradition and applied to the construction of ancient sacred structures. We shall describe Brunes's hypothetical reconstruction of this geometry.

3 The Ancient Geometry of Tons Brunes

In ancient times it was an important problem to find a way to create a square or rectangle with the same area or circumference as a given circle—"squaring the circle," as it was known. Since the circle symbolized the celestial sphere while a square or rectangle oriented with its sides perpendicular to the compass directions of north, east, south, and west symbolized the Earth, the squaring of the circle could be thought to symbolically bring "heaven down to earth." Brunes demonstrates one way in which ancient geometers may have attempted to solve this problem using only compass and straightedge (we now know that this cannot be done exactly). To square the circle with respect to circumference Brunes first considers a geometric construction which he refers to as a "sacred cut."

To create the sacred cut of a side of a unit square, place the point of your compass at a vertex and draw an arc through the center of the square as shown in Figure 2. This cuts the side down by a factor of $1/\sqrt{2}$. In Figure 3 arc AB and the diagonal CD of the half square are approximately equal. In

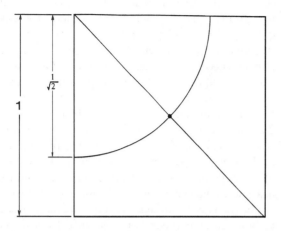

Figure 2: The sacred cut.

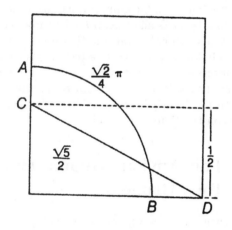

Figure 3: Comparison of lengths.

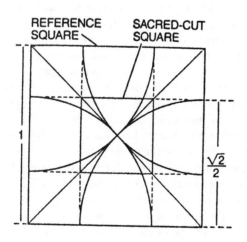

Figure 4: Four sacred cuts within a square.

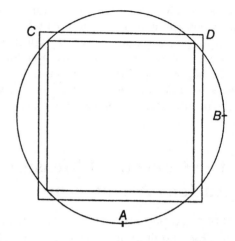

Figure 5: The circumference of the circle is approximately equal to the perimeter of the outer square.

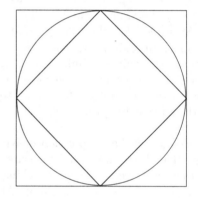

Figure 6: The ad quadratum square.

fact,

$$AB = \pi\sqrt{2}/4 = 1.1107\ldots \qquad \text{while}$$
$$CD = \sqrt{5}/2 = 1.1118\ldots.$$

In Figure 4 four sacred cuts AB are placed into a square. In Figure 5 the four sacred cuts form a circle whose circumference is equal to the perimeter of a square with edge CD to within 1.6%.

In Figure 6, we see that a circle is drawn that is tangent to an outer square (inscribed circle) and touching the vertices of an inner square (circumscribed circle). This square-within-a-square, called an "ad quadratum" square, was much used in ancient geometry and architecture [19]. The area of the inner square is obviously half the area of the outer square. In a sequence of circles and squares inscribed within each other each square is 1/2 the area of the preceding square. Figure 7 shows a sequence of ad quadratum squares which are shaded to form a logarithmic spiral known as the Baravelle

Figure 7: The Baravelle spiral.

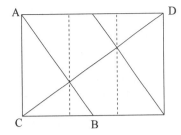

Figure 9: Trisection of the diagonal of a rectangle.

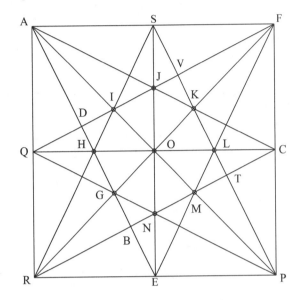

Figure 10: Construction of the Brunes star.

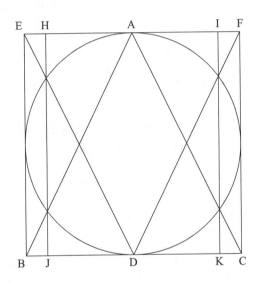

Figure 8: Approximate squaring of the circle.

the circle equals 1/2 and

$$\text{Area of circle} = \pi(1/2)^2 = .7853\ldots$$
$$\text{Area of rectangle} = 4/5 = .80,$$

an error of 1.8%.

In Figure 9 we show that for an arbitrary rectangle the line AB from a vertex to the center of the opposite side cuts the diagonal CD at the 1/3 point. We now use this geometrical property to describe the structure of the Brunes star of Figure 1. Take the circumscribing square and subdivide it by placing perpendicular axes within it, as shown in Figure 10. This divides the outer square into four overlapping rectangular half-squares. Place two diagonals into each of the four half squares and add the two diagonals of the outer square. Notice that the resulting diagram (shown in Figure 10) is the Brunes star.

We now see that this star contains all the information needed to get good approximations to squaring the circle in both circumference and area. Also hidden within the Brunes star are numerous

spiral. It is easy to construct and with color makes an interesting design.

Another geometric structure used by ancient geometers was the upward-pointing triangle ABC in Figure 8 which also has half the area of the circumscribing square $BCFE$. If the downward-pointing triangle DEF is constructed, then rectangle $JKIH$, formed by the vertical lines through the intersection points of the upward- and downward-pointing triangles and the circle, has approximately the same area as the circle. It can be determined (not shown here) that the width of this rectangle is 4/5 of the diameter of the circle. Taking the square to have length equal to 1 unit, then the radius of

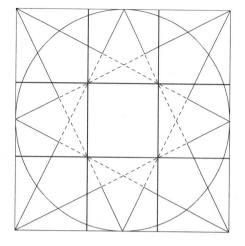

Figure 11: The 3×3 grid of subsquares determined by the Brunes star.

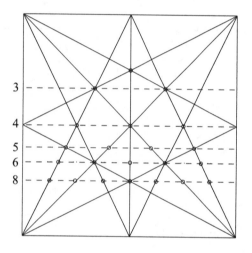

Figure 12: Division of a line into 3, 4, 5, 6, or 8 equal parts.

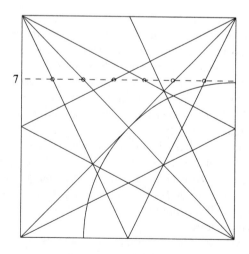

Figure 13: Division of a line into seven approximately equal parts.

3,4,5-right triangles. For example, triangle ABC is a 3,4,5-right triangle because,

$$\tan \frac{1}{2}C = \frac{AQ}{QC} = \frac{1}{2}.$$

Therefore using the trigonometry identity,

$$\tan C = \frac{2 \tan \frac{1}{2}C}{(1 - \tan^2 \frac{1}{2}C)},$$

it follows that,

$$\frac{AB}{BC} = \tan C = \frac{1}{(1 - \frac{1}{4})} = \frac{4}{3}. \tag{1}$$

If the Brunes star with all of its construction lines depicted in Figure 10 is placed on each face of a cube, it can be shown that the vertices of all of the six Archimedean solids and two Platonic solids (cube and octahedron) related to the cubic system of symmetry as well as the tetrahedron coincide with the points of intersection of the construction lines [10]. The Brunes star also succeeds in providing the geometrical basis for dividing an arbitrary length into any number of equal sublengths without the use of measure.

4   Equidivision of Lengths: A Study in Perspective

Figure 10 contains the construction points with which to subdivide lengths into 3 and 4 equal parts without the need of a standard measure, i.e., points I and M divide diagonal AP into thirds (see Figure 9) while H, O, and L divide QC in quarters. The central cross of Figure 10 is therefore subdivided by the central irregular octagon $GHIJKLMN$ into four equal parts and the diagonals into three equal parts in a similar way. Points I, K, G, and M then provide the points that subdivide the outer square into a 3×3 grid of subsquares, as shown in Figure 11. Figure 12 indicates how the Brunes star divides a line segment into 3, 4, 5, and 8 equal parts, while a sacred cut drawn from a vertex of the outer square in Figure 13 defines the level that partitions a line into seven parts which are approximately the same; the error is within 2%. In similar ways, Brunes has shown that the Brunes star can be used to equipartition a line into between 1 and 12 parts in a manner which does not require a standard measure, but only a length of stretched rope.

This equipartitioning property of the Brunes star has its roots in another ancient geometric construction [15] which was first related to me by Michael

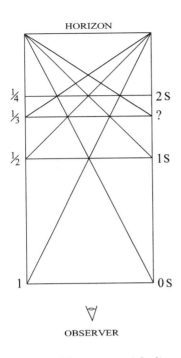

Figure 14: Double square with diagonals.

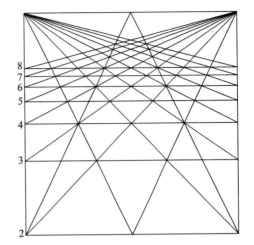

Figure 15: Square with diagonals.

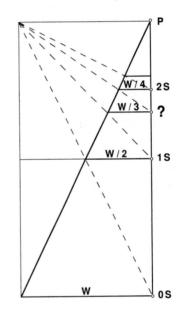

Figure 16: Railroad tracks in perspective. The point P is the central vanishing point.

Porter, a Professor of Architecture at Pratt Institute. In Figure 14 the outer square of the Brunes star has been extended to a double square. The principal diagonals of the double square divide the width of the upper square into two equal parts. The principal diagonals intersect the two diagonals of the upper square at the trisection points of the width. At the same time, the trisected width intersects the long side of the double square at the 1/3 point. Continuing one more step, the two diagonals of the 1/3-rectangle intersect the principal diagonal at points which divide the width into four equal parts. This width also divides the long side of the double square at the 1/4 point. This construction may be continued to subdivide a line segment into any number of equal parts, as in Figure 15 with eight subdivisions.

As is often the case with mathematics, a diagram set up to demonstrate one concept is shown to have a deeper structure. We could also view Figures 14 and 15 as a pair of railroad tracks in perspective receding to the horizon line. The diagonal and the right side of the double square play the role of the railroad tracks as shown in Figure 16. If the observer is at an arbitrary location in the foreground, then the distance between the tracks appears half as great as at the base of the double square at some measured distance in the direction of the horizon referred to as a "standard distance," or $1S$. At a distance from the observer of $2S$ the distance be-

tween the tracks appears to be 1/4 as large as the base width. In a similar manner, the tracks appear to be 1/8 as wide at $3S$ (not shown). How many standard units S make the tracks appear 1/3 as wide? To answer this question requires us to analyze the pattern in greater depth. Table 1 shows the relation between apparent width between the railroad tracks L and the receding distance D (in units of S) towards the horizon. The receding distance is also expressed in terms of logarithms to the base 2. In other words, the relation between D and L in Table 1 can be expressed by the formula:

$$D \text{ (in units of } S) = \log_2 1/L.$$

It is clear from Table 1 that for $L = 1/3$, $1/L = 3$ and

$$? = \log_2 3 = \log_{10} 3 / \log_{10} 2 = 1.58S.$$

The author has further examined the projective transformation that gives rise to Figures 14, 15, and 16 and has related it to the series of overtones resulting from plucking the string of a monochord or other stringed instrument [14].

Apparent width (L)	Receding distance (D)
$1 = 1/2^0$	$0 = \log_2 1$
$1/2 = 1/2^1$	$1 = \log_2 2$
$1/3 = 1/2^?$	$? = \log_2 3$
$1/4 = 1/2^2$	$2 = \log_2 4$
$1/8 = 1/2^3$	$3 = \log_2 8$

Table 1

5 The 3,4,5-Triangle in Sacred Geometry and Architecture

We showed in Section 3 that triangle ABC in Figure 10 is a 3,4,5-right triangle. The 3,4,5-right triangle was called the Egyptian triangle by Vitruvius, the architect of the Emperor Augustus, and had great significance in the construction of the pyramid of Cheops [9, 17]. In cap. 56, Plutarch [17] described this triangle as the symbol of the Egyptian trinity, associated with the three significant Egyptian deities:

$$3 \iff \text{Osiris}$$
$$4 \iff \text{Isis}$$
$$5 \iff \text{Horus}$$

The key to understanding the geometry of the Brunes diagram lies in its construction. But how did ancient architects construct the star diagram? This diagram is easy to construct if one begins with a square, but it is not an easy matter to construct a large square if one has only a length of rope and some stakes to work with. However the entire diagram can equally well be constructed beginning with the 3,4,5-right triangle. The 3,4,5-right triangle can be constructed from a loop of rope with 12 knots, as shown in Figure 18. The 12 sectors of the circle shown in Figure 17 could also have represented the 12 regions of the zodiac visited by the sun during the course of the year, as viewed from a geocentric standpoint.

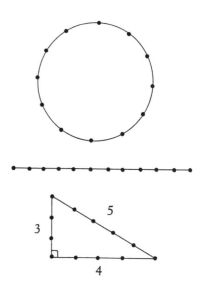

Figure 17: Ropes.

Figure 18: Rope used to create the Brunes star in Figure 19.

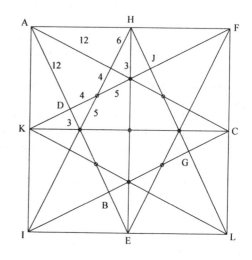

Figure 19: Segment lengths of the Brunes star.

I have created a videotape of a group of students constructing this star on an open field using four lengths of 50-foot clothesline anchored by camping stakes [11]. To construct the Brunes star begin with four lengths of rope each length divided into 12 equal sections by 12 knots as shown in Figure 17. Although the rope is shown stretched out in a line, the ends are connected so that it forms a loop. Four such loops—$ADBGCA$ (see Figure 19), $FJDBEF$, $IBGJHI$, and $LGJDKL$—are stretched into four 3,4,5-right triangles, each providing one vertex of the outer square of Figure 10. The right angles of these 3,4,5-triangles are located at the vertices of the inner square $DBGJ$. We have succeeded in constructing the outer square $AILF$ along with the midpoints of its sides $HKEC$. Now that the outer square has been formed, we can stand back and observe the harmony of this figure. In order to better appreciate its geometry, we must make a brief digression and consider the geometry of the 3,4,5-right triangle. From Equation 1 it follows that triangle ABC is a 3,4,5-right triangle. All other right triangles in Figure 10 are either 3,4,5-right triangles or fragments of a 3,4,5-triangle obtained by bisecting its acute angles. In Figure 19 the dimensions of the sublengths are indicated. These may be gotten from Figure 10 by assigning each segment of the string a length of 6 units. The properties of 3,4,5-triangles given by Equation 1 can also be used to verify these lengths. Figure 19 shows the star diagram to have 3,4,5-right triangles at four different scales. Referring to vertex labels of Figure 10,

$$
\begin{array}{ccccccc}
\triangle ABC & : & 18 & : & 24 & : & 30 \\
\triangle ADJ & : & 9 & : & 12 & : & 15 \\
\triangle QDG & : & 6 & : & 8 & : & 10 \\
\triangle DHI & : & 3 & : & 4 & : & 5
\end{array}
$$

So we see that the star diagram is entirely harmonized by the 3,4,5-right triangle.

As we previously mentioned, Brunes used these principles of geometry to show how many of the structures of antiquity might have been proportioned [10]. He subsumed the principles of this geometry into a series of 21 diagrams (not shown) related to the star diagram and the sacred cut [11]. He claims that each step in the creation of a plan for one of the ancient structures follows one or another of these diagrams. We illustrate the result of Brunes's analysis for the Temple of Ceres by the gulf of Salerno in Southern Italy built by Greek colonists during the period from 550–450 B.C. Brunes has reconstructed his analysis from the ruins of this temple. Although Brunes obtained close fits between key lines of the elevation and plan (not shown) of these structures, his constructions require an initial "reference circle" the choice of which is quite arbitrary as shown in Figures 21 and 22. Despite the close fits between Brunes's diagrams and the actual temple, one never knows the degree to which

Figure 20: The Temple of Ceres.

Figure 21: Analysis of the Temple of Ceres by Brunes using the sacred cut.

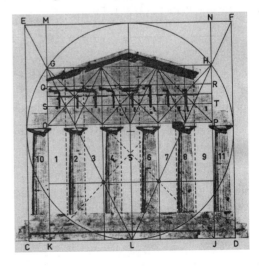

Figure 22: Analysis of the Temple of Ceres by Brunes using triangles.

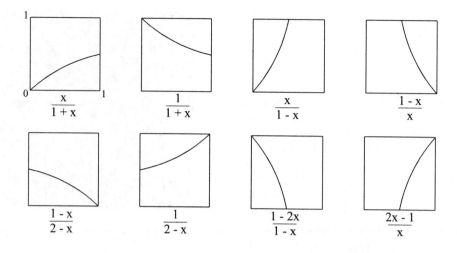

Figure 23: Line segments used to generate the generalized Brunes star.

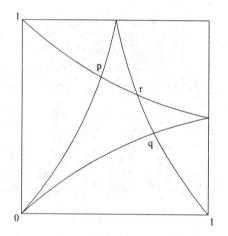

Figure 24: Intersection points: $p = (1/\theta, 1/\sqrt{2})$, $q = (1/\sqrt{2}, 1/\theta)$, $r = (1/\tau, 1/\tau)$.

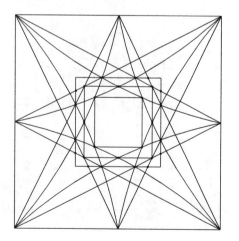

Figure 25: The generalized Brunes star.

they have been forced by his imagination. In my

opinion, it is unlikely that this method was actually used as described by Brunes. Nevertheless, the simplicity and harmony of Brunes's diagrams make it plausible that they could have been used in some unspecified manner as a tool for temple design.

6   A Generalized Brunes Star

Gary Adamson [1] has generalized the Brunes star by replacing the eight line segments that make up the diagonals of the four half-squares by a segment of an hyperbola juxtaposed in eight different orientations within a unit square, as shown in Figure 23. Four of these hyperbolas intersect as shown in Figure 24 at three characteristic points p, q, r with coordinates:

$$p = (.414..., .707...) = \left(\frac{1}{\theta}, \frac{1}{\sqrt{2}}\right) \text{ where } \theta = 1 + \sqrt{2}$$

$$q = (.707..., .414...) = \left(\frac{1}{\sqrt{2}}, \frac{1}{\theta}\right)$$

$$r = (.618..., .618...) = \left(\frac{1}{\tau}, \frac{1}{\tau}\right) \text{ where } \tau = (1 + \sqrt{5})/2$$

Therefore, the key numbers of the ancient Roman system of proportions $\sqrt{2}$ and θ (also referred to in modern dynamical systems theory as the "silver mean") [13], and the golden mean τ are represented in a single diagram. The generalized Brunes star is shown in Figure 25. The points of intersection lie on the edges of the three inner squares. The edge length of the innermost square is τ^{-3}, the middle square is $1/3$, and the outer square is $\theta^{-1} = \sqrt{2} - 1$.

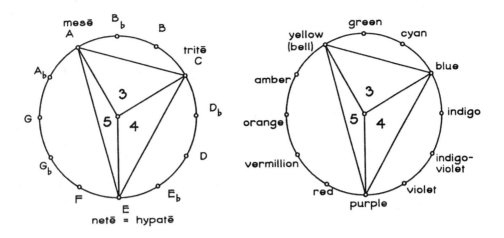

Figure 26: The equal-tempered chromatic scale and the color wheel.

7 What Pleases the Ear Should Please the Eye

We have seen that 3,4,5-triangles pervade the Brunes star. Not all relationships involving the numbers 3, 4, and 5 refer to the 3,4,5-right triangle. Such relationships play a major role in the structure of the musical scale and make a surprise appearance in the structure of the color spectrum of light which could be thought of as a kind of "musical scale" for the eye. Elsewhere I have shown that the ancient musical scale, in which tones are associated with the ratio of string lengths, is organized by a 3,4,5-relationship between the tones [14]. Also, if we regard the 12 sectors of the circle as tones of the equal-tempered chromatic scale, we see in Figure 26 that a subdivision of the tonal circle into 3, 4, and 5 semitones gives rise to the tones A,C,E of the musical A minor triad [6]. The association between tones and number ratios led the architects of the Italian Renaissance to build a system of architectural proportions based on the musical scale [14].

According to the leading architect of that period, Leon Battista Alberti [20]: "The numbers by which the agreement of sounds affect our ears with delight are the very same which please our eyes and our minds. We shall therefore borrow all our rules for harmonic relations from the musicians to whom this kind of numbers is well known and wherein Nature shows herself most excellent and complete." Eberhart [6] has made the observation that the wavelengths of visible light occur over a range between 380 mμ (millimicrons; 1 m$\mu = 10^{-7}$cm) in the ultraviolet range to about twice that amount in the

infrared, or a visual "octave." He states,

When the colors of visible light are spread out in such a way that equal differences in wavelength take equal amounts of space, it stands out that blue and yellow occupy relatively narrow bands while violet, green, and red are broad [see Figure 27]. Observe that the distance from the ultraviolet threshold to blue to yellow to the infrared threshold is very closely 4 : 3 : 5 of that spectral "octave", i.e., $383.333\ldots \times 2^{4/12} = 483$ mμ (mid blue) and $383.333\ldots \times 2^{7/12} = 574.333\ldots$ mμ (mid yellow). This means that if we subjectively identify the two thresholds of ultraviolet and infrared, as is commonly done in making color wheels, calling both extremes simply "purple," then the narrow bands of "blue" and "yellow" have approximate centers lying at corners of the same triangle with 3,4,5 proportions as the A minor triad [see Figure 26].

Eberhart's observation adds some additional substance to the Renaissance credo that what pleases the ear also pleases the eye.

8 Conclusion

According to Plato, the nature of things and the structure of the universe lay in the study of music, astronomy, geometry and numbers, the so-called quadrivium. Built into sacred structures would be not only a coherent geometrical order but also a sense of the cosmic order in terms of the cycles of

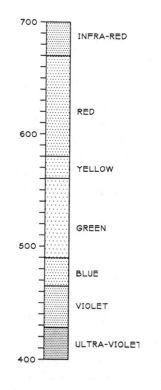

Figure 27: Color Scale.

the sun and the moon and the harmonies of the musical scale. The Brunes star with its ability to approximately square the circle, its equipartioning properties, its relationship to 3,4,5-triangles, and its ability to be generalized to a geometrical figure exhibiting the golden and silver mean makes it a plausible tool for use of the builders of ancient sacred structures.

References

[1] G. Adamson, 1994. Private communication.

[2] B. Artmann. The cloisters of Hauterive. *The Mathematical Intelligencer*, 13(2):44–49, 1991.

[3] Boethius. *"Boethius" Geometrie II*. F. Steiner, Weisbaden, 1970. Translated by Menso Folkerts.

[4] T. Brunes. *The Secrets of Ancient Geometry— And Its Uses*. Rhodos, Copenhagen, 1967.

[5] R. A. Schwaller di Lubicz. *The Temple of Man*. Inner Traditions Press, 1996. Translated by R. and D. Lawlor.

[6] S. Eberhart. *Mathematics Through the Liberal Arts: A Human Approach*. U. of Calif. at Northridge, 1994. Lecture notes.

[7] M. Ghyka. *The Geometry of Art and Life*. Dover, New York, 1978.

[8] J. Hambridge. *The Fundamental Principles of Dynamic Symmetry as They are Expressed in Nature and Art*. Gloucester Art, Alberquerque, 1979.

[9] J. Kappraff. *Connections: The Geometric Bridge between Art and Science*. McGraw-Hill, New York, 1991.

[10] J. Kappraff. Secrets of ancient geometry: An introduction to the geometry of Tons Brunes. Unpublished monograph, 1992.

[11] J. Kappraff. *The Mathematics of Design*. A set of 11 videotapes produced by the Center for Distance Learning and the Media Center of NJIT, 1993. Videotape 4.

[12] J. Kappraff. Linking the musical proportions of Renaissance, the Modulor, and Roman systems of proportions. *Space Structures*, 11(1 and 2), 1996.

[13] J. Kappraff. Musical proportions at the basis of systems of architectural proportions. In K. Williams, editor, *Nexus '96*. Edizioni Dell' Erba, Fuccechio, 1996.

[14] J. Kappraff. *Mathematics beyond Measure: A Random Walk through Nature, Myth, and Number*. Plenum Press, New York, 1998.

[15] H. Kayser. *Harmonical Studies: Harmonical analysis of a proportion study in a sketchbook of the medieval Gothic architect*. Villard de Honnecourt, 1946. Its title is "Ein Harmonikaler Teilungskanon" (A Harmonical Division-Canon).

[16] A. Tyng. Geometric extensions of consciousness. *Zodiac*, 19, 1969.

[17] H. F. Verheyen. The icosahedral design of the Great Pyramid. In I. Hargittai, editor, *Fivefold Symmetry*. World Scientific, Singapore, 1992.

[18] D. J. Watts. Private communication.

[19] D. J. Watts and C. Watts. A Roman apartment complex. *Scientific American*, 255(6):132–140, December 1986.

[20] R. Wittkower. *Architecture in the Age of Humanism*. Norton, New York, 1971.

Part 2

Vedic Civilization

Square Roots in the Śulba Sūtras

Dedicated to Sri Chandrasekharendra Sarasvati who died in his hundredth year while this paper was being written.

David W. Henderson*
Department of Mathematics, Cornell University
Ithaca, NY, 14853–7901, USA
dwh2@cornell.edu

Abstract

In this paper I will present a method for finding the numerical value of square roots that was inspired by the *Śulba Sūtras* which are Sanskrit texts written by the Vedic Hindu scholars before 600 BC. This method works for many numbers. It will produce values to any desired degree of accuracy and is more efficient (in the sense of requiring fewer calculations for the same accuracy) than the divide-and-average method commonly taught today.

1 Introduction

Several Sanskrit texts collectively called the *Śulba Sūtras* (see [1, 4, 7]) were written by the Vedic Hindus starting before 600 BC and are thought (see, for example, [6]) to be compilations of oral wisdom which may go back to 2000 BC. These texts have prescriptions for building fire altars, or *Agni*. However, contained in the *Śulba Sūtras* are sections which constitute a geometry textbook detailing the geometry necessary for designing and constructing the altars. As far as I have been able to determine these are the oldest geometry (or even mathematics) textbooks in existence. It is apparently the oldest applied geometry text. It was known in the *Śulba Sūtras* (for example, Sūtra i.52 of Baudhāyana's *Śulba Sūtras*) that the diagonal of a square is the side of another square with two times the area of the first square, as we can see in Figure 1.

Thus, if we consider the side of the original square to be one unit, then the diagonal is the side (or root)

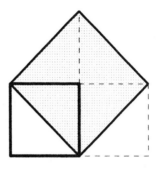

Figure 1: Dvi-karaṇī.

of a square of area two, or simply the square root of 2, that is $\sqrt{2}$. The Sanskrit word for this length is *dvi-karaṇī* or, literally, "that which produces 2". The *Śulba Sūtras*[1] contain the following prescription for finding the length of the diagonal of a square:

> Increase the length [of the side] by its third and this third by its own fourth less the thirty-fourth part of that fourth. The increased length is a small amount in excess (*saviśeṣa*).[2]

Thus the above passage from the *Śulba Sūtras* gives the approximation:

$$\sqrt{2} \approx 1 + \frac{1}{3} + \frac{1}{4} \cdot \frac{1}{3} - \frac{1}{34} \cdot \frac{1}{4} \cdot \frac{1}{3}.$$

I use $\approx$ instead of $=$, indicating that the Vedic Hindus were aware that the length they prescribed is a little too long (saviśeṣa). In fact my calculator

---

*This article grew out of researches which were started during my January, 1990, visit to the Sankaracharya Mutt in Konchipuram, Tamilnadu, India, where I was given access to the Mutt's library. I thank Sri Chandrasekharendra Sarasvati, the Sankaracharya, and all the people of the Mutt for their generous hospitality, inspiration, and blessings.

[1] *Baudhāyana Śulba Sūtras*, i.61–2. *Āpastamba Śulba Sūtras*, i.6. *Kātyāyana Śulba Sūtras*, II.13.

[2] This last sentence is translated by some authors as "The increased length is called saviśeṣa". I follow the translation of "saviśeṣa" given in [1, pp. 196–202]; see also [3] where the word is translated as "a special quantity in excess".

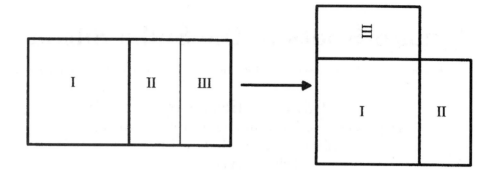

Figure 2: A rectangle as the difference of two squares.

gives

$$\sqrt{2} \approx 1.4142135\ldots$$

and the *Śulba Sūtras* value expressed in decimals is

$$\sqrt{2} \approx 1.4142156\ldots.$$

So the question arises—how did the Vedic Hindus obtain such an accurate numerical value? Unfortunately, there is nothing that survives which records how they arrived at this saviśeṣa. There have been several speculations (see [1, pp. 196–202] and [3, pp. 234–236]) as to how this value was obtained, but no one as far as I can determine has noticed that there is a step-by-step method (based on geometric techniques in the *Śulba Sūtras*) that will not only obtain the approximation

$$\sqrt{2} \approx 1 + \frac{1}{3} + \frac{1}{4}\cdot\frac{1}{3} - \frac{1}{34}\cdot\frac{1}{4}\cdot\frac{1}{3},$$

but can also be continued indefinitely to obtain as accurate an approximation as one wishes. This method will in one more step obtain

$$\sqrt{2} \approx 1 + \frac{1}{3} + \frac{1}{4}\cdot\frac{1}{3} - \frac{1}{34}\cdot\frac{1}{4}\cdot\frac{1}{3} - \frac{1}{1154}\cdot\frac{1}{34}\cdot\frac{1}{4}\cdot\frac{1}{3},$$

where the only numerical computation needed is $1154 = 2[(34)(17) - 1]$ and, moreover, the method shows that the square of this approximation is less than 2 by exactly

$$\frac{1}{(1154\cdot 34\cdot 4\cdot 3)^2} = \frac{1}{221,682,772,224}.$$

The interested reader can check that this approximation is accurate to eleven decimal places. The object of the remainder of this paper is a discussion of this method and related topics from the *Śulba Sūtras*.

2   Bricks and Units of Length

In the *Śulba Sūtras* the *agni* are described as being constructed of bricks of various sizes. Mentioned often are square bricks of side 1 *prādeśa* (span of a hand) on a side. Each *prādeśa* was equal to 12 *aṅgula* (finger width) and one *aṅgula* was equal to 34 sesame seeds laid together with their broadest faces touching [4, Śulba Sūtras, i.3–7]. The estimate given there of 2/3 inch for an *aṅgula* results in an approximation of about 8 inches for the span of a hand. Accordingly, the diagonal of a *prādeśa* brick had length:

1 prādeśa+4 aṅgula+1 aṅgula−1 sesame thickness.

I do not believe it is purely by chance that these units come out this nicely. Notice that this length is too large by roughly one-thousandth of the thickness of a sesame seed. Presumably there was no need for more accuracy in the building of altars!

3   Dissecting Rectangles and $A^2 + B^2 = C^2$

None of the surviving *Śulba Sūtras* tell how they found the saviśeṣa. However, in Baudhāyana's *Śulba Sūtras* the description of the saviśeṣa is the content of Sūtras i.61–62 and in Sūtra i.52 he gives the constructions depicted in Figure 1. Moreover in Sūtra i.54 he gives a method for constructing geometrically the square which has the same area as any given rectangle. If N is any number then a rectangle of sides N and 1 has the same area as a square with side equal to the square root of N. Thus Sūtra i.54 gives a construction of the square root of N as a length. So let us see if this hints at a method for

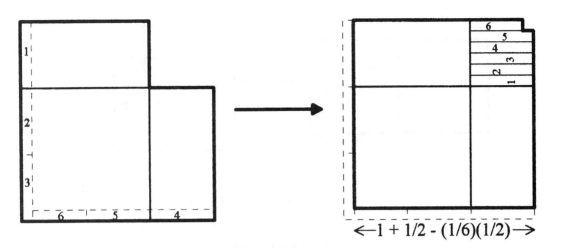

Figure 3: Second stage in construction of saviśeṣa.

finding numerical approximations of square roots. The first step of Baudhāyana's geometric process (Sūtra i.54) is:

> If you wish to turn a rectangle into a square, take the shorter side of the rectangle for the side of a square, divide the remainder into two parts and, inverting, join those two parts to two sides of the square.

See Figure 2. This process changes the rectangle into a figure with the same area which is a large square with a small square cut out of its corner.

In Sūtra i.51 Baudhāyana had previously shown how to construct a square which has the same area as the difference of two squares. In addition, Sūtra i.50 describes how to construct a square which is equal to the sum of two squares. Sūtras i.50, i.51, and i.52 are related directly to Sūtra i.48 which states:

> The diagonal of a rectangle produces by itself both the areas which the two sides of the rectangle produce separately.

This Sūtra i.48 is a clear statement of what was later to be called the "Pythagorean Theorem" (Pythagoras lived about 500 BC). In addition, Baudhāyana lists the following examples of integral sides and diagonal for rectangles (what we now call "Pythagorean Triples"):

$$(3,4,5), \ (5,12,13), \ (7,24,25), \ (8,15,17),$$
$$(9,12,15), \ (12,35,37), \ (15,36,39)$$

which the Śulba Sūtras used in its various methods for constructing right angles. (For further discussions of the geometric methods in the Śulba Sūtras, see [1], [2], [3], and [5].)

4 Construction of the Saviśeṣa for the Square Root of Two

If we apply Sūtra i.54 to the union of two squares each with sides of 1 prādeśa we get a square with side $1\frac{1}{2}$ prādeśa from which a square of side $\frac{1}{2}$ prādeśa had been removed. See Figure 2. Now we can attempt to take a strip from the left and bottom of the large square—the strips are to be just thin enough that they will fill in the little removed square. The pieces filling in the little square will have length $\frac{1}{2}$ and six of these lengths will fit along the bottom and left of the large square. The reader can then see in Figure 3 that strips of thickness $\left(\frac{1}{6}\right)\left(\frac{1}{2}\right)$ prādeśa ($=$ 1 aṅgula) will (almost) work:

There is still a little square left out of the upper right corner because the thin strips overlapped in the lower left corner. Notice that

$$\left(1 + \frac{1}{2} - \frac{1}{6} \cdot \frac{1}{2}\right) \text{prādeśa}$$
$$= 1 \text{ prādeśa} + 5 \text{ aṅgulas}$$
$$= \left(1 + \frac{1}{3} + \frac{1}{4} \cdot \frac{1}{3}\right) \text{prādeśa}.$$

We can get directly to $1 + \frac{1}{3} + \frac{1}{4} \cdot \frac{1}{3}$ by considering the dissection in Figure 4.

We now have that two square prādeśas are equal to a large square minus a small square. The large

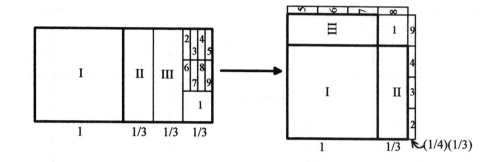

Figure 4: Alternate second stage.

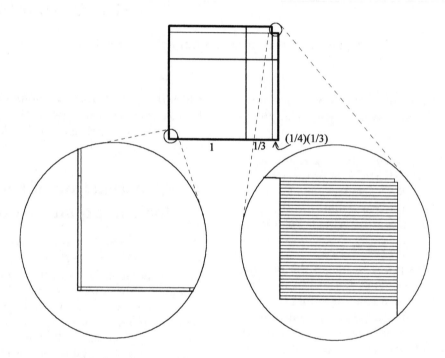

Figure 5: Third stage in construction of saviśeṣa.

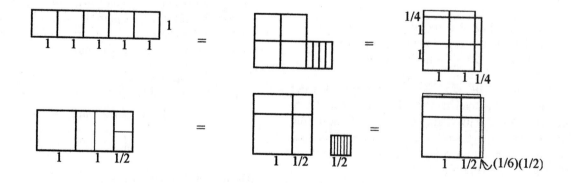

Figure 6: First stage for 5 and $2 - 1/2$.

square has side equal to 1 *prādeśa* plus 1/3 of a *prādeśa* plus 1/4 of 1/3 of a *prādeśa*, or 1 *prādeśa* and 5 *aṅgulas* and the small square has side of 1 *aṅgula*. To make this into a single square we may attempt to remove a thin strip from the left side and the bottom just thin enough that the strips will fill in the little square. Since these two thin strips will have length 1 *prādeśa* and 5 *aṅgulas* or 17 *aṅgulas* we may cut each into 17 rectangular pieces each 1 *aṅgula* long. If these are stacked up they will fill the little square if the thickness of the strips is 1/34 of an *aṅgula* (or $\frac{1}{34} \cdot \frac{1}{4} \cdot \frac{1}{3}$ *prādeśas*). Without a microscope we will now see the two square *prādeśas* as being equal in area to the square with side $1 + \frac{1}{3} + \frac{1}{4} \cdot \frac{1}{3} - \frac{1}{34} \cdot \frac{1}{4} \cdot \frac{1}{3}$ *prādeśas*. But with a microscope we see that the strips overlap in the lower left corner and thus that there is a tiny square of side $\frac{1}{34} \cdot \frac{1}{4} \cdot \frac{1}{3}$ still left out. See Figure 5.

Thus $1 + \frac{1}{3} + \frac{1}{4} \cdot \frac{1}{3} - \frac{1}{34} \cdot \frac{1}{4} \cdot \frac{1}{3}$ is still a little in excess. We can now perform the same procedure again by removing a very very thin strip from the left and bottom edges and then cutting them into $\frac{1}{34} \cdot \frac{1}{4} \cdot \frac{1}{3}$ *prādeśas* lengths in order to fill in the left out square. If w is twice the number of $\frac{1}{34} \cdot \frac{1}{4} \cdot \frac{1}{3}$ lengths in $1 + \frac{1}{3} + \frac{1}{4} \cdot \frac{1}{3} - \frac{1}{34} \cdot \frac{1}{4} \cdot \frac{1}{3}$ *prādeśas*, then the strips we remove must have width $\frac{1}{w} \cdot (\frac{1}{34} \cdot \frac{1}{4} \cdot \frac{1}{3})$ *prādeśas*. We can calculate w easily because we already noted that there were 17 segments of length $\frac{1}{4} \cdot \frac{1}{3}$ in the length $1 + \frac{1}{3} + \frac{1}{4} \cdot \frac{1}{3}$ and each of these segments was divided into 34 pieces and then one of these pieces was removed. Thus $w = 2[34(17) - 1] = 1154$ and

$$\sqrt{2} \approx$$
$$1 + \frac{1}{3} + \frac{1}{4} \cdot \frac{1}{3} - \frac{1}{34} \cdot \frac{1}{4} \cdot \frac{1}{3} - \frac{1}{1154} \cdot \frac{1}{34} \cdot \frac{1}{4} \cdot \frac{1}{3}$$

with error expressed by

$$2 \cdot 1 =$$
$$\left(1 + \frac{1}{3} + \frac{1}{4} \cdot \frac{1}{3} - \frac{1}{34} \cdot \frac{1}{4} \cdot \frac{1}{3} - \frac{1}{1154} \cdot \frac{1}{34} \cdot \frac{1}{4} \cdot \frac{1}{3}\right)^2$$
$$- \left(\frac{1}{1154} \cdot \frac{1}{34} \cdot \frac{1}{4} \cdot \frac{1}{3}\right)^2.$$

I write "2 · 1" instead of "2" to remind us that for Baudhāyana (and, in fact, for most mathematicians up until near the end of the 19th century) that $\sqrt{2}$ denoted the side (a *length*) of a square with *area* 2. If we again follow the same procedure of removing a very thin strip from the left and bottom edges and cutting them into $\frac{1}{1154} \cdot \frac{1}{34} \cdot \frac{1}{4} \cdot \frac{1}{3}$ length pieces, then the reader can check that the number of such pieces must be

$$2[1154(1154/2) - 1] = (1154)^2 - 2 = 1{,}331{,}714$$

and thus that the next approximation (saviśeṣa) is

$$\sqrt{2} \approx 1 + \frac{1}{3} + \frac{1}{4} \cdot \frac{1}{3}$$
$$- \frac{1}{34} \cdot \frac{1}{4} \cdot \frac{1}{3} - \frac{1}{1154} \cdot \frac{1}{34} \cdot \frac{1}{4} \cdot \frac{1}{3}$$
$$- \frac{1}{1{,}331{,}714} \cdot \frac{1}{1154} \cdot \frac{1}{34} \cdot \frac{1}{4} \cdot \frac{1}{3}.$$

The difference between $2 \cdot 1$ and the square of this saviśeṣa is

$$\left(\frac{1}{1{,}331{,}714} \cdot \frac{1}{1154} \cdot \frac{1}{34} \cdot \frac{1}{4} \cdot \frac{1}{3}\right)^2.$$

This method will work for any number N which you can first express as the area of the difference of two squares, $N \cdot 1 = A^2 - B^2$, where the side A is an integral multiple of the side B. For example,

$$5 \cdot 1 = \left(2 + \frac{1}{4}\right)^2 - \left(\frac{1}{4}\right)^2,$$

$$7 \cdot 1 = \left(2 + \frac{2}{3}\right)^2 - \left(\frac{1}{3}\right)^2,$$

$$10 \cdot 1 = \left(3 + \frac{1}{6}\right)^2 - \left(\frac{1}{6}\right)^2,$$

$$12 \cdot 1 = \left(3 + \frac{1}{2}\right)^2 - \left(\frac{1}{2}\right)^2,$$

and $\left(2 + \frac{1}{2}\right) \cdot 1 = \left(1 + \frac{1}{2} + \frac{1}{6} \cdot \frac{1}{2}\right)^2 - \left(\frac{1}{6} \cdot \frac{1}{2}\right)^2.$

I find that the easiest way for me to see that these expressions are valid is to represent them geometrically in a way that would also have been natural for Baudhāyana. To illustrate, see Figure 6. Figures 3 and 4 give other examples. The reader should try out this method to see how easy it is to find saviśeṣa for the square roots of other numbers, for example, 3, 11, $2\frac{3}{4}$.

5 Fractions in the Śulba Sūtras

You have probably noticed that all the fractions above are expressed as unit fractions, but this is not always the case in the Baudhāyana's *Śulba Sūtras*. For example, in Sūtra i.69 he discusses how to find a length which is an approximation to the diagonal of a square whose side is the "third part of" 8 *prakramas* (which equals 240 *aṅgulas*). He describes the construction:

D&A—calculator	D&A—Fractions	Baudhāyana's Method
$a_1 = 1.416666667$	$17/12$	$1 + (1/3) + (1/4)(1/3)$
$a_2 = \frac{1}{2}(a_1 + (2/a_1))$ $= 1.414215686$	$\frac{1}{2}[(17/12) + 2(12/17)]$ $= (577/408)$	$k_2 = 2[(3 \cdot 4) + 4 + 1] = 34$ $c_2 = -(1/34)(1/4)(1/3)$
$a_3 = \frac{1}{2}(a_2 + (2/a_2))$ $= 1.414213562$	$\frac{1}{2}[(577/408) + 2(408/577)]$ $= (665857/470832)$	$k_3 = (34)^2 - 2 = 1154$ $c_3 = -(1/1154)(1/34)(1/4)(1/3)$
$a_4 = \frac{1}{2}(a_3 + (2/a_3))$ $= 1.414213562$	$\frac{1}{2}[(665857/470832) + 2(470832/665857)]$ $= (886731088897/627013566048)$	$k_4 = (1154)^2 - 2 = 1331714$ $c_4 = -(1/1331714)c_3$

Table 1: Comparison of Baudhayana with D&A.

> ... increase the measure [the 8 *prakramas*]
> by its fifth, divide the whole into five parts
> and make a mark at the end of two parts.

In more modern notation if we let D equal 8 *prakramas*, then this gives the approximation of the diagonal of a square with side $(1/3)D$ as

$$\frac{2}{5}\left(D + \frac{1}{5}D\right).$$

This is equivalent to $\sqrt{2}$ being approximated by 1.44.

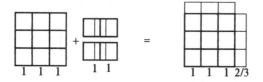

Figure 7: First stage for 13.

If you attempt to find the saviśeṣas for other square roots you will find it convenient to use non-unit fractions. For example, by starting with Figure 7 you can make slight modifications in the above method to find

$$13 \cdot 1 \approx \left(3 + \frac{2}{3} - \frac{1}{11} \cdot \frac{2}{3} - \frac{1}{120} \cdot \frac{1}{11} \cdot \frac{2}{3}\right)^2 - \left(\frac{1}{120} \cdot \frac{1}{11} \cdot \frac{2}{3}\right)^2.$$

6 Comparing with the Divide-and-Average (D&A) Method

Today the most efficient method usually taught to find square roots is called "divide-and-average". It is also sometimes called Newton's method. If you wish to find the square root of N then you start with an initial approximation a_0 and then take as the next approximation the average of a_0 and N/a_0. In general, if a_n is the nth approximation of the square root of N, then $a_{n+1} = \frac{1}{2}(a_n + (N/a_n))$. The interested reader can check that if you start with $[1 + \frac{1}{3} + \frac{1}{12}] = \frac{17}{12} = 1.416666666667$ as your first approximation of $\sqrt{N}$, then the succeeding approximations are numerically the same as those given by Baudhāyana's geometric method.

However, Baudhāyana's method uses significantly fewer computations (in addition, of course, to the drawings either on paper or in one's mind). For example, look at Table 1 which compares the methods for the first four approximations. For Baudhāyana's method at the nth stage let k_n denote the number of thin pieces added into the missing square and let c_n denote the correction term that is added.

Notice that the (10-digit) calculator reaches its maximum accuracy at the third stage. At this stage the Baudhāyana method obtained more accuracy (it can be checked that it is accurate to 12-digits) and the only computation required was $(34)^2 - 2 = 1154$ which can easily be accomplished by hand. Baudhāyana's approximations are numerically identical to those attained in the D&A method using fractions, but again with significantly fewer computations. Of course, Baudhāyana's method has this efficiency only if you do not change Baudhāyana's representation of the approximation into decimals or into standard fractions. At the fourth stage the Baudhāyana method is accurate to less than

$$2[(1331714^2 - 2)(1331714)(1154)(34)(4)(3)]^{-1}$$

or roughly 24-digit accuracy with the only calculation needed being $(1154)^2 - 2 = 1331714$.

Notice that in Baudhāyana's fourth representation of the saviśeṣas for the square root of 2,

$$\sqrt{2} \approx 1 + \frac{1}{3} + \frac{1}{4}\cdot\frac{1}{3} - \frac{1}{34}\cdot\frac{1}{4}\cdot\frac{1}{3}$$
$$- \frac{1}{1154}\cdot\frac{1}{34}\cdot\frac{1}{4}\cdot\frac{1}{3}$$
$$- \frac{1}{1,331,714}\cdot\frac{1}{1154}\cdot\frac{1}{34}\cdot\frac{1}{4}\cdot\frac{1}{3},$$

the unit is first divided into 3 parts and then each of these parts into 4 parts and then each of these parts into 1154 parts and each of these parts into 133174 parts. Notice the similarity of this to standard USA linear measure, where a mile is divided into 8 furlongs and a furlong into 220 yards and a yard into 3 feet and a foot into 12 inches. Other traditional systems of units work similarly except for the metric system where the division is always by 10. Also, when some carpenters I know have a measurement of $2\frac{7}{16}$ inches, they are likely to work with it as $2 + \frac{1}{2} - \frac{1}{8}\cdot\frac{1}{2}$, or 2 inches plus a half inch minus an eighth of that half—this is a clearer image to hold onto and work with. From Baudhāyana's approximation it is easier to have an image of the length of $\sqrt{2}$ than it is from the D&A's (886731088897/627013566048).

7   Conclusions

Baudhāyana's method can not come even close to the D&A method in terms of ease of use with a computer and its applicability to finding the square root of any number. However, the *Śulba Sūtras* contain many powerful techniques, which, in specific situations have a power and efficiency that is missing in more general techniques. Numerical computations with the decimal system in either fixed point or floating point form has many well-known problems (see, for example, [8]). Perhaps we will be able to learn something from the (apparently) first applied geometry text in the world and devise computational procedures that combine geometry and numerical techniques.

References

[1] P. Datta. *The Science of the Sulba*. University of Calcutta, 1932.

[2] D. W. Henderson. *Experiencing Geometry on Plane and Sphere*. Prentice-Hall, Upper Saddle River, NJ, 1966.

[3] G. Joseph. *The Crest of the Peacock*. Taurus, London, 1991.

[4] S. Prakash and R. M. Sharma, editors. *Baudhāyana Śulbasūtram*. Ram Swarup Sharma, Bombay, 1968. Thibaut, trans.

[5] J. F. Price. Applied geometry of the Śulba Sūtras. In C. Gorini, editor, *Geometry at Work*, pages 46–55. MAA, Washington, DC, 2000.

[6] A. Seidenberg. The ritual origin of geometry. *Archive for the History of the Exact Sciences*, 1:488–527, 1961.

[7] S. N. Sen and A. K. Bag. *The Śulbasūtras of Baudhāyana, Āpastamba, Kātyāyana, and Mānava*. Indian National Science Academy, New Delhi, 1983.

[8] P. R. Turner. Will the 'real' real arithmetic please stand up? *Notices of AMS*, 34:298–304, April 1991.

Applied Geometry of the Śulba Sūtras

John F. Price*
School of Mathematics
University of New South Wales
Sydney, NSW 2052, Australia
johnp@sherlockinvesting.com

Abstract

The Śulba Sūtras, part of the Vedic literature of India, describe many geometrical properties and constructions such as the classical relationship

$$a^2 + b^2 = c^2$$

between the sides of a right-angle triangle and arithmetical formulas such as calculating the square root of two accurate to five decimal places. Although this article presents some of these constructions, its main purpose is to show how to consider each of the main Śulba Sūtras as a finely crafted, integrated manual for the construction of citis or ceremonial platforms. Certain key words, however, suggest that the applications go far beyond this.

1 Introduction

The Śulba Sūtras are part of the Vedic literature, an enormous body of work consisting of thousands of books covering hundreds of thousands of pages. This section starts with a brief review of the position of the Śulba Sūtras within this literature and then describes some of its significant features. Section 2 is a discussion of some of the geometry found in the Śulba Sūtras while Section 3 is the same for arithmetical constructions. Section 4 looks at some of the applications to the building of citis and Section 5 integrates the different sections and discusses the significance of the applied geometry of the Śulba Sūtras. I am grateful to Ken Hince for assistance with the diagrams and to Tom Egenes for helping me avoid some of the pitfalls of Sanskrit transliteration and translation.

---
*This article was written while the author was in the Department of Mathematics, Maharishi University of Management, Fairfield, IA.

1.1 Vedic Literature and the Śulba Sūtras

To understand the geometry of the Śulba Sūtras and their applications, it is helpful first to understand their place in Vedic knowledge as expressed through the Vedic literature.

Maharishi Mahesh Yogi explains that the Vedic literature is in forty parts consisting of the four Vedas plus six sections of six parts each. These sections are the Vedas, the Vedāṅgas, the Upāṅgas, the Upa-Vedas, the Brāhmaṇas, and the Prātishākhyas. Each of these parts "expresses a specific quality of consciousness" [4, p. 144]. This means that often we have to look beyond the surface meanings of many of the texts to find their deeper significance.

The Śulba Sūtras form part of the Kalpa Sūtras which in turn are a part of the Vedāṅgas. There are four main Śulba Sūtras, the Baudhāyana, the Āpastamba, the Mānava, and the Kātyāyana, and a number of smaller ones. One of the meanings of śulba is "string, cord or rope." The general formats of the main Śulba Sūtras are the same; each starts with sections on geometrical and arithmetical constructions and ends with details of how to build citis which, for the moment, we interpret as ceremonial platforms or altars. The measurements for the geometrical constructions are performed by drawing arcs with different radii and centers using a cord or śulba.

There are numerous translations and references for the Śulba Sūtras. The two that form the basis of this article are [7] (which contains the full texts of the above four Śulba Sūtras in Sanskrit) and [9] (which is a commentary on and English translation of the Baudhāyana Śulba Sūtra). Another useful book is [8] since it contains both a transliteration of the four main Śulba Sūtras into the Roman alphabet and an English translation.

It is timely to be looking at some of the mathematics contained in the Vedic literature because of the renewed understanding brought about by Maharishi of the practical benefits to modern life of this ancient Vedic knowledge. Details and further references can be found in [5].

1.2 Features of the Śulba Sūtras

For me, there are three outstanding features of the Śulba Sūtras: the wholeness and consistency of their geometrical results and constructions, the elegance and beauty of the citis, and the indication that the Sūtras have a much deeper purpose.

1.2.1 Integrated wholeness of the Śulba Sūtras

When each of the main Śulba Sūtras is viewed as a whole, instead of a collection of parts, then a striking level of unity and efficiency becomes apparent. There are exactly the right geometrical constructions to the precise degree of accuracy necessary for the artisans to build the citis. Nothing is redundant. This point is nicely made by David Henderson [2] who argues that the units of measurement used easily lead to the accuracy of the diagonal of one of the main bricks of "roughly one-thousandth of the thickness of a sesame seed."

There is also a remarkable degree of internal consistency such as the way that the 'square to circle,' 'circle to square' and 'square root of two' constructions fit together with an accuracy of 0.0003%. (Details are given in Section 5.) A related discussion in the literature is whether or not the authors of the Sūtras knew their construction of the root of two was an approximation.[1] Viewing the Sūtras as utilitarian construction manuals suggested that they knew that they were describing an approximation but that they achieved what they set out to do, namely to provide the first terms of an expansion of the root of two sufficient to ensure the reciprocity of the 'square to circle' and 'circle to square' constructions.

If mathematicians were asked to write such a manual, it is likely that they would do two things, firstly give the construction procedures to level of accuracy appropriate for the actual constructions, and secondly, for their own enjoyment, show to other mathematical readers that they really understood that they were dealing with approximations. Both these features are observed in the Śulba Sūtras.

---

[1]This discussion hinges on the meaning of *viśeṣa*, which [1] and others take to mean, in this context, a small excess quantity or difference. See also [8, p. 168].

1.2.2 Beauty of the citis

Each of the citis are low platforms consisting of layers of carefully shaped and arranged bricks. Some are quite simple shapes such as a square or a rhombus while others are much more involved such as a falcon in flight with curved wings, a chariot wheel complete with spokes, or a tortoise with extended head and legs. These latter designs are particularly beautiful and elegant depictions of powerful and archetypal symbols, the falcon as the great bird that can soar to heaven, the wheel as the 'wheel of life,' and the tortoise as the representative of stability and perseverance.

1.2.3 Deeper significance

Sanskrit is a rich language full of subtle nuances. Words can have quite different meanings because of their context and, in any case, frequently there is no reasonable English equivalent.

There are a number of key terms in the Śulba Sūtras which, because of their etymology and phonetics, suggest that there is a much deeper significance to the Sūtras. One is the word *citi* introduced above. In the context of the Śulba Sūtras, the usual translation is a type of ceremonial platform but it is close to the word *cit* which means consciousness. Another is *vedi* which is usually translated as the place or area of ground on which the citi is constructed. But since the word *veda* means "pure knowledge, complete knowledge" [4, p. 3], vedi also means an enlightened person, a person "who possesses Veda."

A third is *puruṣa* which is usually translated as a unit of measurement obtained by the height of a man with upstretched arms (Mānava Śulba Sūtra IV, 5) or as 120 *aṅgulas* (Baudhāyana Śulba Sūtra I, 1–21), a measurement based on sizes of certain grains.

However, in [3] *puruṣa* is defined as "the uninvolved witnessing quality of intelligence, the unified . . . self-referral state of intelligence at the basis of all creativity" (p. 109). Thus we could easily infer that a more expanded role of the Śulba Sūtras is as a description of consciousness. Further discussion of this point is given in Section 5.

When these and other examples are combined with the general direction of all the Vedic literature towards describing "qualities of consciousness," we are led to the conclusion that the Śulba Sūtras are describing something much beyond procedures for building brick platforms, no matter how far-reaching their purpose. This theme is referred to again in the concluding section. In this article

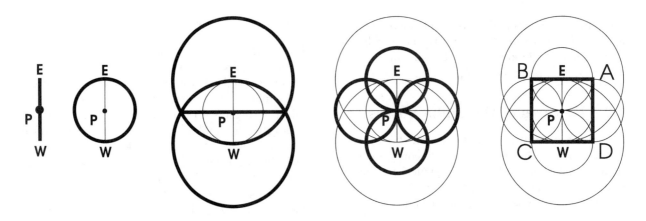

Figure 1: Steps for the construction of a square.

the focus is on the geometrical content of the text, but because of the range of meanings of the key terms, they will generally be left in their original (but transliterated) form. In a later article, I hope to develop some of these deeper themes of the Śulba Sūtras.

2   Geometry

Most of the geometric procedures described in the Śulba Sūtras start with the laying out of a *prācī* which is a line in the east-west direction. This line is then incorporated into the final geometric objects or constructions, generally as a center line or line of symmetry. This section describes some of the main geometric constructions given in the Sūtras.

2.1   Construction of a square with a side of given length

From verses I, 22–28 of the Baudhāyana Śulba Sūtra (BSS)[2], the procedure is to start with a prācī and a center point (line EPW in Figure 1) and, by describing circles with certain centers and radii, construct a square $ABCD$ in which E and W are the midpoints of AB and CD. The steps of the construction are displayed in Figure 1.

2.2   Theorem on the square of the diagonal

Verse I, 48 of BSS states:

---
[2]There are two main numberings of the verses of the BSS, one used by Thibaut [9] and one attributed to A. Bürk in [8] and used there. In this article we follow the numbering used by Thibaut.

The diagonal of a rectangle produces both (areas) which its length and breadth produce separately.

There appears to be no direct mention of areas in this verse. When, however, it is combined with the subsequent one,

This is seen in rectangles with sides three and four, twelve and five, fifteen and eight, seven and twenty-four, twelve and thirty-five, fifteen and thirty-six,

it is clear that it is an equivalent statement to the theorem named in the west after Pythagoras, namely that in a right-angle triangle with sides a, b and c (c the hypotenuse), $a^2 + b^2 = c^2$. Note also that all possible pairs (a, b) are given (except one) which (i) allow an integer solution of $a^2 + b^2 = c^2$ with $1 \leq a \leq 12$ and $a \leq b$, and (ii) are coprime. (The interested reader might like to check this.) This accounts for five of the listed pairs, the last pair $(15, 36)$ being derivable from the earlier pair $(5, 12)$.

Further evidence that Baudhāyana had a clear understanding of this result and its usefulness is provided by the next two constructions.

2.3   A square equal to the sum of two unequal squares

Verse I, 50 of BSS describes the construction of a square with area equal to the sum of the areas of two unequal squares. Suppose that the two given squares are $ABCD$ and $EFGH$ with $AB > EF$. Mark off points J, K on AB and DC with $AJ = DK = EF$ as shown in Figure 2. Then the line AK is the side of a square with area equal to the sum of the areas of $ABCD$ and $EFGH$.

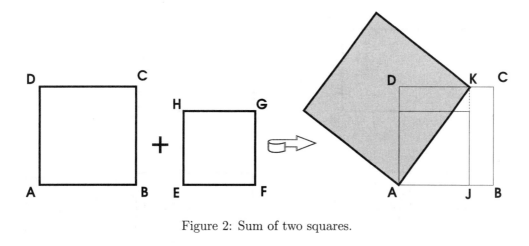

Figure 2: Sum of two squares.

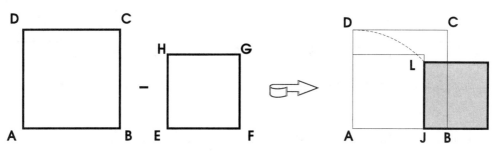

Figure 3: Difference of two squares.

2.4 A square equal to the difference of two squares

The subsequent verse (I, 51) describes the construction of a square with area equal to the difference of two unequal squares. With the notation the same as the preceding example, form an arc DL with center A as shown in Figure 3. Then JL is the side of the required square.

This follows from the facts that $(AJ)^2 + (JL)^2 = (AL)^2$, $AL = AB$, and $AJ = EF$, so that $(JL)^2 = (AB)^2 - (EF)^2$.

2.5 Converting a rectangle into a square

The method of converting a rectangle into a square with the same area described in BSS I, 54 makes use of the previous construction. Start with a rectangle $ABCD$ with $AB > CD$ as shown in Figure 4 and form a square $AEFD$. The excess portion is cut into equal halves and one half is placed on the side of the square. This gives two squares, a larger one $AGJC'$ and a smaller one $FHJB'$; the required square is the difference of these two squares. In Figure 4 the side of this square is GL, where L is

determined by $EL = EB'$.

To see this, denote the sides of the rectangle by $AB = a$ and $AD = b$. Then the side GL satisfies

$$(GL)^2 = (EL)^2 - (EG)^2 = (EB')^2 - (EG)^2$$
$$= \left(b + \frac{a-b}{2}\right)^2 - \left(\frac{a-b}{2}\right)^2 = ab$$

and so the area of the shaded square equals the area of the initial rectangle, as required.

Datta [1] suggests that the steps in this construction of marking off a square, dividing the excess, and rearranging the parts could be the basis of the method described later in the BSS I, 61 (see 3.1) of finding the square root of two. By repeating these steps, he shows that the successive approximations to the square root of two described in I, 61 are obtained. If this is the procedure they used, then this is another example of the tightly knit methods and logic of the Sūtras.

2.6 Converting a square into a circle

Verse I, 58 describes the procedure for constructing a circle with area approximately equal to that of a given square. Start with a square $ABCD$ as in

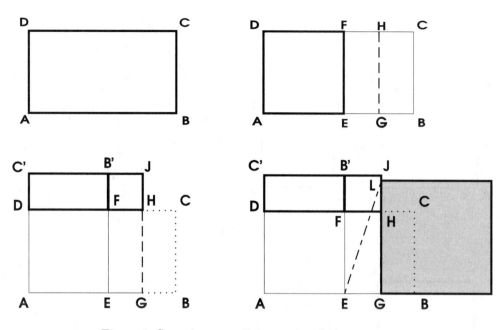

Figure 4: Steps in converting a rectangle to a square.

Figure 5 with center O. Draw an arc DG with center O so that OG is parallel to AD. Suppose that OG intersects DC at the point F. Let H be a point $1/3$ of the distance from F to G. Then OH is the radius of the required circle.

To see what is going on here, let $2a$ be the length of the side of the square $ABCD$ and r the radius of the constructed circle. Then

$$
\begin{aligned}
r &= OH = OF + FH \\
&= OF + \frac{1}{3}(OG - OF) \\
&= a + \frac{1}{3}(a\sqrt{2} - a)
\end{aligned}
$$

and hence

$$ r = \frac{a}{3}\left(2 + \sqrt{2}\right). \tag{1} $$

If we substitute the values $\pi = 3.141593$ and $\sqrt{2} = 1.414214$ we get that the area of the constructed circle is

$$ \text{Area} = \pi r^2 = 4.069011\ldots \times a^2 $$

which is within about 1.7% of the correct value of 4.

In the next verse Baudhāyana describes how to go in the opposite direction, namely from a circle to a square.

2.7   Converting a circle into a square

Thibaut's translation of Verse I, 59 is:

If you wish to turn a circle into a square, divide the diameter into eight parts and one of these eight parts into twenty-nine parts; of these twenty-nine parts remove twenty-eight and moreover the sixth part (of the one part left) less the eighth part (of the sixth part).

In modern mathematical notation, starting with a circle of diameter d, the length of the side of the corresponding square is:

$$ \text{length} = d - \frac{d}{8} + \frac{d}{8 \times 29} $$

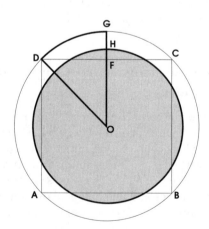

Figure 5: Converting a square to a circle.

$$-\frac{d}{8 \times 29}\left(\frac{1}{6} - \frac{1}{6 \times 8}\right)$$

$$= \frac{9785}{11136} \times d. \qquad (2)$$

Taking $\pi = 3.141593$ as before, we get

$$\frac{\text{Area of square}}{\text{Area of circle}} = \left(\frac{9785}{11136}\right)^2 d^2 \Big/ \frac{\pi}{4}d^2$$

$$= 0.983045\ldots.$$

The correct value is 1 and so, just as in the previous section, the result is accurate to approximately 1.7%. The precision of these accuracies will be looked at in more detail in Section 5 below.

3   Arithmetic

We shall first consider the example in the Baudhāyana Śulba Sūtra of the calculation of the square root of two and then look at what could be the underlying principle.

3.1   The square root of two

Verse I, 61 of BSS writes:

> Increase the measure by a third and this (third) again by its own fourth less its thirty-fourth part; this is the (length of) the diagonal of a square (whose side is the measure).

Using modern notation, the assertion is that if the length of the side of the original square is a, then the length of its diagonal is

$$\text{diagonal} = \left(1 + \frac{1}{3} + \frac{1}{3 \times 4} - \frac{1}{3 \times 4 \times 34}\right) \times a$$

$$= \frac{577}{408} \times a$$

$$= 1.414215\ldots \times a$$

3.2   Discussion

Given that $\sqrt{2} = 1.414213\ldots$, it is not surprising that many commentators have proposed ways that explain the underlying method of achieving such an accurate result. The common thread passing through most of the explanations is to start with the initial approximation

$$\sqrt{2} \approx 1 + \frac{1}{3}$$

and then sequentially estimate and reduce the size of the errors. Let x_1 be the first error so that

$$\sqrt{2} = 1 + \frac{1}{3} + x_1.$$

Squaring both sides and neglecting terms in x_1^2 gives $x_1 = \frac{1}{3 \times 4}$. Now define the next approximation x_2 by

$$\sqrt{2} = 1 + \frac{1}{3} + \frac{1}{3 \times 4} + x_2.$$

Repeating the preceding step gives

$$x_2 = -\frac{1}{3 \times 4 \times 34}$$

which yields the approximation given in the BSS just described.

Another repetition of this step gives the approximation:

$$\sqrt{2} \approx 1 + \frac{1}{3} + \frac{1}{3 \times 4} - \frac{1}{3 \times 4 \times 34}$$

$$- \frac{1}{3 \times 4 \times 34 \times 2 \times 577}$$

$$= \frac{665857}{40832} = 1.414213562374\ldots$$

which is accurate to 13 decimal places.

Another approach has been suggested by [1] which maintains the geometric flavor of the Sūtras. The idea is to use the first steps of the method described in 2.5 to convert a rectangle of size 2×1 into a square of side $\sqrt{2}$. If these steps are repeated the above sequence of approximations is obtained.

Henderson [2] shows just how natural the procedure is and how it can be generalized to a large class of real numbers. He also observes that same sequential approximation to $\sqrt{2}$ can be obtained by applying Newton's method for finding the roots of an algebraic equation. Take $f(x) = x^2 - 2$ with an initial approximation of $x_0 = 4/3$ for its positive root. Newton's method is that the successive approximations are given by

$$x_{n+1} = x_n - \frac{f(x_n)}{f'(x_n)} = \frac{x_n}{2} - \frac{1}{x_n}.$$

It is easily checked that this gives the approximations described in the Śulba Sūtras.

4   Applications to the construction of citis

In this section we look at the applications of the geometrical and arithmetical constructions of the

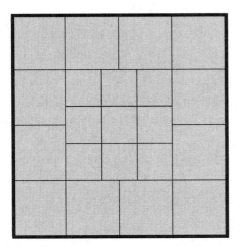

Figure 6: Possible configuration of the gārhaptya citi for the odd and even layers.

Śulba Sūtras to the construction of citis. As stated in Section 1, this is just the first level of applications of the Sūtras.[3]

4.1   General introduction

Each of the citis is constructed from five layers of bricks, the first, third and fifth layers being of the same design, as are the second and fourth. Quite a lengthy sequence of units is used; the two that are referred to the most are the *aṅgula* and the *puruṣa* which were discussed in 1.2.3 above. They are approximately 0.75 inches and 7 feet 6 inches with 120 *aṅgulas* equal to a *puruṣa*.

The heights of each layer are 6.4 *aṅgulas* which is about 4.8 inches and the successive layers are built so that no joins lie along each other. This last requirement is sometimes difficult to achieve and adds to the aesthetics of the finished product as much as to its strength. Generally each layer has 200 bricks with the exception of the gārhaptya citi which has 21 bricks in each layer.

For each design (with the exception of the gārhaptya citi), the citi is first constructed with an area of 7.5 square puruṣas, then with 8.5 square puruṣas, and so on up to 101.5 square puruṣas (BSS II, 1–6). Verse II, 12 explains how these increases in size are to be brought about. If the original citi of 7.5 sq. puruṣas is to be increased by q sq. puruṣas, a square with area one sq. puruṣa has substituted for it a square with area $1 + (2q/15)$ sq. puruṣas. The sides of this square form the unit of the new construction replacing the original puruṣa. Hence

---

[3]Photographs from Kerala, India of citis and associated ceremonies are contained in [6].

the area of the enlarged citi is

$$7.5 \times \left(\sqrt{1 + \frac{2q}{15}} \right)^2 = 7.5 + q \text{ sq. puruṣas,}$$

as required.

Many of the designs described in the BSS have a number of variations. For simplicity not all the variations will be described and we shall choose one of them usually without commenting that there are other possibilities.

4.2   The gārhaptya citi

This citi, described in BSS II, 66–69, is one vyāyāma square and consists of 21 bricks on each level. (A vyāyāma is 96 *aṅgulas* or about 6 feet; BSS I, 21). Three types of bricks are used: one-sixth, one-fourth and one-third of a vyāyāma. The odd layers consist of 9 bricks of the first type and 12 of the second, while the even layers consist of 6 bricks of the third type and 16 of the first. One possible arrangement is shown in Figure 6.

4.3   The śyena citi

This is a particularly beautiful depiction of a falcon (*śyena*) in flight, its construction steps being described in BSS III, 62–104. See Figures 7 and 8.

4.4   The rathacakra citi

This citi is in the shape of a chariot wheel (*rathacakra*) with nave (or center), spokes and felly (or rim), its construction being described in BSS III, 187–214. It requires seven types of bricks for the

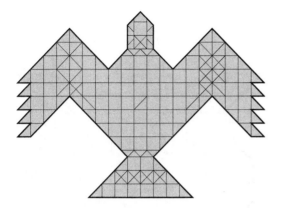

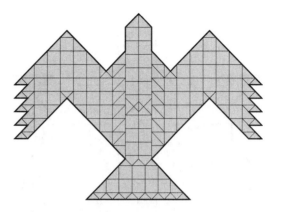

Figure 7: The śyena citi: layers 1, 3, and 5.   Figure 8: The śyena citi: layers 2 and 4.

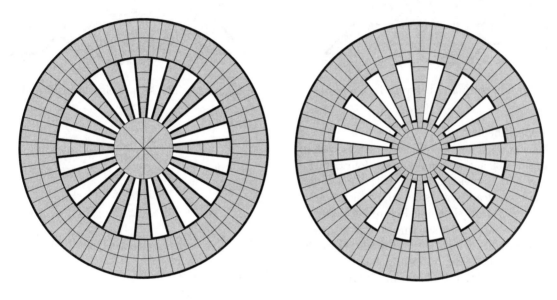

Figure 9: Design of the rathacakra citi.

odd layers and nine types for the even layers. There seems to be some flexibility about the final design, Figure 9 being one possibility.

The initial calculations for determining the different parts of the wheel are in terms of square bricks each of area 1/30 square puruṣas. Since the final area is required to be 7.5 square puruṣas, the number of bricks is $7.5 \times 30 = 225$. The nave of the wheel consists of 16 of these bricks, the spokes 64 and the rim 145, making 225 in all.

The spaces between the spokes are equal in area to the spokes and so, if these spaces are included, the overall area is $225 + 64 = 289$ bricks. (Notice that these numbers satisfy $15^2 + 8^2 = 17^2$ and are one right-angle-triangle triples described above.) Hence the radius of the outer rim of the wheel is equal to the radius of a circle equal in area to a square of side 17 bricks so the methods described in subsection 2.6 could be used to construct this circle. Similarly, the inner radius of the rim is the radius of a circle equal in area to a square with side the square root of $16 + 64 + 64 = 144$, namely 12. Finally, the radius of the nave comes from a square of side 4.

4.5   The kūrma citi

Another fascinating series of constructions are in the form of a tortoise (*kūrma*). In the BSS there are two types of these constructions, one is described as having twisted limbs (*vakrāṅgāśca*) and the other as having rounded limbs (*parimaṇḍalāśca*). Figures 10

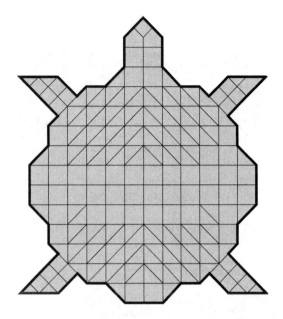

Figure 10: The kūrma citi: layers 1, 3, and 5.

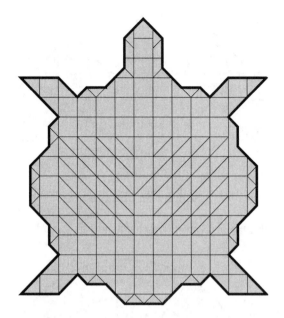

Figure 11: The kūrma citi: layers 2 and 4.

and 11 depict the first type.

The construction for the odd layers starts with a square of side 300 aṅgulas and then the four corners are removed by isosceles triangles with equal sides of 30 aṅgulas. Head, legs, sides and tail are now added with the result shown in Figure 10.

For the even layers, the starting step is a square of side 270 aṅgulas which is offset from the basic square for the odd layers by 15 aṅgulas. The plan of the final construction is shown in Figure 11.

5 Discussion

Having listed some of the constructions of the Śulba Sūtras, we are now in a position to look at how they fit together. A key example is the duality of the 'circle-square' constructions. As explained in 3.1, verse I, 61 of BSS gives a value of the square root of two as $\frac{577}{408}$. Using this value, there is a remarkable duality between the 'square to circle' and 'circle to square' results.

Suppose that we start with a square of side 2. By equation (1), the diameter d of the corresponding circle is

$$
\begin{aligned}
d &= \frac{2}{3}\left(2+\sqrt{2}\right) \\
  &= \frac{2}{3}\left(2+\frac{577}{408}\right).
\end{aligned}
$$

Now use equation (2) to get the side of the square

as

$$
\begin{aligned}
\text{side} &= \frac{2}{3}\left(2+\frac{577}{408}\right)\frac{9785}{11136} \\
&= \frac{13,630,505}{6,815,232} \\
&= 2+\frac{41}{6,815,232} = 2.000006,
\end{aligned}
$$

an accuracy of 0.0003%.

A second example is the use made of the 'difference of two squares' construction to construct a square with area equal to a given rectangle. Also, as explained in Section 2, construction may well form the basis of the method of finding the root of two.

Other examples are the conversion of squares to circles for the construction of citis in circular shapes such as the rathacakra citi, or the methods used to increase the size of the citis by scaling up all the dimensions. These and other similar results show the level of integration and completeness of the body of results in the Śulba Sūtras.

In the opening section, several examples were given of key words in the Śulba Sūtras that either had alternative deeper meanings or were related to such words. The study of the meaning of words in Sanskrit is a large and technical field founded on the work of Pāṇini. Because of the technical nature of the area, drawing the specific conclusion that the Sūtras were also intended to be dealing with the field of consciousness would require considerably more work. But there is an example that explicitly connects the construction of the citis with consciousness, namely Verse II, 81 of BSS. It reads that after having constructed a citi for the third time, then a *chandaścit* is to be constructed. The word *chandas* means mantra or mantras which are the 'structures of pure knowledge, the sounds of the Veda' [4, p. 3]. Commentators interpret this verse as indicating that the fourth and later constructions are to be carried out on the level of consciousness with mantras replacing the actual bricks. (See, for example, [8, 9].)

As with all the Vedic literature, the Śulba Sūtras can be read and interpreted on many levels. At the very least, they provide a fascinating chapter in the growth of geometrical and arithmetical knowledge and its application to the design and construction of complex brick platforms. But there are many indications, some of which have been pointed out above, that they are also a description, or perhaps a map, of the structure and qualities of the field of consciousness.

References

[1] B. Datta. *The Science of the Śulba.* Calcutta University Press, Calcutta, 1932.

[2] David W. Henderson. Square roots in the śulba sūtras. In C. Gorini, editor, *Geometry at Work*, pages 39–45. MAA, Washington, DC, 2000.

[3] Maharishi Mahesh Yogi. *Maharishi's Absolute Theory of Government.* MVU Press, Vlodrop, the Netherlands, 1993.

[4] Maharishi Mahesh Yogi. *Vedic Knowledge for Everyone.* MVU Press, Vlodrop, the Netherlands, 1994.

[5] MERU. *Scientific Research on Maharishi's Transcendental Meditation and TM Sidhi Programme: Collected Papers, Volumes 1–5*, 1977.

[6] Ajit Mookerjee. *Ritual Art of India.* Thames and Hudson, London, 1985.

[7] Satya Prakash and Usha Jyotishmati. *The Śulba Sūtras: Texts on the Vedic Geometry.* Ratna Kumari Svadhyaya Sansthana, Allahabad, India, 1979.

[8] S. N. Sen and A. K. Bag. *The Śulbasūtras of Baudhāyana, Āpastamba, Kātyāyana and Mānava.* Indian National Science Academy, New Delhi, India, 1983.

[9] George F. W. Thibaut. *Mathematics in the Making in Ancient India.* K. P. Bagchi & Co, Calcutta, India, 1984. This is a reprint of two articles by Thibaut: *On the Śulvasūtras*, J. Asiatic Soc. Bengal, 1875 and *Baudhāyana Śulva Sūtram*, Paṇḍit, 1874/5–1877.

Part 3

The Classroom

Ethnomathematics for the Geometry Curriculum

Marcia Ascher
524 Highland Road
Ithaca, New York 14850

1 Introduction

Ethnomathematics—the study of the mathematical ideas of traditional peoples—can make a unique contribution to the mathematics curriculum. In general, among mathematical ideas are ideas involving number, logic, spatial configuration and, more significant, the combination or organization of these into systems and structures. Mathematical ideas can occur in diverse contexts in any culture but, particularly in traditional cultures where there is no specific category called "mathematics", the ideas may be found embedded in contexts we categorize as, for example, art, navigation, record keeping, games, religion, or kinship. Here our specific concern is topics of interest in the teaching and learning of geometry.

Introducing ideas from traditional cultures into the classroom broadens the students' views in several ways. One major contribution of ethnomathematics is that students learn about other peoples and can come to appreciate their ideas. Another is that students meet with mathematical ideas as they arise in human contexts rather than as artifacts of textbooks and mathematics courses. And, when students confront expressions of ideas different from their own, they often come to see their own more clearly.

Three diverse suggestions for incorporating ethnomathematics into the collegiate mathematics curriculum are offered below. There are, of course, other possibilities but these provide a beginning.

2 Conceptions of Space

Of major importance is that cultures, in fact, differ in how they conceive of, discuss, and organize the space around them. In Western culture, space is viewed as having three dimensions, being uniform, continuous, and infinite. There is a focus on points, lines, and surfaces and a belief that these can be used to separate space into parts. The world we create around us is dominated by lines, flat surfaces, right angles, and rectangular solids. Our view of space led to and is reinforced by Euclidean geometry. An important issue in the history of mathematics is the relationship between Euclidean geometry and the truth about space. One of the reasons people believed that Euclidean geometry was the true representation of space was because they could visualize no other.

This issue can be addressed in the geometry curriculum by having students read Chapter 5 in *Ethnomathematics: A Multicultural View of Mathematical Ideas* [1]. The discussion there of the Navajo view of space/time (the two are inextricably combined and so cannot be described separately) will contrast with many of the most fundamental ideas in Western culture and so place the Western ideas in bold relief [1, pages 128–132]. For example, to most of our students Cartesian coordinates are quite familiar and, if they learn about polar and spherical coordinates, these latter often seem less straightforward and harder for them to visualize. For cultures in which circles rather than straight lines and rectangles are the dominant mode of conceptualization, the situation might be quite the reverse. And, to those in cultures where change and motion are an integral part of their spatial concepts, differential geometry might be the most obvious of all. The reading of contrasting statements—one by a pair of contemporary American mathematicians and the other by an Oglala Sioux—may stimulate an interesting class discussion on the cultural role and meaning of geometric forms [1, pages 124–125]. Still referring to this chapter, there is a discussion of the Inuit and their descriptors of place and renderings of reality. Perspective drawing, an often used application of geometric ideas, is intimately related to the Euclidean way of seeing in Western culture. The example of the Inuit will serve to highlight that this mode of representation is neither natural nor necessary; those who see space differently use modes of depictions in keeping with their spatial ideas [1, pages 132–140].

The suggestion that students be assigned this reading is different in kind from most suggestions that are made to instructors of geometry. It differs in that the instructor must forego the role of presenter and explicator of a topic with students as receptors. In this assignment, there are no constructions, proofs, or calculations. What is intended is that the instructor, who has greater sophistication in both geometry and reading, serves as discussion leader while the students discuss what they have read and the thoughts that it has stimulated. My own experience has been that students do react to the reading and welcome an opportunity to talk about it.

3   Symmetric Strip Patterns

Another suggestion for using ethnomathematics in the geometry curriculum is the investigation of strip patterns and their symmetry groups. There are numerous writings about this subject and it has been included by many instructors in a variety of courses (e.g., [8, 9]). However, since the creation of strip patterns is ubiquitous, examples can be drawn from a wide array of cultures. Realistic patterns are far more varied, complex, and visually appealing than those usually supplied in mathematics texts and, through their use, students see different expressions of a mathematical abstraction as they arise in meaningful human settings.

In support of the instructor's presentation, as an introduction to the analysis of strip patterns with one or two colors, Chapter 6 of [1] can be read by students. The reading includes strip patterns created by the Maori, the indigenous people of New Zealand, and by the Inca of sixteenth-century South America. The strips are analyzed mathematically but are also discussed within their cultural frameworks. A valuable resource for the instructor is the book *Symmetries of Culture: Theory and Practice of Plane Pattern Analysis* [10]. Not only does this book have numerous examples but, in addition, an appendix delves more deeply into the mathematical background by including proof that there are only four rigid motions of the plane and proof that there are only seven one-color symmetry groups for a strip.

The topic of strip patterns is attractive in that it can be presented at different levels of mathematical sophistication and expanded or contracted depending on the interest of the instructor and the time one wishes to allot. Student assignments that I have used include:

1) selection of a basic design unit and, with it, the creation of strip patterns representing the seven possible types (drawn large enough to be seen by everyone when held up in class);

2) analyzing sets of strip patterns drawn from other cultures; and

3) doing library research to locate a set of at least five strip patterns from one other culture (including no more than one simple translation), and then writing a paper including the cultural context of the strips as well as a mathematical analysis of them.

In my experience, the topic has provoked lively interest among students and has also increased their awareness of geometric aspects of decoration in their own environment. Although, at first, the library and writing portions of the last assignment meet some resistance, it is, to me, an important integrative contribution to the student's overall educational experience. And, some of the strips that the students locate raise a most important point: as is the case when any mathematical construct is being applied to a realistic situation, the students are faced with judgements as to the fit of the mathematical model.

4   Geometric Game Boards

A third and final suggestion is somewhat lighter than the first two and may be most appropriate for classes including future school teachers. It concerns a specific expression of geometric ideas by the herders of Mongolia. Among the Mongolian herders there is a long tradition of playing board games of strategy. The games were played primarily in association with New Year festivities in order to help assure a good year. Some of the board games are the same as those found widely dispersed throughout the world. These are in the category generally referred to as three-in-a-row games. However, the Mongolian herders have created more complicated versions of the games by geometrically elaborating the common game boards [2].

The common three-in-a-row games are played on the boards shown in Figure 1. The rules of the game differ slightly in different cultures; only the Mongolian rules are described here. In the games, each of two players has a set of markers distinguishable from the opponent's markers. For the games shown,

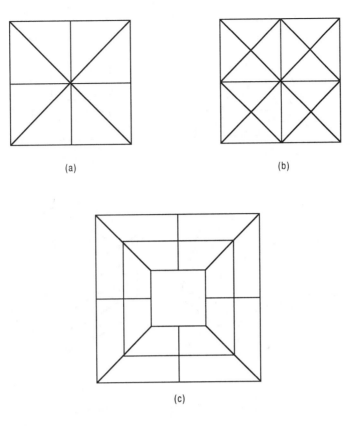

Figure 1: Boards in common with other cultures.

the number of markers per player are 3, 6, and 11 respectively. Each player, in turn, places a marker on the board with the eventual aim of forming three of his markers into a row and interfering with the opponent's forming of threes-in-a-row. Once all the markers are placed, the players take turns moving, one marker at a time, along a line to an empty point in an attempt to form threes-in-a-row. During this phase of the game, each time a player forms a row of three, he takes one of his opponent's markers off the board. Markers that are in a row of three cannot be taken. A player wins when his opponent is blocked from moving or has only two markers left. The strategy of play must consider the potential formation of rows of three but also the potential of the markers to move out of the rows to form new rows.

The board in Figure 1c is made up of squares, that is, of three regular 4-sided polygons. The Mongolian herders have geometrically elaborated this into a family of board games. The family of boards, shown in Figure 2, is made up of three regular n-sided polygons for $n = 3, 4, 5, 6$. (The rules for the games remain the same with $3n - 1$ markers per player.) Some simple classroom exercises involving the boards could be, for example, to find, in terms

of n, the total number of marker positions on each board, the number of different threes-in-a-row that are possible or, for each point on a board, the number of different threes-in-a-row that it can be part of. And, of course, the boards can be the basis for any other questions one usually asks about regular n-sided polygons.

Another set of elaborated boards created by the Mongolians is shown in Figure 3. These combine and extend the nesting of squares and the rotation of squares used on the common boards. For a construction process, I hypothesize that a square was constructed, the midpoints of its sides were found and connected, then the midpoints of its sides were found and connected, and so on. For the boards in Figure 3, the finding and connection of midpoints was repeated n times where $n = 4$ and $n = 6$. Then lines were drawn connecting corresponding midpoints and corresponding vertices. Other construction processes are possible, such as beginning with a set of nested squares and then connecting the midpoints of the sides of the squares to form the rotated squares. This, however, is less plausible since the original set of nested squares would be required to have specific relative spacing and specific relative side lengths. Assuming the hypothesized construc-

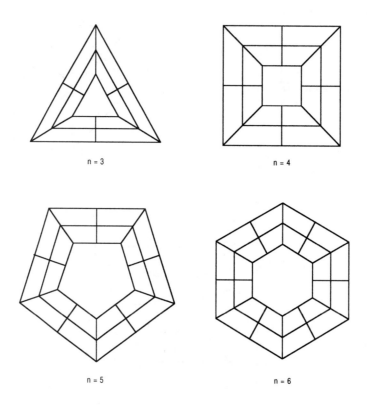

Figure 2: Boards with n-sided regular polygons [5, 7].

tion method, some questions, then, are to find, in terms of n, the resulting relative spacing and relative side lengths of the nested squares. Further, the numerical questions posed for the polygonal board games also apply to these boards but become somewhat more difficult, particularly if n is permitted to be odd as well as even. And, a wide variety of conjectures can be made and proved about the numerous triangles and subsquares.

The Mongolian board games can also serve to remind students that games of strategy played on geometric configurations have been long-time favorites in many cultures. In most cases the geometric forms and their properties are intimately linked with the rules of the games and with developing winning strategies.

5   Conclusion

The suggestions presented above are based primarily on my own investigations in ethnomathematics. Other writing in ethnomathematics, for instance [3, 4, 11], can provide further ideas for the geometry instructor. The world view presented in the geometry curriculum can be even further enriched and expanded by the inclusion of geometric ideas as they were expressed and elaborated in Indian,

Chinese, and Arab cultures (see, for example, [6]).

The three suggestions presented in this paper differ in the mathematical ideas they involve as well as in conceptual depth and cultural scope. The last is the most limited and straightforward; it is a specific geometric idea expressed in a specific context within a particular culture. However, like most mathematical ideas occurring in realistic settings, the geometric idea is not isolated from other ideas; here it is linked to the logic and strategy of a board game. The study of symmetric strips patterns is quite different because it deals with a type of spatial ordering that is widespread throughout many cultures and arises in different contexts from culture to culture or even within one culture. The strips are amenable to formal mathematical description and analysis, keeping in mind that our mode of analysis does not necessarily have any correspondence to the ideas of the strips' creators. The analysis, however, increases our insight into the spatial relationships within the strips and increases our ability to see structural similarities within cultural and stylistic differences. The first example, that of conceptions of space, is by far the most philosophical. A culture's conception of space is not a limited idea but rather a deep and significant part of a culture's world view. Each of us is so imbued with the spatial concepts of our own culture that it is difficult to visualize that of any other. Attempting to do so, however, is particularly appropriate for a mathematics curriculum in which students are being taught to deal with and visualize a variety of humanly constructed abstract systems.

References

[1] M. Ascher. *Ethnomathematics: A Multicultural View of Mathematical Ideas.* Chapman & Hall, New York, 1991.

[2] M. Ascher. Elaborating board games: An example from Mongolia. MS. submitted for publication, 1996.

[3] M. Ascher and U. D'Ambrosio, editors. *Special Issue on Ethnomathematics in Mathematics Education*, volume 14, number 2, of *For the Learning of Mathematics*. 1994.

[4] P. Gerdes and G. Bulafo. *Sipatsi: Technology, Art, and Geometry in Inhambane.* Institute Superior Pedagógico, Mozambique, 1994.

[5] I. Kabzińska-Stawarz. *Games of Mongolian Shepherds.* Institute of the History of Material Culture, Polish Academy of Sciences, Warsaw, 1991.

[6] V. J. Katz. *A History of Mathematics: An Introduction.* HarperCollins College Publishers, New York, 1993.

[7] A. Popova. Analyse formelle et classification des jeux de calculs Mongols. *Études Mongoles*, 5:7–60, 1974.

[8] D. Schattschneider. In black and white: How to create perfectly colored symmetric patterns. *Computers and Mathematics with Applications*, 12B:673–695, 1986.

[9] T. Sibley. Bridging the two cultures: Mathematics appreciation through symmetry. *Humanistic Mathematics Network Newsletter*, 2, 1988.

[10] D. K. Washburn and D. W. Crowe. *Symmetries of Culture: Theory and Practice of Plane Pattern Analysis.* University of Washington Press, Seattle, 1988.

[11] C. Zaslavsky. *Africa Counts: Number and Pattern in African Culture.* Lawrence Hill Books, New York, 1979.

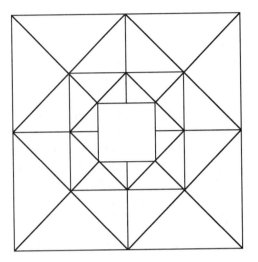

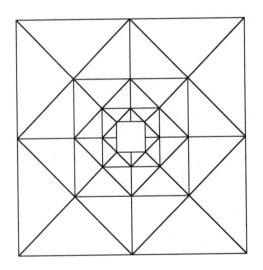

Figure 3: Boards with rotated squares [5, 7].

Education with Fascination: Teaching Descriptive Geometry with Applications

Marina V. Pokrovskaya

Associate Professor of Mathematics

Moscow State Technological University (named after Bauman)

1 Introduction

This article is devoted to the experience of teaching applications of descriptive geometry—or projective geometry—to geometry students.

The saying "Drawing is the international language of engineering and descriptive geometry is the grammar of that language" is old but true. Actually, a drawing is the designer's idea put down in lines. It is a guide to the worker and is the means of communication between technicians. Almost all man-made things around us as well as the tools used for their production were made using drawings done following the rules of descriptive geometry.

Gaspard Monge, the French geometer and a founder of the science of descriptive geometry, wrote that descriptive geometry has two goals. The first goal is to provide a method of representing objects with three dimensions—length, width, and height—on a sheet of paper having two dimensions only—just length and width. The second goal of descriptive geometry is to deduce from the exact description of the objects everything that follows from their shapes and respective positions.[1]

Nikolay Rynin, who made great contributions to the advance of the Russian school of descriptive geometry, said that the study of descriptive geometry is the best method for the development of the intuitive ability of the human spirit. He called this ability imagination and said that it could hardly be explained by means of the exact sciences. It became a step towards another human ability—creativity; no great discoveries could be made without it [1].

Without awareness of the fundamental basis of the graphical language of geometry, it is not possible to communicate with technically educated people, especially nowadays when international contacts are being spread and promoted. However, despite the appearance of the powerful means of computer graphics today, the level of basic knowledge in the field of geometry tends to be less. To my mind, one of the reasons for this tendency is in the overly abstract presentation of the subject. It is very difficult for many students to understand the internal logic of that science and to think in 3-dimensional space. Moreover, our critically thinking students want to realize the usefulness of learning this very abstract theory for their future engineering practice.

One of the ways of teaching a course in descriptive geometry properly is illustrated by a course using applications. We should support an interest in learning the subject through solving problems that give a clear correlation between abstract geometrical figures and situations on one side and real objects and real situations on the other side.

2 Teaching Descriptive Geometry

I would like to take an opportunity to share my experience of this teaching method with my colleagues and readers. I teach descriptive geometry at the Department of Engineering Graphics of Moscow State Technical University, where descriptive geometry and drawing have been studied since the last century (MSTU, which was founded in 1830, was originally called Moscow Higher Technical School).

The Russian method of technical education, in which theoretical knowledge and practical skills are both an integral part of the students' work, is a tra-

---

[1]Cet art a deux objets principaux. Le premier est de representer avec exactitude, sur des dessins qui n'ont que deux dimensions, les objets qui en ont trois... Le second objet de la descriptive geometrie est de deduire de la description exacte des corps tout ce qui suit nécessairement de leurs formes et de leurs positions respectives.

dition well-known all over the world. This method of education originated in Moscow Higher Technical School during the last century and won a number of honours, prizes, and medals at the World Exhibitions in Paris, Boston, and Philadelphia. Later, it was accepted as a background for teaching by many American technical institutes.

Continuing that tradition, I realize my subject as a kind of triptych. Its central part is the theory of descriptive geometry [3]; one leaf is its history; another leaf is its numerous applications. They complement each other well to fully represent the science.

To draw attention to the historical roots of this science, I have worked out a course entitled "History of Graphics" see [2]. It includes the stories about the wonderful world of prehistorical drawings which are painted on the walls of the ancient caves and rocks of Africa, Spain, and the Caucasus. The course enlightens students with material about Egyptian culture and drawings, the famous Greek mathematicians, and the Roman cartographers. There are stories about the origin of several branches of descriptive geometry: perspective, axonometric projections, orthographic projections, and projections with numerical scales. Students can also learn about the graphic culture of the Middle Ages and about the great icon painters of Early Russian Art who used reverse perspective.

The astonishing lives of the Frenchman Gaspard Monge, the German Albrecht Dürer, the Spaniard Betancourt, the Russian engineers—Peter Romanov (better known all over the world as the Russian Tzar Peter the Great) and Ivan Kulibin—are discussed here as well as many names, data, and facts which might not be as widely known. The course is illustrated with slides of authentic drawings.

3 Applications of Descriptive Geometry

On the other hand, I select a block of tasks covering all of the topics of the course to show applications of descriptive geometry. These are presented during the lectures and seminars or recommended for students' research work. Here are some requirements for selection of these assignments for the students.

1. They are to be varied, interesting, and rather modern.

2. The geometrical essence of the problems must be clearly revealed and they must not require special engineering data. It should be taken into account that we are dealing with first-year students.

3. Each of the problems must promote 3-dimensional thinking.

4. Various theoretical means are to be used for solving the problems to stimulate the creative ability of the students.

One can see geometrical shapes in the things around us: architectural constructions, the patterns of highways, bridges, TV-towers, cars, ships, airplanes, spaceships, etc. Descriptive geometry helps one to visualize data in other sciences, including physics, chemistry, and biology. Painters, sculptors, and landscape artists use the rules of perspective in their work. Geometry exists not only around us, but inside us as well. Geometrical ideas and philosophy have walked hand in hand since ancient times. For example, Plato studied the regular polyhedrons that were identified at that time with the elements Fire, Water, Air, and Earth.

Here are several examples of applied problems that students can work on.

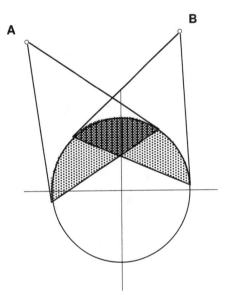

Figure 1: Simultaneous surveying by sputniks.

Problem 1. (See Figure 1). Find the zone of simultaneous surveying by two sputniks located at points A and B (For the topic "A tangency of surfaces").

Problem 2. (See Figure 2). A bullet made a hole in a window at the point A and stuck in the wall at the point B. Find the point C where the killer was on the site behind the house (for the topic "The traces of a straight line").

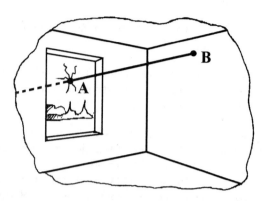

Figure 2: Traces of a straight line.

Problem 3. (See Figure 3). Construct the projections of a ray reflected from a given prism (for the topic "The measurement of angles").

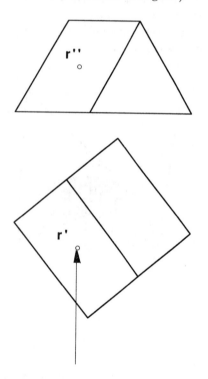

Figure 3: Prism reflections.

Problem 4. (See Figure 4). Find the depth between a bore-hole A and an oil-bed (for the topic "Measurement of distances").

I believe the most effective way to teach applications of geometry to geometry students is through the students' scientific research work. I have had

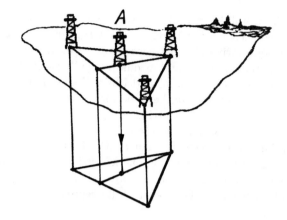

Figure 4: Measurement of distances.

wide experience of this kind of work with the students of our university and each year I help my students to prepare reports for our annual student conference.

One such conference was entitled "Geometry Around Us." It was an attempt to apply geometry to many diversified areas of the sciences, engineering, and fine arts. Two dozen first-year students took part in that conference. Their reports were illustrated with posters made in the traditional manual manner, with wooden and electronic models, and with computer graphics. Some of their topics were as follows: "Geometry in Bionics," "Geometry in Electronics," "Geometry and Philosophy," "The Transformation of Cupola Shapes of Ancient Russian Churches," "The Reverse Perspective in Russian Icon Painting," and "Geometric Poetry." To deepen the knowledge of our Russian students in foreign languages, we decided that they should present their reports in English, German, and French.

On the other hand, another goal, namely, to apply the diversified sciences for solving geometry problems, was the topic of another conference arranged by a special group of students.

4   Projects with Exceptional Students

Three years ago, a group of students with hearing loss began studying at the Educational Center for the Deaf at the Moscow State Technical University under the supervision of the Rochester Institute of Technology. The first year was very hard but fruitful both for the students and for us, their teachers. Six young men and one young woman completed original research on a complex application of the fundamental sciences to an investigation of a certain specific problem. This work was done in addition to the usual credit courses.

Each of the students dealt with one of the aspects of the problem and each aspect was based on one of the fundamental sciences. The topic of their research was the properties of a torus and its sections; the title of their conference was "That Wonderful Many-Faced Torus." The torus (also called a ring, annular torus, or a closed torus) is one of the most striking geometrical figures. Its surface is formed by the revolution of a circle or its arc about some axis. The kind of torus depends on the relationship between the radius r of the circle and the distance R from its center to the axis.

The sections of the torus by a plane parallel to the axis are called Cassini Ovals. The kinds of sections depend on the p-parameter (the distance between the cutting plane and the torus axis). For example, for $p = R + r$, it is a single point; for $p = R - r$, it is a Bernoulli lemniscate; and for $p = 0$, it is two circles.

It is interesting to note that the names of these curves are associated with the names of great mathematicians. The Cassini Ovals were discovered by the 17th-century astronomer Jean Dominique Cassini when he was trying to determine the orbit of the Earth. The first time the lemniscate equation was mentioned was in the article "Acta Eruditorum" (1694) written by Jacob Bernoulli.

The generation of the torus surfaces and their sections can be represented in different ways. They can be described by mathematical equations, and there are also several means of graphical representation. One of them is a traditional method of descriptive geometry using auxiliary lines. The lemniscate can be constructed with the help of a hinged mechanism (see Figure 5). This method uses the property that each point M of the lemniscate satisfies the equation $MA \times MB = a^2$, where A and B are the focuses of the lemniscate.

The projections (horizontal, frontal, and profile) and axonometric views of each of the toruses can be visualized by the students with the help of AutoCAD (See Figure 6). A program using Pascal was written to show the construction of the torus sections dynamically. This program enables one to indicate the location of the cutting plane on a horizontal projection and represent the projections of the superimposed sections. The whole set of the removed sections forms the framework of a surface.

The construction of the planes tangent to any surface is very important in engineering. From the

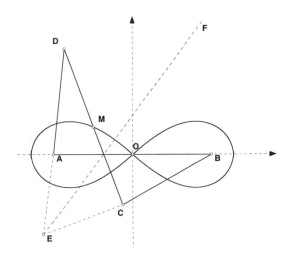

Figure 5: Lemniscate constructed by a mechanism.

point view of differential geometry, there are three kinds of points: elliptical, parabolic and hyperbolic. If the tangent plane at a point has only a single point of intersection with the surface near the point of tangency, this point is *elliptical*. If the point is a saddle point, the point of tangency is *hyperbolic*. If a tangent plane is tangent to all the points on a curve on the surface, the points of the curve are called *parabolic*. The surface of a ring is unique because there are points of these three kinds on it.

The generation of the Cassini Ovals is a geometrical illustration of one of the dialectic laws, namely, the law of transition from quantity into quality. By changing the quantitative parameter p of the cutting plane, the section figures change their quality. They are as follows: a point, a closed convex curve, a closed concave curve, and a lemniscate. Furthermore, a qualitative leap occurs when one curve is broken into two curves.

Torus surfaces are widely applied in engineering. For example, they are used in wheels, flywheels, and elements of water pipes. Scientists predict that space stations may look like a ring one day. We have given here only some fragments illustrating several properties of that wonderful geometrical figure. So, several fundamental sciences were involved in that research. Moreover, two versions of that report were made by the students, one in Russian and one in English.

From the point of view of method, such an approach increases students' interest in scientific observation, helps them to realize contacts between the fundamental sciences, and cultivates interest in creative research and work with scientific literature.

As for me, the psychological aspect of that work

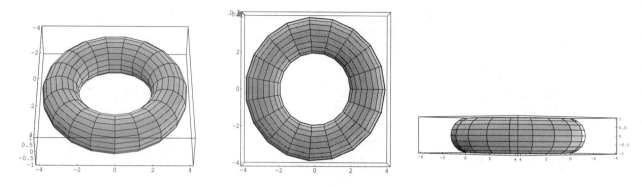

Figure 6: Three views of the torus.

was the most important. I remember those students with hearing loss at the beginning of their education in MSTU, their uneasiness that September, and their difficulties in their communication with other students. And I liked very much helping them to overcome those psychological barriers.

By means of work like this, the students were united as a team for joint research. This approach inspired mutual understanding, responsibility, and competition; it taught the students initial skills of making public presentations; and it gave them the experience of fluent communication to an audience.

Our next conference was devoted to "The Help Project." The main goal of that project was to support people with physical disabilities with the help of the students. We had several aims: teaching the disabled the ABC's of computer practice including computer graphics, providing them with new occupational skills and habits to adapt to their future professional activity, and arousing their interests in life and improving their spirits. Our first practical step in that direction was setting up a computer class in the Department of Spinal Diseases of Moscow District Hospital No. 6.

In addition, work like this can initiate the professional creativity of our healthy students, provide them with teaching experience, help them to realize their social responsibility, and make them more compassionate and sympathetic.

I do hope that our joint efforts will help the students to overcome different and difficult barriers in education and continue successfully their penetration into the wonderful world of science.

References

[1] Н. А. Рынин. Значение начертательной геометрии и сравнительная оценка главнейших ее медогов. Спб., 1907. (N. A. Rynin, *The Importance of Descriptive Geometry and a Comparative Appraisal of its Main Methods*, S. Pb.)

[2] С. А. Фролов и М. В. Покровская. В поисках начала: Рассказы о начертательной геометрии. Вышэйшая школа, Минск, 1985. (S. A. Frolov and M. V. Pokrovskaya, *In Search of the Origin: Stories of Descriptive Geometry*, Higher School Publishing House, Minsk.)

[3] С. А. Фролов и М. В. Покровская. Начертательная геометрия: что это такое? Вышэйшая школа, Минск, 1986. (S. A. Frolov and M. V. Pokrovskaya, *Descriptive Geometry: What is It?*, Higher School Publishing House, Minsk.)

Part 4

Engineering

Making Measurements on Curved Surfaces

James Casey
Department of Mechanical Engineering
University of California at Berkeley
Berkeley, CA 94720
jcasey@newton.berkeley.edu

1 Introduction

All around us, in natural forms, art, architecture, engineering, and consumer products, we find surfaces whose geometry is vastly richer than the Euclidean geometry we learn in high school. To treat surface geometry analytically requires rather advanced tools, but it is possible to gain much insight into the subject by performing measurements on the surfaces of physical objects. Beem [1] suggested some activities along these lines, and Henderson [6] has made ample use of physical models. Casey [3, 4] has described a series of experiments that are designed to bring modern geometrical concepts within the reach of students who might not yet be in a position to study differential geometry formally.

In the present paper, as an illustration, a method of estimating the Gaussian curvature of surfaces from surface measurements is described.

2 Gaussian Curvature and Parallel Transport

The concept of intrinsic, or Gaussian, curvature of a surface is best understood in terms of Gauss's original description, introduced in his celebrated paper, "General Investigations of Curved Surfaces", presented in Latin in 1827 (an English translation is readily available: [5]).

Consider a piece $\mathcal{S}$ of surface, of the type illustrated in Figure 1a: you may think of the piece as being obtainable by deforming an open disk of elastic material; no tearing, cutting, puncturing, joining, or overlapping of the material is allowed. Suppose, additionally, that at every point of the piece, a tangent plane exists, and that these tangent planes change in a smooth manner as one traverses $\mathcal{S}$. Construct a continuous unit normal vector field on $\mathcal{S}$, and let $\mathbf{N}$ denote the unit normal at a generic

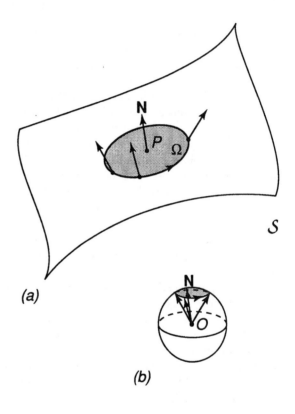

(a)

(b)

Figure 1: Gaussian curvature.

point P of $\mathcal{S}$.

Next, take a unit sphere centered at an arbitrary point O in space (Figure 1b). Translate the vector $\mathbf{N}$ so that its tail is at O; its tip will then touch the sphere at a unique point. This assignment of points of the unit sphere to points of $\mathcal{S}$ is called the *Gauss map*.

Consider now a simple closed curve on $\mathcal{S}$ that encloses a region Ω about P. Let a be the area of Ω. Through the Gauss map, a region of the sphere is made to correspond to Ω. Let the area of the latter region be $\bar{a}$. As one traverses the boundary of Ω

in the counterclockwise direction, if the boundary of the corresponding portion of the sphere is traversed also in the counterclockwise direction, then $\bar{a}$ is reckoned as positive, and if the boundary of the portion of the sphere is traversed in the clockwise direction, then $\bar{a}$ is reckoned as negative. $\bar{a}$ is the *total curvature* of Ω.

The *Gaussian curvature* at P is defined by the limit

$$K = \lim_{a \to 0} \frac{\bar{a}}{a}. \tag{1}$$

For the Euclidean plane, $K = 0$ everywhere, and likewise for a cylinder. For a sphere of radius R, $K = 1/R^2$. For a saddle-shaped surface, K is negative. On a torus, K is positive on the very outside, is negative on the very inside, and is zero on two circles in between. Gauss proved that K is an intrinsic quantity: it can be determined from measurements carried out on the surface itself, without any reference to the manner in which the surface occupies space. Hence, if the surface is bent without stretching any line in it, K will be unaltered. Thus, cylinders and cones, and any other surface that can be locally developed on a plane—or flattened—have zero Gaussian curvature. Again, if you cut a piece out of a thin rubber ball of radius R, and bend it into any shape, its Gaussian curvature is still $1/R^2$. K is a pointwise measure of the non-Euclideanness of a surface.

Euclidean parallelism breaks down for surfaces having nonzero Gaussian curvature. However, a beautiful generalization of the concept was introduced by Tullio Levi-Civita in 1917 [7]. It may be described in physical terms as follows:

(*i*) Let AB be any curve on the piece of surface S in Figure 1, and let $\mathbf{v}$ be any vector lying in the tangent plane at A (Figure 2).

(*ii*) Cut a narrow strip of squared paper in a shape that covers AB (you may have to tape some strips end to end to do so). Trace AB and $\mathbf{v}$ on the strip.

(*iii*) Lay the strip flat on a plane and translate $\mathbf{v}$ along it by drawing Euclidean parallels.

(*iv*) Lay the strip along AB again and secure it with tape. You now have a set of vectors lying in tangent planes to S along AB.

By taking different paths from A to B, it can be readily seen that parallel transport is path-dependent in general. This phenomenon is called *holonomy*.

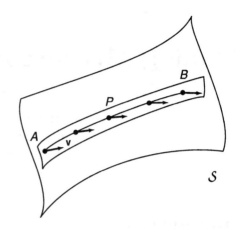

Figure 2: Parallel transport.

Suppose that we parallel transport a vector around the closed curve that forms the boundary of the region Ω in Figure 1. As a result of holonomy, the vector will have rotated through some angle ψ when we return to the starting position. It can be shown that the holonomy angle is equal to the total curvature of Ω (see Spivak [9, Vol. III, pp. 390–391]:

$$\psi = \bar{a} = \int_{\Omega} K\, da. \tag{2}$$

3 Estimating the Gaussian Curvature from Surface Measurements

For a sufficiently small region Ω, Equation (2) may be approximated by

$$\psi \approx Ka. \tag{3}$$

A measurement of the holonomy angle together with a measurement of the area a then furnish an approximation to the Gaussian curvature at P. I tried out this scheme on a honeydew melon. The circumference of the melon along the line EW (its "equator") was 18 in, and the equatorial cross-section was close to being circular. The polar circumference was 20.5 in. See the photograph in Figure 5.

Procedure

(1) From squared paper, cut out a strip $ABCD$ of the type shown in Figure 3. There is a cut at A. (I used a 3×3 inch strip.) At A, draw a vector $\mathbf{v}$ lying along AB, and translate

it around the square $ABCD$: it will always lie along horizontal lines, and in particular will be perpendicular to DA as one approaches the cut at A.

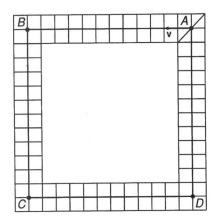

Figure 3: A strip of squared paper.

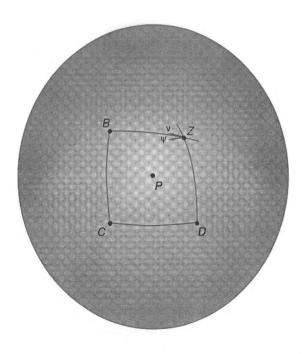

Figure 4: Holonomy on a melon.

(2) Place the vertex C on the melon and tape it. Gently run your finger along CB and apply tape at B. Do the same for CD. Carefully run your finger along DA and tape the end of the strip at A. Do the same for BA. The line $ABCD$ in Figure 3 will now appear as illustrated in Figure 4. The segments BA and DA meet at some new point Z, and the vector $\mathbf{v}$ experiences a holonomy angle ψ.

(3) Measure the angle BZD using a plastic protractor. (Alternatively, place a narrower strip along BZ, and another along DZ, tape them together at Z, and then remove them and measure the angle between them on a plane surface.)

(4) Cut strips of paper one square in width and carefully fill the region enclosed by $ZBCDZ$ with them. (This is a tedious process, but it is worth doing carefully.) Count the squares and fractions thereof to estimate the area of the region.

(5) Use the approximate expression (3) to estimate the Gaussian curvature near the center of the region.

I did two tests of the above type on a region of the melon (See Figure 5). For Test 1, I found that

$$\psi \approx 116° - 90° = 0.454 \text{ rad},$$
$$a \approx 148 \times (0.2)^2 \text{ in}^2 = 5.92 \text{ in}^2, \qquad (4)$$

and hence
$$K \approx 0.077/\text{ in}^2, \quad \frac{1}{K} \approx 13.0 \text{ in}^2. \qquad (5)$$

For Test 2, I got

$$\psi \approx 123.5° - 90° = 0.585 \text{ rad},$$
$$a \approx 189.5 \times (0.2)^2 \text{ in}^2 = 7.58 \text{ in}^2, \qquad (6)$$

and hence
$$K \approx 0.077/\text{in}^2, \quad \frac{1}{K} \approx 13.0 \text{ in}^2. \qquad (7)$$

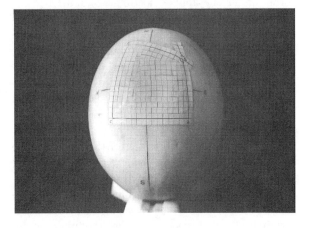

Figure 5: Photograph of the melon.

I also cut the melon through a polar plane and tried to estimate the curvature of the resulting curve, by measuring the change in direction of the tangent vector and dividing by the length of arc. (I traced the curve on cardboard and used string, a ruler, and a protractor.) This yielded an approximate radius of curvature $R_2 \approx 4.125$ in/1.02 rad ≈ 4.04 in. For the radius of the equatorial section, we have the approximate value $R_1 \approx 18$ in/$2\pi \approx 2.86$ in. I cut the two halves of the melon through their equatorial planes, and held the quarters together so I could trace the equatorial section on cardboard. I estimated a radius of curvature of $R_1 \approx 3.0625$ in/1.13 rad ≈ 2.71 in, which is pretty close to the previously estimated value of 2.86 in.

If our values for R_1 and R_2 were exact, their product would be equal to the reciprocal of the Gaussian curvature. We see, however, that $R_1 R_2 \approx 10.9$ in^2, which is significantly less than the value of 13.0 in^2 that we found for $1/K$ using surface measurements. The difference is largely due to the difficulty of drawing tangents to a noncircular curve with sufficient accuracy.

4 How Do Thin Structures Carry Loads?

Besides being of obvious interest to mathematicians, surface geometry is also of importance to engineers and scientists. For example, in designing thin-walled structures such as aircraft fuselages and wings, fuel tanks, and domes, engineers have to analyze how loads are supported by and distributed throughout a structure, and how the structure deforms under the applied loading. Similarly, in order to understand how such biological structures as blood vessels, the heart wall, the bladder, and cell walls function mechanically, bioengineers calculate the stresses (loads per unit area) and the deformations that occur in these structures.

Engineers model thin structures as mathematical surfaces that are capable of deforming under loads. The loads are carried primarily by some combination of bending and stretching of the surface. The relationship between the changes in geometry and the forces required to effect them are studied experimentally. Using these studies as a basis, a theory of the mechanical behavior of shell structures is formulated. Shell theory furnishes the fundamental equations that make structural calculations possible.

I would like to close this paper with some activities that may help students develop their intuitions about how surfaces respond to applied loads.

4.1 Bending

In bending a mathematical surface, no line is to experience a change in length. For thin physical objects undergoing bending, fibers on the convex side of the object are stretched a little, while fibres on the concave side are compressed a little: on some surface in between, called the *middle surface*, no stretching takes place. Here are some examples:

(a) Take a steel or plastic ruler, hold it partially over the edge of a table and press on the overhanging end. Watch how its radius of curvature changes as you apply more force. Apply equal force at different projecting lengths to study the relationship between torque (force times distance) and curvature.

(b) Take a wooden yardstick, two pieces of string, and a spring balance. Tie a piece of string to each end of the yardstick, and attach the spring balance to one of the strings. Press on the yardstick to cause it to bow out. Tie the other string to the spring balance and measure the load required to keep the yardstick buckled. For more details, see Casey [2].

(c) Study how a thin-walled cardboard cylinder deforms as you press it radially against a table. Cut the cylinder longitudinally and now study its mechanical behavior.

(d) Make a band from a narrow strip of paper and turn the band inside out.

(e) Take a retractable steel measuring tape and pull out a few feet of it. Notice that the tape is curved across its width. Hold the tape with the convex side facing you. Now gradually apply torques at the ends of the extended piece: you induce longitudinal curvature. As you increase the torques, the tape will suddenly buckle. The longitudinal curvature will occur over a much smaller portion of the tape but will be larger than before, while the transverse curvature will be eliminated over that portion.(For a discussion of this and other interesting buckling phenomena, see Sewell [8].)

4.2 Stretching

In the following examples, stretching of the materials occurs:

(a) Tie one end of a spring balance to an immobile support. Hook a rubber band on the other end. Measure the force required to cause different changes in length.

(b) Take a piece of rubber sheet (exercising material is ideal) and place it on the curved surface of a solid object, such as a steel bowl. Pull on opposite sides of the sheet as you hold it against the curved surface.

(c) Blow up a balloon or tube. How does the pressure vary with inflation?

4.3 Bending and Stretching

In the following examples, the material is both bent and stretched:

(a) Press on opposite sides of an inflated balloon or tube.

(b) Press on a stretched rubber band.

(c) Stretch a rubber sheet on a curved cardboard surface and bend the cardboard.

References

[1] J. R. Beem. Measurement on surfaces. *1976 Handbook of the National Council of Teachers of Mathematics*, pages 156–162, 1976.

[2] J. Casey. The elasticity of wood. *The Physics Teacher*, 31:286–288, 1993.

[3] J. Casey. Using a surface triangle to explore curvature. *Mathematics Teacher*, 87:69–77, 1994.

[4] J. Casey. *Exploring Curvature*. Vieweg, 1996.

[5] K. F. Gauss. *General Investigations of Curved Surfaces*. Translated by A. Hiltebeitel and J. Morehead, Raven Press, 1965.

[6] D. W. Henderson. *Exploring Geometry on Plane and Sphere*. Prentice-Hall, 1996.

[7] T. Levi-Civita. *The Absolute Differential Calculus*. Blackie and Son Limited, 1926. Reprinted by Dover Publications, 1977.

[8] M. J. Sewell. Demonstrations of buckling by bifurcation and snapping. In *Elasticity— Mathematical Methods and Applications (The Ian N. Sneddon 70th Birthday Volume)*, pages 315–345. John Wiley and Sons, 1990.

[9] M. Spivak. *A Comprehensive Introduction to Differential Geometry*, volume 3. Publish or Perish, 1970.

Mathematics to the Aid of Surgeons

Ramin Shahidi

Assistant Professor of Neurosurgery

Stanford University Medical Center

1 Introduction

Historically, the methods of calculus have developed in close connection with physical applications ranging from Newton's laws of motion to Einstein's theory of relativity. In recent years these methods have played a vital role in the development of robotics and computer sciences as well. Whether it is a robotic vehicle with automatic navigation on the surface of Mars trying to find the best path from point A to point B while avoiding obstacles, or it is a supercomputer trying to use artificial intelligence to learn a new language, all such tasks require mathematical methods. This mathematics can come in the form of artistic impressions such as fractals or significant complex theoretical solutions to aid daily engineering problems.

One of the most recent applications of mathematics has flourished in the form of computer vision. Computer vision is the basic science of making computers see. What we mean by "seeing" is not the complex form of vision as we humans know it, but the basic functions of distinguishing objects from each other by detecting their colors, edges, and their spatial relationships such as depth and orientation. In this section, we discuss some applications of elementary trigonometry to complex decision problems that arise in the tracking of surgical instrumentation (surgical devices, surgical microscopes, etc.) in the operating room, which is a form of computer vision.

Let us start by asking why we need to track instruments in the operating room. To answer this question, we can start with the two-dimensional case shown in Figure 1.

Assume that Figure 1 represents a cross section of a patient's head with an abnormality (lesion) shown in black. Let's call the coordinate system that is used to calibrate the patient's head the *surgical coordinates* or, as it is known to surgeons, *stereotactic space*. Two points (P_0 and P_1) on a given surgical device are calibrated with respect to a predefined stereotactic space. Using the simple equation of

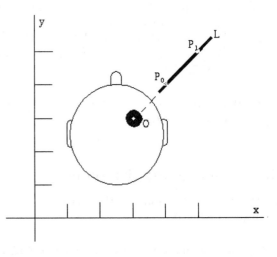

Figure 1: A two-dimensional illustration of surgical trajectory approaching a tumor.

a line, the spatial relationship between the surgical instruments (the line passing between the two points P_0 and P_1) and the center of the abnormality (point O) can be defined. If an imaginary line is drawn extending the surgical device's trajectory (line L), this line should hit the center of the lesion. In summary, tracking the patient's head and surgical instrumentation in the operating room would give surgeons the ability to locate the abnormality within the body prior to an incision. This type of procedure is known as *stereotactic surgery*.

In order for such an implementation to occur, the following are required:

1. Obtaining the patient's imaging data.

2. Tracking a moving pointer in the surgical space using a computer vision approach.

3. Superimposing the patient's volumetric data onto the actual physical anatomy.

Even though these tasks seem to be very complicated, we will show that all of them are derived from simple mathematical methodologies.

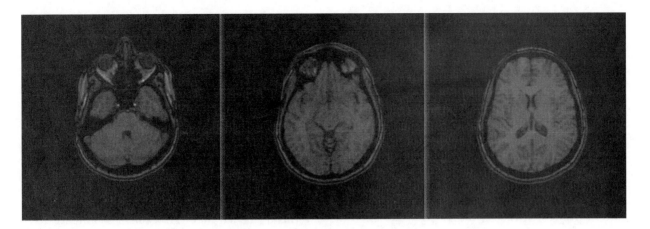

Figure 2: Cross sections of the brain.

2 Obtaining Patient's Imaging Data

A patient's data can be obtained through Magnetic Resonance Imaging (MRI) or Computed Tomography (CT); MRI and CT are also known as imaging machinery. Using imaging machinery, the internal anatomy of a patient's body can be visualized. Cross sections of a patient's anatomy can be obtained in the form of a series of given slices with gray-scale values representing different tissue types. Figure 2 shows three MRI cross sections of the author's brain.

Once a series of these spatially contiguous and aligned 2D axial slices is obtained, they can be reconstructed into a three-dimensional object. This three-dimensional model should resemble the patient's physical anatomy. One popular technique of presenting 3D structures is surface rendering. These kinds of surface display algorithms are based on conventional surface models, which represent a surface by gatherings of millions of small surface patches. These small surface patches or *polygons* are simply created by passing a small plane through any given three points on the surface of the object (Figure 3). The accumulation of all these small patches creates the surface of the object in whole. Figure 4 illustrates an example of such a three-dimensional model. In this figure the surface representations of the brain, the vascular structures, and the surface of the face are shown using polygon meshing.

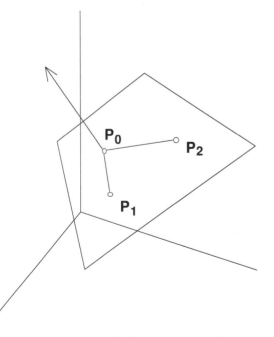

Figure 3: Plane through three points and normal vector

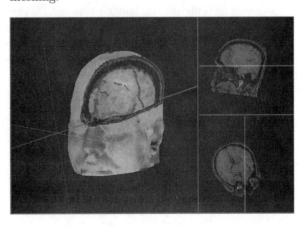

Figure 4: Three-dimensional model of the patient's head.

3   Defining Surgical Coordinates Using Computer Vision

The heart of this project involves the development of a technique which provides surgeons with the ability to define the coordinates of a given point in a surgical space. The basic principle involved in the recovery of depth by using two sets of camera images is triangulation. In this technique, the location of the object must be determined using features from two images that correspond to some physical feature in space. Then, provided that the positions of the centers of projection, the effective focal length, the orientation of the optical axis, and the sampling interval of each camera are all known, the depth can be reconstructed using triangulation. Based upon this correspondence problem, a particular paradigm can be constructed which depends upon the specific matching features used, the number of cameras used, the positioning of the cameras, and the scene domain. We can calculate the depth of a point from a pair of cameras mounted on a platform. These cameras can translate horizontally and vertically, and rotate left/right and up/down. The focal lengths of the cameras and their viewing angles can also be adjusted and are known. Further, any pair of cameras can verge by rotating towards each other. The maximum vergence angle is approximately six degrees.

As it is shown in Figure 5, we model each lens as a pinhole, assuming that up to the first order all lines of sight intersect at a unique lens center. The distance between the centers O_L and O_R of the two lenses is specified as b, and the focal length (from the lens center to the image plane) as f. The frames L and R are associated with the simulated left and right camera frames, with origins at the lens centers, O_L on the left and O_R on the right. The Z axes Z_L and Z_R are along the optical axes, with positive direction pointing toward the scene; the X axes X_L and X_R are perpendicular to the Z axes and also point toward the scene. There is also a defined reference frame W with the origin O_W at the midpoint of the baseline (with Z-axis normal to the baseline and X-axis pointing along the baseline toward O_R). The vergence angle is the angle Θ between the baseline $O_L O_R$ and the camera's X axis X_L.

The goal of this reconstruction algorithm is to identify the spatial position of the point P with reference to the frame W as $P = (X_W, Y_W, Z_W)$. The input to this algorithm is the 2D projections of the point P to the L and R frames as (X_L, Y_L) and (X_R, Y_R). The primary goal is to determine the depth (Z_W) as closely as possible to its actual value in space.

In Figure 5, suppose that we rotate the left camera about the left lens center by the angle Δ, and that we rotate the right camera about the right lens center by the angle Δ'. Kortokov, a 20th century Russian scientist, has implemented an analytical relationship between range, disparity, and vergence angle [4]. Using this method, the coordinates of the point $P = (X_W, Y_W, Z_W)$ can be expressed in terms of the left and right camera coordinates.

The distance normal to the baseline, from the baseline to the object point, is Z_W. It is convenient to measure distances not from the baseline, but from a plane attached to the platform that supports the cameras, and parallel to the baseline. We accomplish this by writing $Z = Z_W + Z_0$, where Z_0 is the baseline offset (distance from the baseline to O_L or O_R). Using this algorithm, the 3D coordinates of one point and in that manner a series of points can be calculated as follows.

Let Δ be the angle that the axis X_L parallel to the right image plane makes with the ray PO_L from the point P to the left lens at O_L and let Δ' be the angle that the axis X_R parallel to the right image plane makes with the ray from the point P to the right lens at O_R. Then

$$
\begin{aligned}
\sin(\Theta + \Delta) &= Z_L / \overline{O_L P} \\
\cos(\Theta + \Delta) &= X_L / \overline{O_L P} \\
\sin(\Delta) &= Z_W / \overline{O_L P} \\
\cos(\Delta) &= \frac{\left(\frac{b}{2} + X_W\right)}{\overline{O_L P}}
\end{aligned}
$$

from which we get

$$
\begin{bmatrix} X_L \\ Y_L \\ Z_L \end{bmatrix} = \begin{bmatrix} (X_W + \frac{b}{2})\cos\Theta - (Z_W \sin\Theta) \\ Y_W \\ (X_W + \frac{b}{2})\sin\Theta + (Z_W \cos\Theta) \end{bmatrix}.
$$

Similarly on the right side,

$$
\begin{aligned}
\sin(\Theta' + \Delta') &= Z_R / \overline{O_R P} \\
\cos(\Theta' + \Delta') &= X_R / \overline{O_R P} \\
\sin(\Delta') &= Z_W / \overline{O_R P} \\
\cos(\Delta') &= \frac{\left(\frac{b}{2} - X_W\right)}{\overline{O_R P}}
\end{aligned}
$$

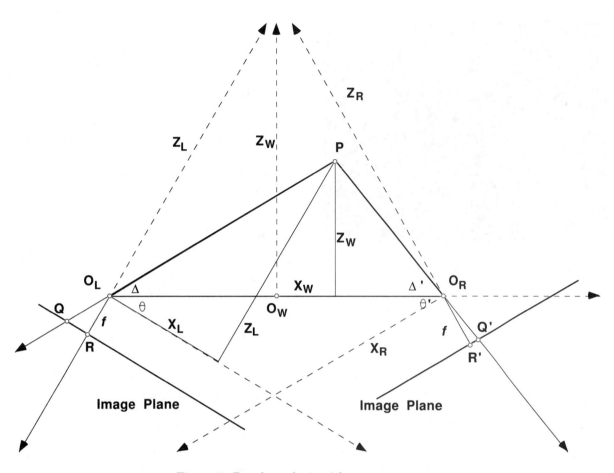

Figure 5: Depth analysis with stereo cameras.

from which we get

$$\begin{bmatrix} X_R \\ Y_R \\ Z_R \end{bmatrix} = \begin{bmatrix} \left(X_W - \frac{b}{2}\right)\cos\Theta' + (Z_W\sin\Theta') \\ Y_W \\ \left(-X_W + \frac{b}{2}\right)\sin\Theta' + (Z_W\cos\Theta') \end{bmatrix}.$$

Using these values, we are able to obtain the lengths of segments QR on the left and $Q'R'$ on the right:

$$QR = f\frac{\left(X_W + \frac{b}{2}\right)\cos\Theta - (Z_W\sin\Theta)}{\left(X_W + \frac{b}{2}\right)\sin\Theta + (Z_W\cos\Theta)}$$

$$Q'R' = f\frac{\left(X_W - \frac{b}{2}\right)\cos\Theta' + (Z_W\sin\Theta')}{\left(-X_W + \frac{b}{2}\right)\sin\Theta' + (Z_W\cos\Theta')}.$$

Solving each of these equations for X_W and then eliminating X_W allows us to obtain the following value for Z_W:

$$Z_W = \frac{b}{\dfrac{f\sin\Theta + X_L\cos\Theta}{f\cos\Theta - X_L\sin\Theta} + \dfrac{f\sin\Theta' - X_R\cos\Theta'}{f\cos\Theta' + X_R\sin\Theta'}}.$$

In Figure 1, since we can define the spatial coordinates of points P_0 and P_1 using this method and

since the distance of the tip of the pointer from either P_0 or P_1 is known, then the coordinates of the tip of the pointer in our surgical space can also be calculated.

4   Superimposing the Actual Anatomy on the 3D Model

So far we have managed to track a pointer in our surgical space or stereotactic space. At this point we have to superimpose the patient's actual anatomy on the patient's 3D model. To do this, we rely on the mathematical fact that if we can register four points (which are not in the same plane) in any two volumes, then we have registered the volumes. In order to accomplish this, the surgeon would place the tip of the pointer at a known anatomical landmark of the patient's actual anatomy, e.g., the tip of the nose. The position of the tip of the pointer's model in the 3D space would be defined to correspond with the same anatomical landmark (the tip of the nose on the patient's 3D model). This

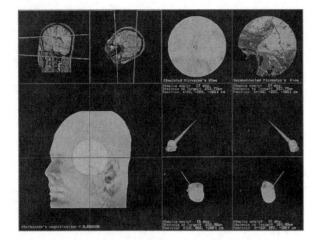

Figure 6: Simulation of the surgical setup. The reconstructed microscope's view is shown in the two top right pictures and the four camera views are shown on the bottom right.

will be repeated four times for four known anatomical landmarks. This procedure would superimpose the patient's 3D model with the patient's physical anatomy, so as we move the pointer around the patient's head, the model of the pointer in the computer will be moving around the patient's 3D model with the same orientation.

Looking back at Figure 1, we can compute the angle of the pointer's trajectory by using the equation of the line through points P_0 and P_1. Then if we rotate the 3D data set in an opposite angle, the viewing projection will be the same as the surgical pointer's angle of attack (its bird's-eye view). If we assume that all of the critical organs are interactively or automatically delineated from imaging modalities and that different trajectories are related to different possible entry points, then the target can be generated and the most appropriate trajectory selected. This will allow the surgeon to continually investigate the outcome of each step of the surgery. It will also allow the surgeon to "fly" through and around the surgical site in order to quantitatively determine the best solution among all available scenarios. Figure 6 illustrates the simulation of such a setup.

5   Conclusion

Simple geometric methods are necessary to transform the kind of data recorded using MRI and CT into models and images that can be of use to the surgeon. Using the mathematical methods mentioned in this paper, we have been able to develop a volumetric image navigation system for stereotactic surgery. This system integrates many recent developments in computational visualization, medical image analysis, and virtual reality, all of which are elegantly driven by simple geometric models. The real-world implementation of this model requires a set of stereo cameras and a laser targeting system. Our research group at Stanford University Medical School Image Guidance Labs is currently working to transform this theoretical model into real-life surgical practice.

References

[1] N. Alvertos, D. Brzakovic, and R. C. Gonzalez. Camera geometries for image matching in 3D machine vision. *IEEE PAMI*, 11(9), 1989.

[2] U. R. Dhond and J. K. Aggarwal. Structure from stereo—a review. *IEEE PAMI*, 19(6), 1989.

[3] J. Flifla, R. Collerec, A. Bouliou, and J. L. Coatrieux. Three dimensional positioning from multimodality medical imaging. In *Proc. 10th IEEE Conf.*, pages 422–423, New Orleans, 1988.

[4] E. Kortokov, K. Henriksen, and R. Kories. Stereo ranging with verging cameras. *IEEE PAMI*, 12(12), December 1990.

[5] D. F. Rogers. *Procedural Elements For Computer Graphics*. McGraw-Hill, New York, 1985.

[6] D. F. Rogers and J. A. Adams. *Mathematical Elements For Computer Graphics*. McGraw-Hill, New York, 1990.

[7] P. K. Sahoo, S. Soltani, A. K. Wong, and Y. C. Chen. A survey of thresholding techniques. *Computer Vision Graph Image Proc*, 41:233–260, 1988.

[8] R. Shahidi, R. Mezrich, and D. Silver. Proposed simulation of volumetric image navigation using a surgical microscope. *Jour. Image Guided Surgery*, 1(5), 1995.

[9] E. A. Spiegel, H. T. Wycis, M. Marks, and A. J. Lee. Stereotaxic apparatus for operations on the human brain. *Science*, 106:349–350, 1947.

The Geometry of Frameworks: Rigidity, Mechanisms, and CAD

Brigitte Servatius
Worcester Polytechnic Institute

Abstract

This paper is an introduction to the study of constrained geometric structures. The motion and rigidity of frameworks of rods and joints is examined, and connections are drawn to other constrained structures that are useful in many applications, in particular to Computer Aided Design (CAD).

1 Geometry of Frameworks

Mathematical applications, beyond increasing our understanding of the world, often refocus our attention on the underlying mathematics, shifting our point of view and deepening our understanding of a familiar abstract relationship. The most mundane theorems may be transformed by this process of redirection and reinterpretation.

THEOREM 1 (SSS) *If the lengths of corresponding sides of two triangles are equal, then the triangles are congruent.*

Let us consider this theorem in the context of two physical triangles constructed of straight rods joined at their endpoints, where the second triangle represents the result of forces and stresses acting on the first. We may interpret the theorem as stating that a triangular framework retains its structural integrity as long as the individual rods do, and as long as the joints don't break apart. This means that a triangular framework made of rigid members

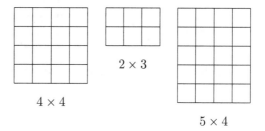

Figure 2: Deforming a square.

is rigid even if each individual joint allows twisting, as does a pin joint, see Figure 1.

There is no SSSS theorem for quadrilaterals, as we see in Figure 2. Even if the joints of a simple rod structure do fix angles, for analysis it is often preferable to idealize them as pin joints (or ball joints in three dimensions). This allows for the fact that any angle-fixing joint is prone to failure due to twisting, since each of the rods it connects acts as a lever with respect to it.

With a mathematical model for a framework we would like to be able to determine which frameworks are, like the triangle, rigid, so that they would be candidates for the supporting structure of a building or a bridge. On the other hand, if the framework is flexible like the square, or like the frameworks of mechanical machines or DNA molecules, we would like to know how it moves.

A nice example of a plane framework is an $n \times m$ grid of squares; see Figure 3. Since, as we have seen, a single square made of rigid rods is not rigid in the plane, we expect a grid of squares to allow even

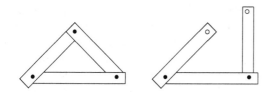

Figure 1: Breaking a triangle.

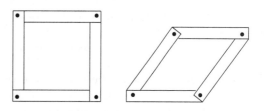

4 × 4

2 × 3

5 × 4

Figure 3: Grids.

more complicated deformations. An easy measure of the deformability of a framework is its *degree of freedom*. A point in the plane has two degrees of freedom, say its x and y coordinate, while a more complicated geometric object may require three values to specify it among its congruence class and to allow for the possibility of rotation.

An $n \times m$ grid has $(n+1)(m+1)$ vertices, each of which has two degrees of freedom, but the grid as a whole has three degrees of freedom (two translations and one rotation) which are regarded as trivial since they correspond to rigid motions of the entire grid. This gives $(n+1)(m+1) - 3 = 2nm + 2n + 2m - 1$ degrees of freedom which must be constrained in order for the grid to be rigid. The edges, each fixing the distance between two vertices, give $n(m+1) + m(n+1)$ constraints, so we expect that at least $n + m - 1$ additional constraints are required to rigidify the grid.

A rigid bracing with exactly $n + m - 1$ braces is achieved by adding diagonal braces in all the squares of the first row and column of the grid, since the braced squares then are triangulated, hence rigid, and it is easy to see that the whole braced grid is then rigid.

As might be expected, it is not sufficient to match the degree of freedom with the number of constraints and, indeed, not all choices of $n + m - 1$ braces will rigidify an $n \times m$ grid, see Figure 4. Analyzing which braces are required to rigidify a grid may be reduced to a simple combinatorial exercise. Since a square deforms to a rhombus, any deformation of a grid leaves all the vertical edges in row i parallel, and likewise with the horizontal edges in column j. Thus, a brace in square (i, j) forces the vertical rods in row i to remain perpendicular to the horizontal rods in column j under any deformation. Define a graph, called the *brace graph*, on the set of rows and columns of the $n \times m$ grid, and declare row i adjacent to column j if there is a brace in square (i, j).

Figure 4: A braced grid with a motion, and its brace graph.

THEOREM 2 *An $n \times m$ braced grid is rigid if and only if its brace graph is connected.*

PROOF: If the brace graph is connected then there is a path from every row or column to column 1. A simple induction on the length of the path shows that, under any deformation, the rods in every column are constrained to be parallel to those in column 1, and the rods in every row are constrained to be perpendicular to the rods in column 1, hence the grid is rigid.

If, on the other hand, the brace graph is disconnected, then make the rods in the rows and columns connected in the brace graph to row 1 vertical and horizontal respectively, and assign to those not connected to row 1 two other perpendicular directions. This gives a deformation of the grid, see Figure 4. □

If we brace a square with cables instead of rods, then, since cables can buckle but not stretch, the constraints become inequalities, namely that the distance between the endpoints of the cable is at most the length of the diagonal. In other words, the constraint is that the angles where the cables are attached are not acute, and the rigidity of the grid may be analyzed by a directed graph. For an elementary treatment of grids see [16]. Also, an analogous development may be attempted for grids of cubes in space, and other natural generalizations, see [15].

A framework which deforms, while not desirable for the structures of civil engineering, is fundamental in mechanical engineering, robotics, and some branches of organic chemistry. From the standpoint of machinery, the simplest and most useful type of framework has its deformations governed by a single parameter which is associated with time by a driving force, so that the behavior of the framework is predictable. A 1-parameter framework is called a *mechanism*. Even more complicated programmable machines may be analyzed as an assembly of mechanisms. The deformations of a mechanism are called its *motions*.

One of the classical problems in this area was to find a mechanism one point of which traces out a straight line, analogous to the compass which traces out a circle. The mechanism of James Watt is pictured in Figure 5, in which two vertices are "grounded" to indicate that their positions are to be held fixed. As the mechanism moves, points p_1 and p_2 are constrained to move on circles centered at the grounds, and the point q traces out a curve which is straight to third order at its inflection point. This line-approximating engine is so mechanically simple that one finds it in almost universal use even though a mechanism drawing perfect lines requires just twice as many bars. Since the requirement is

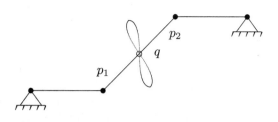

Figure 5: The Watt mechanism.

to transform circular motion about a "ground" vertex into straight line motion, we will make use of the fact that inversion in a circle transforms the collection of circles and lines into itself. Recall that inversion in the circle $|z| = R$ in the complex plane is given by the linear fractional transformation $w = -R^2/z$, see [1]. We may calculate the altitude of the triangle in Figure 6 in two different ways to get

$$b^2 - (\frac{y+x}{2})^2 = a^2 - (\frac{y-x}{2})^2$$

which gives $xy = b^2 - a^2$, so the points P and Q are inversions of each other in the circle of radius $\sqrt{b^2 - a^2}$ and center O.

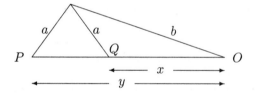

Figure 6: Triangle showing inversion of P and Q.

Inversion in this circle sends every circle passing through O into a circle passing through infinity, that is, a straight line. If, therefore, we fix O, force P, Q, and O to lie on a straight line, and force Q to move in a circle passing through O, then P will move along a straight line, which is accomplished by the mechanism of Peaucellier, see Figure 7.

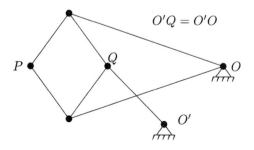

Figure 7: Peaucellier's mechanism.

A more general idea than looking at curves traced out by mechanisms is to look at the *configuration space* of a framework. We may regard a framework with n vertices in $\mathbb{R}^d$ as a single point in $\mathbb{R}^{dn}$ by simply listing all the coordinates of all the vertices as one vector. The configuration space of a framework is the subset of $\mathbb{R}^{dn}$ consisting of the framework together with all its deformations. Since there are only distance constraints between the vertices, the Pythagorean Theorem implies that the configuration space is the space of solution of a system of quadratic equations.

A true mechanism will have a one-dimensional configuration space; for instance, the configuration space of the Watt mechanism is homeomorphic to a circle, with the homeomorphism given by $(\alpha, \beta)/\sqrt{\alpha^2 + \beta^2}$, where α is the counterclockwise angle the leftmost bar makes with the horizontal, and β the clockwise angle of the rightmost bar, $-\pi < \alpha, \beta < \pi$. The configuration space of a planar two-bar robot arm is a torus; of a pentagon in the plane, it is a 5-holed torus. For some topological aspects see [20]. The configuration spaces of frameworks have long been used as examples of manifolds; however, Figure 8 shows a recently discovered curiosity: a mechanism in the plane whose configuration space contains a cusp; see [4]. A consequence of this is that any differentiable motion of this mechanism must have zero velocity in the cusp position.

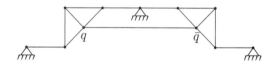

Figure 8: A plane mechanism in a cusp configuration.

2 Rigidity and Strong Rigidity

Up until now we have been content with an intuitive definition of framework, but it is time for a precise definition. A *framework* $\mathcal{F} = (G, \mathbf{p})$ is a graph $G = (V, E)$ together with an embedding $\mathbf{p}$ of V into Euclidean space, in practice usually two- or three-dimensional. By this definition, a framework is a mathematical model of a physical framework in a fixed position in space. A *motion* of the framework $\mathcal{F} = (G, \mathbf{p})$ is a continuous one-parameter family $\mathbf{p}(t)$ of embeddings of V so that $\mathbf{p}(0) = \mathbf{p}$ and so that for all t the distance between points corresponding to adjacent vertices is constant. We say a framework is *rigid* if the only motions which it admits arise from Euclidean congruences.

Consider a quadrilateral with one diagonal in the plane. If we regard this framework as being embedded in 3-space, then it is not rigid since the diagonal rod acts as a hinge between the rigid triangles (illustrating the fact that the rigidity of a framework depends upon the dimension of the space in which it is embedded). If we hold one triangle fixed and turn the other on the hinge through 180°, the framework again lies in a plane, but now it is non-congruent to its original form, see Figure 9, and each of these two forms lies in a different connected component of the configuration space of the plane framework. Nevertheless, both forms of the quadrilateral with one diagonal are rigid in the plane, since there are no non-trivial motions in the plane which deform them. We say that the rectangle with one diagonal is not strongly rigid. A framework is *strongly rigid* if the underlying graph, together with specified edge lengths, determines the congruence class of the framework.

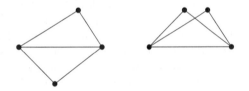

Figure 9: Embedding the quadrilateral.

Obviously any strongly rigid framework is rigid; however, testing for strong rigidity is very delicate, and the complexity is daunting even for moderate numbers of vertices, see [3].

For a comprehensive history and background of rigidity see [9, 21].

3   Infinitesimal Rigidity and Parallel Drawings

Suppose there is a differentiable motion, $\mathbf{p}(t)$, of a framework. Let $\{\mathbf{p}_i', \mathbf{p}_j'\}$ denote the initial velocities of the endpoints of edge (i, j). Since the distance between $\mathbf{p}_i$ and $\mathbf{p}_j$ is held fixed during the motion, the components of $\mathbf{p}_i'$ and $\mathbf{p}_j'$ in the direction parallel to the edge must be equal, i.e.,

$$(\mathbf{p}_i - \mathbf{p}_j) \cdot \mathbf{p}_i' = (\mathbf{p}_i - \mathbf{p}_j) \cdot \mathbf{p}_j' \text{ for } (i, j) \in E. \quad (1)$$

This condition gives us a system of linear equations whose variables are the coordinates of the vectors $\mathbf{p}_i'$, where i ranges over V, with one equation for

each edge in E. A solution to this linear system of equations 1 is called an *infinitesimal motion*, or a *flex*. If a framework has a differentiable motion, then the instantaneous velocities of the vertices form a flex. We denote a flex in a diagram by drawing the initial velocity vectors at each vertex, see Figure 10. If the only infinitesimal motions are *trivial*, that is, they arise from infinitesimal translations or rotations of $\mathbb{R}^d$, then we say that the framework is *infinitesimally rigid*.

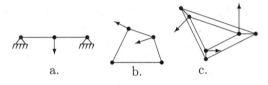

a.   b.   c.

Figure 10: Infinitesimal motions.

If a framework is infinitesimally rigid, then it is rigid, see for example [2]. On the other hand, since not every flex need be realized as the initial velocities of an actual motion of the framework, a rigid graph is not necessarily infinitesimally rigid. For instance, the framework of Figure 10a is clearly rigid, yet it has a non-trivial infinitesimal motion. We shall see later that the rigid framework of Figure 10c is also not infinitesimally rigid.

One computational advantage of infinitesimal rigidity is that the set of length constraints, a system of quadratic equations, is replaced by a system of linear equations. There is also a greater engineering utility to infinitesimally rigid structures. The vulnerability of Figure 10a to sagging is typical of rigid, but non-infinitesimally rigid, frameworks.

Infinitesimal motions may not correspond to real motions, and so it is often difficult to make use of our geometric intuition in detecting them. An old engineering trick is to transform the rigidity problem into one of parallel redrawing, see [21]. Two frameworks on the same graph are *parallel drawings* if corresponding edges are parallel. Translations and dilations trivially result in parallel drawings, however any infinitesimal motion can be transformed into a parallel drawing by rotating each of the velocity vectors 90° and adding them to the position vectors, see Figure 11. Conversely, any parallel redrawing corresponds to an infinitesimal motion. In the plane framework of Figure 12a, the segments which join the two triangles may be extended to intersect at a single point, so the inner triangle may be dilated with respect to the point of intersection producing the parallel drawing Figure 12b, which corresponds to the infinitesimal motion of Figure 12c. This infinitesimal motion does

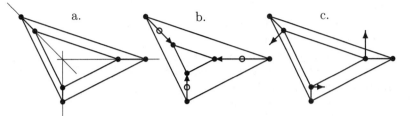

Figure 12: Constructing an infinitesimal motion.

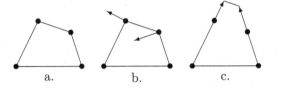

Figure 11: Infinitesimal motions and parallel redrawings.

not correspond to a real motion, and in fact this framework can be shown to be rigid.

4 Generic Rigidity

If a small perturbation in the embedding of a framework alters its rigidity properties, then such a framework is in some sense a singularity and, consequently, often a poor design choice. We call a framework *generic* if the vertices are embedded so that we can "wiggle" the vertices a little bit without altering any of the framework's rigidity properties. Most embeddings are generic, and, in fact, generic embeddings form an open dense subset in the space of all embeddings.

Since the rigidity of a generic framework does not depend on the particular embedding, we may speak of the rigidity of the graph itself. A graph G is called *generically rigid* (in dimension d) if there is a generic embedding of G in $\mathbb{R}^d$ which is rigid. A non-generic framework may have more or less rigidity than a generically embedded one. In Figures 13a and b we see two frameworks on the same graph, with the first being rigid since the path with three edges is pulled taut. The second embedding is generic and flexible. In Figures 13c and d we

again have two frameworks on the same graph, but now the "singular" framework is flexible, since we can hold the lower triangle fixed and simultaneously rotate the corresponding vertices of the upper triangle about them. On the other hand, the generic embedding, Figure 13d, is rigid.

If G is generically rigid, then almost all embeddings of G will yield a rigid framework. While generic rigidity is purely a combinatorial concept, no combinatorial characterization is known for dimensions 3 and up. The characterization question in dimension 3, in particular, is one of the most compelling open problems in geometry today. For dimension 1, rigidity on a line, generic rigidity is equivalent to ordinary connectivity. For two-dimensional frameworks, we have the following theorem.

THEOREM 3 **Laman, 1970.** *A graph $G = (V, E)$ is generically rigid if and only if there is a subset F of edges so that $|F| = 2|V| - 3$ and $|F'| \leq 2|V(F')| - 3$ for all subsets $F' \subseteq F$, where $V(F')$ denotes the set of vertices which are endpoints of F'.*

The conditions of Laman's Theorem justify the intuitive ideas of constraint and degrees of freedom with which we began. The condition that $|F| = 2|V| - 3$ insures that G has enough edges to constrain the two degrees of freedom of each vertex leaving aside the three which arise from isometries. The condition that $|F'| \leq 2|V(F')| - 3$ insures that no subset of these constraints is used wastefully to overbrace some subset of vertices. This combinatorial form of rigidity is also of interest just as a concept of graph theory—the one-dimensional version of this theorem states that every connected graph has a spanning tree.

As was observed in [12], the edge sets F which satisfy the conditions of Laman's Theorem form the bases of a matroid, so many of the questions of generic rigidity in two and higher dimensions are stated in terms of the rigidity matroid, see [8, 10]. For instance, matroid algorithms can be used to test for rigidity in dimension 2 in polynomial time even

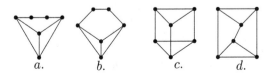

Figure 13: Generic and non-generic frameworks.

though checking the condition of Laman's Theorem directly requires testing all subsets of edges, see [7, 5]. See [14] for background on matroids.

Laman's theorem may also be reworded to apply to parallel drawings in the plane. Given a generic drawing of a graph G in the plane, if G satisfies Laman's condition, then all parallel redrawings are related by similarity. If G is not generically rigid, then there definitely are non-similar parallel redrawings.

5   Applications to CAD

In computer aided design (CAD), the object is to specify and represent a complicated design using a standard collection of geometric objects such as points, circular arcs, and line segments. The design is specified by requiring the objects to satisfy certain constraints. For example, certain point-line incidences, segment lengths, or angles may be prescribed. The basic design problems are: realizability (does there exist a design satisfying the constraint?), (local) uniqueness (do the constraints determine the congruence class of the design?), and constraint independence (are all the constraints necessary?).

Theoretically, just as with rigidity of frameworks, the constraints may be written as a system of algebraic equations whose variables are the coordinates or other parameters of the geometric objects, see [13, 18]. The rank of the Jacobian of the system of constraints is used to answer some of the basic design questions. Even these linear computations may be slow or unstable because of degeneracies; for example, the simple cusp framework in Figure 8 crashed our CAD software. Since typical CAD designs use hundreds of objects, avoiding degeneracies to achieve computational stability is useful and may be accomplished by using generic parameters.

For plane length designs, where the objects are points and the constraints specify distances between certain pairs of points, rigidity theory provides a purely combinatorial solution to generic design questions. Parallel drawings serve the same role for direction designs. For designs with both direction and length constraints, see [17], an analogue of Laman's theorem is also valid.

Purely combinatorial characterizations are available in this setting as well: A set of direction and length constraints is independent and achieves local uniqueness if and only if there is a decomposition of the edges of the underlying graph into two spanning trees, such that no two proper subtrees of the

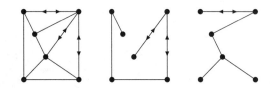

Figure 14: Decomposing a direction length design.

same constraint type have the same span. Figure 14 shows a two tree decomposition of a direction length design in which the edges which correspond to direction constraints are indicated by arrows in their interior.

Other combinations of CAD motivated constraint systems have not yet been satisfactorily analyzed. Observe that a direction constraint corresponds to an equation among the coordinates of two points, and the same is true for a length constraint. An angle constraint, by contrast, involves three points. A theory of designs mixing lengths and angles would be very interesting for map making; so far, however, none has been formulated.

References

[1] L. Alfors, (1966). *Complex Analysis*, McGraw-Hill.

[2] R. Connelly, (1980). *The rigidity of certain cabled frameworks and the second-order rigidity of arbitrary triangulated convex surfaces*, Advances in Math. **37**, 272–298.

[3] R. Connelly, (1990). *On generic global rigidity*, in "Applied Geometry and Discrete Mathematics", DIMACS Ser. Discrete Math. Theoret. Comput. Sci. **4**, 147–155.

[4] R. Connelly and H. Servatius, (1994). *Higher-order rigidity, what is the proper definition?* J. Discrete and Computational Geometry **11**, 2, 193-199.

[5] H. Crapo, (1993). *On the generic rigidity of structures in the plane*, Advances in Applied Math.

[6] H. Crapo and W. Whiteley, (1993). *The Geometry of Rigid Structures*, Encyclopedia of Math., Cambridge University Press.

[7] J. Edmonds, (1965). *Minimum partition of a matroid into independent subsets*, J. Res. Nat. Bur. Standards B **69**, 67–72.

[8] Jack E. Graver, (1991). *Rigidity Matroids,* SIAM J. Discrete Math. **4**, 355-368.

[9] J. Graver, B. Servatius and H. Servatius (1993). *Combinatorial Rigidity,* Graduate Studies in Math., AMS.

[10] J. Graver, B. Servatius and H. Servatius (1993). *Abstract rigidity matroids,* Jerusalem Combinatorics '93, Contemporary Math. **178**, 145–152.

[11] G. Laman, (1970). *On Graphs and rigidity of plane skeletal structures,* J. Engrg. Math. **4**, 331–340.

[12] L. Lovász and Y. Yemini, (1982). *On Generic rigidity in the plane,* SIAM J. Alg. Disc. Methods **3**, 91–98.

[13] J. C. Owen, (1991). *Algebraic solutions for geometry from dimensional constraints,* Symposium on Solid Modeling Foundations and CAD/CAM applications, ACM Press.

[14] J. Oxley (1966). *Matriod Theory,* Oxford University Press.

[15] A. Recski, (1988). *Bracing cubic grids—a necessary condition,* Discrete Math. **73**, 199–206.

[16] B. Servatius, (1995). *Graphs, digraphs, and the rigidity of grids,* The UMAP Journal, **16**, No. 1, 43–70.

[17] B. Servatius and W. Whiteley, (1995). *Constraining plane configurations in CAD: Combinatorics of directions and lengths.*

[18] K. Sugihara, (1985). *Detection of structural inconsistency in systems of equations with degrees of freedom,* J. Disc Appl. Math. **10**, 297–312.

[19] T. S. Tay, (1993). *A new proof of Laman's theorem,* Graphs and combinatorics, **9**, No. 4, 365–348.

[20] W. P. Thurston and J. R. Weeks, (1984). *The mathematics of three–dimensional manifolds,* Scientific American **251** 1, 108–120.

[21] W. Whiteley, (1996). *Some Matroids from Discrete Applied Geometry,* in Matroid Theory, J. Bonin, J. Oxley and B. Servatius, eds., Contemporary Math., AMS.

Geometry and Geographical Information Systems

George Nagy

Department of Electrical, Computer, and Systems Engineering

Rensselaer Polytechnic Institute

Troy, NY

nagy@ecse.rpi.edu

Abstract

Most Geographic Information Systems (GIS) make use only of elementary geometric structures, but their intrinsic geometric simplicity is balanced by the need to manipulate large numbers of spatial entities efficiently and accurately. The emphasis in GIS is on robust geometric algorithms rather than on theorems. In GIS, little importance is attached to worst-case complexity analyses, randomized algorithms, and high-dimensional domains. The compilation of realistic geographic databases is beyond individual reach, therefore I provide pointers to several public-domain data sets. Then I single out three aspects of GIS for more extensive discussion: triangulations and proximity diagrams, geometric visibility, and map projections. I close with a list of additional GIS applications of computational geometry.

1   Introduction

Geometry and geography are daughters of Gaea who have grown apart but are now getting reacquainted. This paper presents a sample of geometric concepts that arise in the context of Geographic Information Systems. Because computerized information systems usually operate on large databases (typically millions of coordinates), the emphasis is on algorithms and procedures rather than on theorems and proofs. With few exceptions, only simple Euclidean constructs are invoked, but most of the tools required to manipulate them efficiently were recently developed in computational geometry.

Geographic Information Systems (GIS) are computerized means of storing, analyzing, and displaying data about the earth, its features, the distribution of life on it, and whatever affects human activity. The broad term GIS came into vogue in the mid-sixties, and may include the data as well as the methodology. GIS are designed to answer queries like: "How many towns with a population of over 10,000 are within 50 kilometers of a railroad line in the state of Nebraska?" or "What fraction of Illinois will be flooded if Lake Huron rises by 20 meters?" or "Which areas of Washington County are both in the floodplain and near railroad lines?"

The design of GIS draws on conventional database management systems (DBMS), operations research, numerical cartography, and computer graphics. Because towns may be represented as points, rivers as curves, and political boundaries as polygons, the essential relationships are geometric in nature. Only a small fraction of the computer code that constitutes a GIS may deal with geometric constructs (most of the code is probably buried in the graphical user interface or GUI), but the two- and three-dimensional geometric (and topological) relations are the conceptual core.

Two aspects differentiate a GIS from an ordinary DBMS such as might be used for payroll or class schedules. First, spatial coordinates do not support a global ordering: there is no unique sorting sequence. Second, spatial entities are not readily viewed as a discrete collection of objects: a continuous view of the world is essential.

Among popular applications of GIS are cadasters (property records); planning and management of transportation infrastructure and operations; mineral exploration; environmental monitoring; flood control and hydro power; retail marketing; the location of manufacturing, service and distribution facilities; and agriculture, pasture and timber management. The underlying spatial framework may be obtained from digitized maps, digital elevation models, and satellite images.

Many commercial GIS systems are available,

ranging from simple digital atlases costing a few hundred dollars to elaborate multi-workstation networks for over $100,000. Among the best known ones are Arc/Info, Intergraph, TNT, ERDAS, and Smallworld. Systems designed for utility applications (internal and external plant, conduits, piping, power lines) increasingly resemble computer-aided design (CAD) systems, while those intended for making use of satellite data have complex-image processing components.

Advances in GIS research are presented at conferences.[1] and in three or four specialized journals.[2] Some useful introductory texts are [18, 39, 44, 53, 56] The Federal Geographic Data Committee (http://www.fgdc.gov/) promotes standards and publishes a newsletter. The National Center for Geographic Information and Analysis (NCGIA) at U.C. Santa Barbara distributes a curriculum and teaching materials. Many universities offer a course in computational geometry and several excellent texts are available [9, 29, 35, 40].

In the rest of this chapter, I will provide some sources of geographic data for study and experimentation, and discuss three components of GIS that have both intrinsic interest and solid geometrical foundations.

2   Geodata

In most countries, government organizations collect and disseminate large and medium scale maps, related geographic information, and topographic data in digital form [43]. In the United States, the principal players are the US Geological Survey (USGS), the National Imagery and Mapping Agency (NIMA), the National Oceanographic and Atmospheric Agency (NOAA), the Soil Conservation Service (SCS), and the National Aeronautics and Space Agency (NASA). Mapmakers, like Rand-McNally, Hammond, and Simon & Schuster, distribute highway and city maps in digital form. Specialized firms, like MapInfo and Intergraph, offer software, directory data, and detailed address-location information for computerized mapping and analysis of demographic data for marketing studies.

The USGS maintains an index of digital data on the World Wide Web at http://www.usgs.gov/. Among data sets they distribute are elevation values for the United States recorded at 3 arc-second intervals. Each file contains over one million points and corresponds to a 15-minute quadrangle in the 1:50,000 topographic map series. The NIMA ETOPO5 dataset contains lower resolution (5 arc-minute) topography including bathymetry, for the entire world. A catalog of other elevation data may be obtained from http://www.geo.ed.ac.uk/home/ded.html.

The staggering amount of satellite imagery collected since the launch of the first Landsat (ERTS-A) in 1972 can be obtained (for a price) from the EOSAT Company in Lanham, MD. Accurate registration and rectification of the data, taking into account differences in surface elevation, requires complex geometric transformations and interpolation, and huge computing resources. Several commercial GIS allow the superposition of recent satellite images on digitized topographic maps. This technique is vital in monitoring changes such as coastal erosion, deforestation, desertification, overgrazing, floods, and forest fires. (Examples of beneficent changes are harder to come by.) Landsat images can be found at http://edcwww.cr.usgs.gov/Earthshots1.00/.

Most of these databases are built on a quasi-rectangular grid, such as latitudes and longitudes, or state-plane coordinates. Some of the older statewide GIS are based on 1 km × 1 km cells. Maps are also digitized into rectangular pixel arrays or bitmaps.

Techniques used for accelerating the processing of grid data include the Uniform Grid method for line intersections that takes advantage of the homogeneous distribution of the data [22], quad trees [46], and R-trees [24]. These techniques are all designed to speed up query processing by taking advantage of the locality of most entities of interest, i.e., their small size relative to the extent under consideration. Alternative methods, described in the next section, explicitly partition the entire extent during a preprocessing step, before any queries are processed. These non-grid-based tessellation methods are probably of greater interest to students of geometry.

3   Triangulated Networks and Proximity Diagrams

Aside from the rectangular grids discussed above, the most common data structures used in GIS are

---

[1]The International Symposium on Spatial Data Handling (IGU), the International Symposium on Large Spatial Databases (ACM Sigmod), the International Conference on GIS and Environmental Modeling (NCGIA), the National GeoData Forum, GIS/LIS, and the European Conference on GIS (EGIS Foundation) among others.

[2]*Int'l J. of GIS, Cartography and GIS, Cartographica, GIS World.*

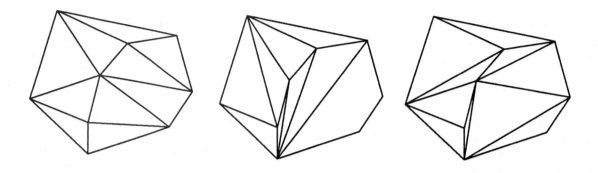

Figure 1: Several different triangulations of a set of nine points enclosed by their convex hull.

based on Delaunay triangles and Voronoi polygons. As we shall see, these two are intimately related.

3.1 Triangulated irregular networks

Instead of simply referring to locations of interest by their coordinates (e.g., latitude and longitude), it is convenient to partition the spatial domain of a geographic information system into mutually exclusive cells. An assembly of triangular cells, called a *Triangulated Irregular Network* (TIN), offers many advantages over more complex cell shapes [14, 19, 38]:

1. Triangles can compactly fill any area: they tessellate the plane.

2. Given a set of points in the plane, simple algorithms exist (try writing one!) for linking them into a TIN [45]:

3. Each cell can be represented by a simple data structure: given three vertices, there is no ambiguity about the edges. (An interesting problem is to devise a data structure where appending a triangle to an edge of the boundary requires adding only the information associated with the new node.)

4. The planar graph formed by the vertices and edges—the triangles are "faces"—has an almost homogeneous structure. On the average the nodes are of degree six, and n points give rise asymptotically to $3n$ edges and $2n$ triangles.

5. When an elevation value is associated with each vertex of a triangulation, the elevation value at any point of the resulting surface can be linearly interpolated from vertex elevations.

Figure 2: Delaunay triangulation of the nine points with their "empty" circumcircles.

This results in a piecewise-linear surface model. In contrast, elevations cannot be linearly interpolated in a rectangular grid model because four points overconstrain a plane.

6. TINs generalize readily to higher dimensions, where they are called *simplicial cell-complexes*. These find uses in solid models of mechanical parts, geological strata, and finite-element solutions of differential equations [4, 5, 49, 52].

Figure 1 shows several different triangulations of a set of nine points enclosed by their convex hull.[3] Among specific types of triangulation are the *minimum weight triangulation*, which minimizes a cost

---

[3]The convex hull is easily visualized as a rubber band around the points in 2-D, and as shrink-wrap in 3-D. Writing an efficient algorithm to find the convex hull of a set of points, especially in higher dimensions, is not easy. The computational geometry literature is littered with faulty attempts.

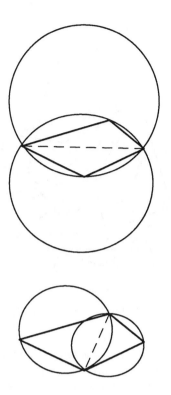

Figure 3: Swapping diagonals to obtain empty circumcircles increases the minimum interior angle.

function defined on the edges, *constrained triangulation*, which requires including some prespecified edges, and *Delaunay triangulation*, discussed below.

3.2 Delaunay Triangulation

In GIS, a particular type of triangulation, called Delaunay Triangulation, is standard. It is named after the author of a paper titled *On The Empty Sphere*. Presented in 1931 to the Soviet Academy of Science and dedicated to George Voronoi, it demonstrated the existence of a triangulation such that the circumcircle of any triangle contains no vertices other than those of its inscribed triangle[4] [7]. Figure 2 shows the Delaunay triangulation of the same nine points with their "empty" circumcircles.

From the point of view of interpolating terrain elevations, the chief merit of Delaunay triangulation is that among all possible triangulations, it maximizes the minimum interior angle of each triangle. This tends to equalize the lengths of the sides of the triangles, resulting in more robust interpolation. The minimax angle property also allows constructing a Delaunay triangulation from any arbitrary triangulation by simply swapping the diagonals of any

[4]If the set of points is in general position, i.e., no three points are collinear.

quadrilateral (formed by two triangles that share a side) if the swap increases the smallest of the six interior angles (Figure 3).

Another unexpected property of Delaunay triangulation (that does not hold for arbitrary triangulations) is a partial ordering with respect to any "viewpoint" in the plane [10, 15]. The triangles can be assigned consecutive integer labels in such a way that an arbitrary point in any triangle can be joined to the viewpoint by a line segment that passes *only through triangles with a lower number*. In the Delaunay triangulation on the top of Figure 4 the viewpoint is one of the vertices (circled) of the triangulation. The numbering shown is consistent with the partial order. For instance, any point in triangle #8 can be joined to the viewpoint with a straight line segment either through triangles #5 and #2, or through triangles #4 and #1. In the bottom non-Delaunay triangulation of the same points, no such numbering exists. This property of Delaunay triangulation (called *acyclicity*) can be exploited for computing the region of the terrain (represented by a TIN) that is visible from a specified viewpoint; see Section 4.

The connectivity of the Delaunay graph has been extensively investigated. It can be shown, for instance, that it contains the minimum spanning tree, which is the straight-line connection among the

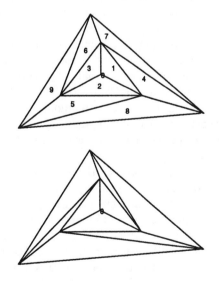

Figure 4: Above: a partial ordering of Delaunay triangles. Any ray from vertex 0 to a point within any triangle passes only through triangles with a lower number. Below: the triangulation is cyclic with respect to the vertex 0: no such numbering is possible.

points that has minimum length. It has also been shown that the Delaunay graph is not necessarily *Hamiltonian* (i.e., contains a cycle passing through all the points), but that it is so with high probability [8].

3.3 Voronoi Diagrams

The Voronoi diagram[5] defined by a set of points is a subdivision of the plane into polygons. Each polygon is associated with one of the given points called its *seed* or *site*, and encloses the convex region of the plane closer to its seed than to any other point. It is therefore convenient to think of the Voronoi diagram in terms of the intersections of the halfplane regions nearest to each of every pair of data points (Figure 5). For a scholarly survey of the properties of the Voronoi diagram, see [1].

Voronoi polygons are sometimes called *Thiessen polygons* in geography and *Dirichlet tessellation* in geometry. They can be used for nearest-point problems: if the seeds represent distributors, then to find out to which distributor a consumer is closest, it is sufficient to determine in which polygon the consumer is located (this is also called the *post-office problem*). Voronoi diagrams are also immediately useful for finding the closest pair among n sites, the *largest empty figure*, and *collision-free path-planning*.

The Voronoi diagram is the *straight-line dual* of the Delaunay triangulation. Each edge in the Voronoi diagram is the perpendicular bisector of an edge of a Delaunay triangle. In fact, some algorithms first construct the Voronoi diagram by a divide-and-conquer method [40], then convert it to the Delaunay triangulation by joining every pair of points that share a Voronoi boundary. "Dynamic algorithms" can modify an existing Voronoi diagram or Delaunay triangulation as points are added to the database.

The *kth-order Voronoi diagram* partitions the plane into regions where every point is nearer to a set of k points than to any other point. The number of Voronoi diagrams of all orders on n points is $O(n^3)$. The *Farthest Point Voronoi Diagram*, useful for locating facilities away from undesirable sites, is the Voronoi diagram of order k, with $k = n - 1$.

To determine the closest points to each road, we need a *Voronoi Line diagram*. The partitions here consist of a mixture of line segments and quadratic curves. Other useful extensions are the *Weighted-Distance Voronoi Diagram*, where the distances are

---

[5]Named after the Russian turn-of-the-century mathematician.

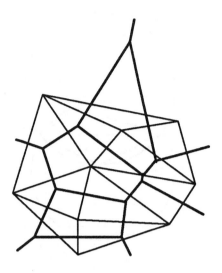

Figure 5: Voronoi diagram of the nine points.

weighted according to the site from which they are measured, and the *Voronoi Diagram with Barriers* [21], where a point cannot be considered near a site if there is an impenetrable barrier (e.g., a river) between them. Current research topics include efficient incremental modification of Voronoi diagrams (when points are added or deleted, or when they are in motion), and Voronoi interpolation.

Naturally, the Voronoi diagram also generalizes to higher dimensions. In 3-D, it is the dual of the Delaunay tetrahedralization. However, the computational resources required to compute and store it rise exponentially with the dimensionality. The maximum number of vertices of the Voronoi diagram for a set of N points in d-dimensional space is of the order of $(d/2)! N^{d/2}$. More efficient algorithms are now available to find the nearest neighbors of query points in the high-dimensional vector spaces found in pattern recognition applications.

4   Terrain Visibility

This section explores the application of geometric algorithms to problems involving lines of sight between points on the earth's surface. One might, for instance, compute from which of several alternative observation sites in San Francisco the largest portion of the Bay would be visible. The most important concept here is the *visible region of a viewpoint*, or that part of the surface that can be "painted" by a searchlight located at the viewpoint.

Geometric visibility is an abstraction based only on the intersection with the terrain of the lines of sight emanating from each viewpoint. Surface attributes, vegetation, atmospheric diffraction, and light intensity are neglected. While visualization shows the *appearance* of the terrain to an observer, geometric visibility is concerned only with the *extent* visible from given observation points. The output of a visualization program is intended for display for human assimilation, but the output of a visibility program can be channeled to another program for further calculation of visibility-based attributes.

Digital elevation terrain models (DEMs) provide an abstract representation (*model*) of the surface of the earth by ignoring all aspects other than topography. Visualization tools, on the other hand, generate a display under some simplified assumptions (*models*) of surface reflectance, illumination, light transmission, and viewing mechanism. For instance, a surface may be visualized using a finite number of colors (that indicate land cover), lambertian reflectance, point-source illumination, and stereographic observation.

Computer-aided visualization of geometric constructs [58] facilitates solving, by inspection, many problems of a geographic nature. Using geometric visibility, however, allows some of these problems to be solved by direct computation instead of inspection. Examples include locating fire towers and microwave transmitters and receivers; siting power lines, pipelines, roads, and rest-stops; navigation and orientation by reference to the horizon; the identification of certain topographic features; and, of course, a host of military emplacement problems. Visibility mapping plays a central role in scenic landscape assessment for establishing jurisdictions, quantifying impacted populations, exploiting sources of energy, and planning transportation corridors [13].

Although the display algorithms that form the core of computer graphics are based on geometric visibility, the application of geometric visibility to terrain models is relatively new.[6] In addition to computer graphics, spatial data processing and topographic analysis, geometric visibility bears on computational geometry, computer vision, and operations research. References to original research on the topics discussed below may be found in a recent survey by the author [31].

---

[6]However, visibility in the plane, or polygon visibility, has been a popular topic in computational geometry. Most of the essential results can be found in [33, 34, 48].

4.1 Basic visibility concepts

For our purposes, a *terrain* is a topographic surface whose elevation above a horizontal datum is a single-valued function of x and y (no overhangs). Two points on such a surface are said to be *mutually visible* if the line segment that joins them does not pass below the surface. The intervisibility of a pair of points is a Boolean function of *four* scalar variables, or a mapping from $[R^2 \times R^2]$ to $\{0, 1\}$.

Given a terrain model on which surface-points, lines and regions can be specified, the intervisibility of the various types of entities is represented by the corresponding Boolean *visibility function* defined on a product space of the entities. Among the nine visibility functions that can be defined among point, line, and region entities, the most useful are the *point-point* and *point-region* visibility functions.

Any visibility function can be represented by a *visibility graph* with arcs that link the nodes corresponding to intervisible entities. The visibility graph for point-point visibility is straightforward because any given point is either visible or invisible from any other point. However, edges and regions may be *partially* visible.

The *point-point visibility* among every pair of data points can be represented by a Boolean array of size N^2, called the *visibility matrix*, or by the corresponding *visibility graph* with N nodes and up to N^2 arcs, where N is the total number of data points. The visibility matrix is symmetric (under the assumption of zero observation height). The row and column sums (projections) of the visibility matrix correspond to the number of data points visible from each data point of the terrain. These *visibility indices* provide useful and relatively compact information about the terrain. In a bowl-shaped terrain, all points are intervisible; on a dome, none are. The highest points don't necessarily have the largest visibility indices [23].

Point-region visibility can be represented by a set of two-dimensional visibility maps showing the vertical projection on the horizontal datum of the visible and invisible parts of the terrain from a specified viewpoint (Figure 6). A visibility map is required for each observation point. In cartographic terms, most viewshed maps, from turn-of-century military conventions to the present, are binary *choropleths* (shaded maps) of visible and invisible zones. The earliest visibility maps were generated by the military using a defilade approach consisting of radial samples of vertical cross-sections derived from topographic contours. The intersection points of the lines of sight were projected back to the original

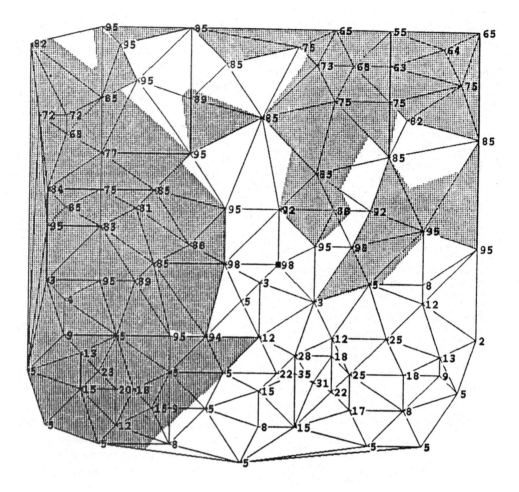

Figure 6: A visibility map. The figure shows the horizontal projection of a terrain on a Delaunay-triangulated irregular network. The dark areas are invisible from the viewpoint near the center, which is marked by a small square.

map, and interpolated. Regions of visibility and invisibility may be nested, as in the case of a mountain peak—that itself contains an invisible crater—which is visible beyond a ridge.

4.2 Visibility Regions

As we have seen, a visibility map is a projection onto the horizontal plane of the 3-D curves (or line segments) that separate visible and invisible regions on a topographic surface. The boundaries of the regions visible from a given viewpoint, projected onto the x-y plane, may be divided into blocking segments and shadow segments. In a sectional view (a vertical section through the viewpoint) of the terrain, such as Figure 7, these segments are just points on the baseline. From the perspective of the viewpoint, a *blocking segment* represents the transition from a visible to an invisible region (again, as

projected onto the x-y plane). An example is the first ridge to the right of the viewpoint (projected onto the baseline). A *shadow segment* represents the transition from invisible to visible. In Figure 7, there are two shadow segments, each corresponding to the "shadow" of the ridge to its left.

Blocking segments typically correspond to ridges and shoulder lines that cross a *line of sight* (i.e., a ray through the viewpoint). Shadow segments correspond to a double projection: the orthogonal (vertical) projection on the horizontal datum of the central projection (from the viewpoint) of a ridge (or shoulder line) onto the terrain on the far side of the ridge.

The boundary of a connected region of visibility or invisibility that does not contain the viewpoint must consist of alternating chains of blocking segments and shadow segments. Any single chain consisting only of blocking segments or only of shadow

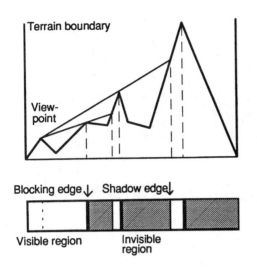

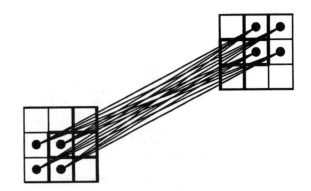

Figure 8: Visibility calculations on a grid. This figure shows the redundancy of computing visibility by following the line of sight from each observation point to each target point. Most grid-based visibility programs eliminate computing some of the redundant intersections of the lines of sight with the edges of the grid cells.

Figure 7: Sectional view: horizons. The viewpoint has three odd right horizons (blocking edges) and three even right horizons (shadow edges) including the terrain boundary. Its only left horizon is the terrain boundary. Terrain segments between odd and even horizons are invisible from the viewpoint, and segments between even and odd horizons are visible.

segments must be a single-valued radial function of the azimuth, and may therefore form a closed curve only if it encloses the viewpoint. Furthermore, along any ray from the viewpoint on the visibility map, blocking and shadow segments must strictly alternate. (But vertical edges and surfaces tangent to a line of sight can give rise to anomalous radial boundaries between visible and invisible regions.)

If the terrain model consists of planar approximations, such as a TIN, then the projections on the horizontal datum of both the visible and invisible regions of an observation point consist of polygonal areas, and each blocking or shadow chain is a piecewise-linear curve. Each blocking segment consists of edges of the triangulation. An edge of the triangulation may be part of a shadow segment only if the plane that contains the corresponding terrain edge and the viewpoint also contains a more proximal terrain edge.

The *horizon* is the set of ridges that corresponds to the blocking segments most distal from the viewpoint. It has been shown that the number of segments comprising the horizon is, in the worst case, only slightly supralinear in the number of terrain edges. The boundaries between visible and invisible regions are sometimes called odd and even order

horizons with respect to the given observation point (Figure 7).

In order to program visibility computations, two questions must be laid at rest. The first question is: What happens beyond the boundary of the terrain, where we have no elevation information? We can assume, for instance, that the terrain is bounded by an infinitely high wall, or that it is surrounded by a flat ocean. Alternatively, we can model the curvature of the earth, which will ensure that visibility from every point is limited, or else simply set an arbitrary limit on the maximum distance from which a point may be visible.

The second question concerns collinear points. Are surfaces tangent to a line of sight visible? How we settle these questions won't have any significant impact on the methods or conclusions that we present, but computer implementation requires unambiguous specifications.

A further assumption may be made with regard to the height of the observer above ground. In most instances, assuming ground-level observation is not realistic. Assuming some *observation height* is essential for some problems, but optional for others.

4.3 Computing visibility

The computation of the visibility matrix on a triangulated irregular network is conceptually straightforward, unless one attempts to exploit the obvious *coherence* in the visibility of neighboring viewpoints [16]. A simple algorithm for computing the complete visibility map on a TIN can be visualized as

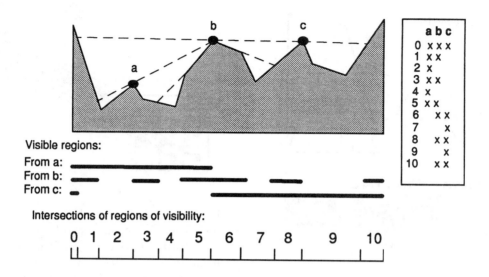

Figure 9: Location of fire towers. Points *a*, *b*, and *c* are candidate locations. The regions visible from each point are shown below the terrain, followed by the ten elementary regions (generated by pairwise intersections) that are each completely visible or completely invisible from all candidate points. Visible elementary regions are marked with an *x* in the table on the right. In this simple example, it is clear that points *a* and *c* are sufficient to see the entire terrain.

a searchlight, located at the viewpoint, which illuminates the terrain in a progressive outward spiral. As the beam is raised, it encounters ridges that cast shadows on the terrain farther from the light. The endpoints of the ridge and shadow segments, which form the boundary between the visible and invisible regions, are recorded. Adjacent viewpoints (vertices of the triangulation) are considered intervisible. With a TIN, a triangle which may cast a shadow on another triangle must be processed first.[7] The visible portion of each triangle is determined by projecting on it the dominant blocking edges between the triangle and the viewpoint. New blocking edges are introduced whenever a partially or fully visible triangle is followed by an invisible triangle.

Grid-based algorithms all compute the intersection of radial lines of sight with the edges of the grid cells that they intersect [42]. The difference between algorithms lies mainly in the choice of rays (Figure 8). There is little loss of accuracy if the number of rays is reduced, provided that the points are weighted to account for the dispersion of the rays [23].

In the next several sections, we examine several specific applications based on visibility computations.

---

[7]The necessary ordering property was mentioned in Section 3.

4.4   Observation points

A *shortest watchtower algorithm* determines the location of the point *with the lowest elevation above the surface* from which an entire polyhedral terrain is visible [47]. Such a point must exist because the terrain elevation is a single-valued function, and therefore entirely visible from any point sufficiently far above it. It is also possible to determine efficiently whether any particular point is visible from a *single observation point* on or above the surface [6].

Finding the location of the *minimum* set of observation points on the surface from which the entire surface is visible (*guard allocation*) is much more time consuming. Topographic applications include the location of fire towers, artillery observers, and radar sites.

On a triangulated terrain, it is customary to restrict consideration to viewpoints located at vertices of the triangulation. First, the area of interest must be partitioned so that each partition is either completely visible or completely invisible from each viewpoint. The required partitions are obtained by successive intersections of the visibility maps. Now finding the smallest number of observation towers can be stated as a *set-covering* (or *facilities-location*) problem of operations research. An example showing a vertical section through the terrain is shown in Figure 9. Variations of the problem include [27]:

a. Find the area visible from a fixed set of observation points.

b. Maximize the area visible from a fixed number of observation points.

c. Given some cost function related to tower height, locate the towers so as to see the entire area at minimum cost.

d. Given some cost function related to tower height, locate the towers that maximize the area visible at a fixed cost.

Landscape analysis is less easy to formalize, but modern scenery analysis distinguishes between superior, normal, and inferior positions relative to local relief. Depending on the application, a commanding vista may be called a *military crest* or a *panorama*.

4.5   Line-of-sight communication

An obvious application of geometric visibility is the location of microwave transceivers for telephone, FM radio, television, and digital data networks. Of course, a realistic solution must take into account the height of the towers, the diffraction from intermediate ridges, and the distance limit imposed by the inverse-square law of electromagnetic propagation.

If the towers are restricted to the vertices of a polygonal terrain, then the only information that is required for line-of-sight computations is the visibility matrix or graph. Finding the minimum number of relay towers necessary for line-of-sight transmission between two transceivers can be formulated as a shortest-path search on the visibility graph. The overall computation can be accelerated by computing dynamically only the portions of the visibility graph that are required at any stage of the shortest-path search.

Now consider the problem of locating relay towers to complete the line-of-sight network between several transceivers (Figure 10). This problem can be solved, under the restriction that the relay towers are located at vertices of the TIN. Here, instead of computing the shortest path, one must find the *Minimum Steiner Tree* (MStT) on the visibility graph. The MStT in this context is the acyclic subgraph of the visibility graph, with the minimum number of (unweighted) "intermediate" edges, that includes all the transceivers. Because the overall computational complexity is the product of the cost

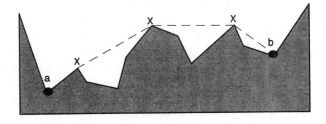

Figure 10: Relay towers for line-of-sight transmitters a and b. The x's indicate the minimum number of transmitters. Although a sectional view is shown here for ease of illustration, in actuality the transceivers and the relays would not be in the same vertical plane.

of computing the visibility graph on the TIN and of the cost of computing the MStT on the visibility graph, the size of the underlying triangulation must be reduced as much as possible. Then a conservative solution is obtained by adding the known bound on the resulting elevation error to the heights of the relay towers [17].

Finally, suppose that identical transmitters are to be located so as to broadcast to a fixed set of receivers. Specifically, it is required to locate the minimum number of transmitters so that each receiver can "see" at least one transmitter. This problem is similar to the fire-tower problem, and can be reduced to set covering on the visibility matrix itself (without intersecting any visibility maps).

4.6   Surface paths

The shortest path from one viewpoint to another along the edges of a triangulated terrain, such that none of the viewpoints traversed is visible from a given observation point, is called a *smuggler's path*. A path on which every vertex is visible is a *scenic path*. We can find such a path (if one exists) by determining either the viewpoints that are visible from the observation point, or those that are not, and applying a standard shortest-path algorithm to the edges that connect them.

Iwamura and his colleagues demonstrate a GIS for interactive planning of scenic paths. Constraints on the path include length, slope, and cost of construction. For any observation point along a candidate path, both a "visual range map" (the representation of a viewshed using radial lines from the viewpoint) and a bird's-eye view of the terrain can be displayed [25].

4.7   Visibility invariants

Visibility functions do not define a terrain uniquely: several different terrains may have the same visibility map. We call these terrains *visibility-equivalent*. To gain some insight into what characterizes this equivalence relation, consider the two sectional views of visibility-equivalent terrains shown in Figure 11. For simplicity, the x-coordinates of the points are uniformly spaced. Under these conditions, the y-coordinates are subject to the following equation:

$$(y_4 - y_1)/(y_4 - y_3) = (y_5 - y_2)/(y_3 - y_2).$$

This equation defines the cross-ratio of four segments, which is invariant under a projective transformation. As a corollary, the visibility functions on a $1\frac{1}{2}$-D terrain are invariant under any projective transformation, including translation and scaling of coordinates, rotations, and affine transformations.

Tools for computing the intervisibility of selected points have long been included in geographic information systems but a recent attempt to compare eight software packages for viewshed determination led to inconsistent results. We expect, however, that the next generation of GIS will offer a number of robust visibility-related application programs.

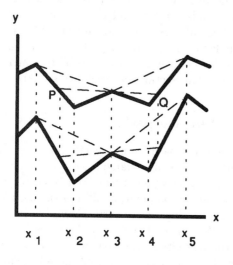

Figure 11: Two visibility-equivalent terrains in section. The locations of shadow edges are identical for every terrain whose elevations satisfy the projective relationship $(y_4 - y_1)/(y_4 - y_3) = (y_5 - y_2)/(y_3 - y_2)$ for each subset of five data points. For example, the shadow edge Q of the observation point P has the same x-coordinate in both terrains. Consequently the intervisibility of any pair of points can be readily determined from the locations of the horizons.

5   Map Projections (Geometry, Geography, and Geodesy)

When the extent of the database under consideration is larger than a city or county, we must allow for the sphericity of the planet. Although for internal computer processing a three-dimensional model for locating objects in terms of spherical coordinates is perfectly adequate, for display and mapping purposes there is no satisfactory alternative to projecting the earth's surface on a 2-D plane.

The ideal projection would preserve relative distances, angles, and areas, but no projection can preserve all of these. The distortions increase with the solid angle subtended at the center of the earth. No projection can show scale correctly throughout the map, but there are usually one or more lines on the map where scale remains true. *Equidistant projections* show true scale between one or two selected points and every other point. *Azimuthal projections*, which preserve the direction of all points with respect to the center, are usually projections on a tangent plane.

The most common mapping projections are the Lambert Conformal and the Transverse Mercator. Both of these are *conformal projections*, and therefore preserve local shape. The former projects the earth's surface on a cone centered about the polar axis with its surface tangent or secant to the extent under consideration. It is suited to regions elongated east to west, like Nepal. Distortions can be minimized by letting the cone cut the surface of the sphere at two parallel circles, and tilting its axis away from that of the earth. The Transverse Mercator is based on a cylinder with its axis perpendicular to the polar axis and its surface tangent to a meridian (longitude) near the center of the projected area. It is suitable for regions whose major dimension is north and south, like Norway. The cylinder of the Oblique Mercator touches the earth's surface along a great circle other than the Equator. While the transformation from Transverse to Oblique is easy for a sphere, it is more difficult for an ellipsoidal model of the Earth. As shown in Figure 12, both cone and cylinder are developed into a flat surface by cutting along a single line segment [51].

State Plane Coordinates (used, for example, by Roads departments) consist of a rectilinear grid defined on a map projection. The origin is placed outside the extent, hence locations can be identified

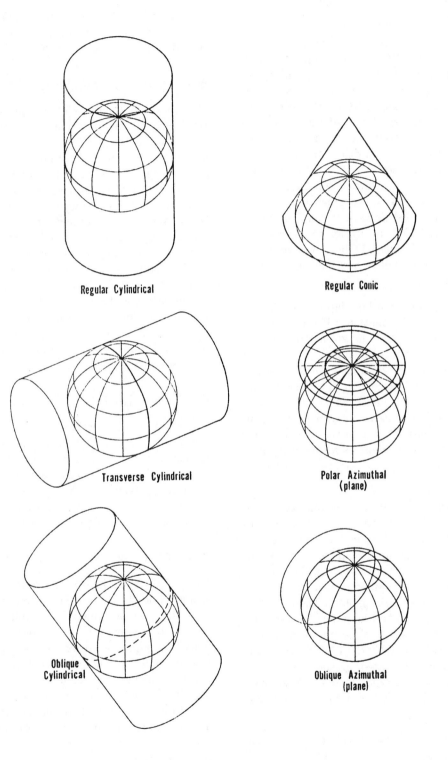

Regular Cylindrical

Regular Conic

Transverse Cylindrical

Polar Azimuthal
(plane)

Oblique
Cylindrical

Oblique Azimuthal
(plane)

Figure 12: Projection of the earth on a cylinder, a cone, or a plane. For some maps, a central projection is used, but in most cases more complex mappings are necessary to obtain desirable properties.

entirely by pairs of positive integers. In large states, the state plane coordinates are based on more than one projection.

The newest projections are designed for satellite mapping. Since 1972 several mapping satellites have been launched into nearly circular sun-synchronous orbits inclined at 99 degrees to the Equator at about 919 km altitude. Because of the double motion of the satellite and the earth's rotation, these satellites track a continuous S-shaped swath (about 185 kms in width). In the case of Landsat, the course does not return to the same point for 251 orbits (18 days). For mapping, a "dynamic" map projection on which the ground-track remains true-to-scale is required. The Space Oblique Mercator (SOM) projection was derived for this purpose from the Hotine Oblique Mercator projection (which Hotine called "rectified skew orthomorphic") [50].

This new projection can be adapted to orbits of any eccentricity and inclination. The central line of the SOM closely follows the groundtrack of the satellite, which is not a great circle but a path of constantly changing curvature. Figure 13 shows two orbits of the SOM projection. The forward and backward projection formulas, which require iteration and numerical integration, can be found in [50]. Prior to the introduction of these projections, the preparation of photo-mosaics for mapping purposes from satellite images was more an art than a science.

The shape of the earth is approximated by an oblate ellipsoid of revolution. The difference of 1/300th between the polar and equatorial radius must be taken into account in maps at a scale of 1:100,000 or larger. Furthermore, slightly different ellipsoidal surfaces are required to provide the best fit at different parts of the globe. Therefore the ellipsoid is used with an "initial point" reference location to provide the sea-level datum for mapping. The Clarke ellipsoid, used for the 1927 North American Datum, was supplanted in the 1983 North American Datum by the World Geodetic System ellipsoid. The new datum is based on satellite tracking data.

The distortions in distance, relative angle, and area can be readily computed and plotted for the simpler projections under the assumption of a spherical earth. They can also be studied by direct measurement on existing maps, making use of the known locations of the intersections of latitudes and longitude lines or ticks.

6 Ranging Further

Here we briefly list some operations that are commonly encountered in processing GIS queries and provide references for additional information.

6.1 Line intersections

Given a set of line segments in the plane, described by their endpoint coordinates, it is often of interest to find and report the location of every intersection among them. A number of fast techniques exist for accomplishing this, including line-sweep methods based on presorting the segments in one dimension [2, 3, 40], and uniform-grid techniques that presort the segments in both dimensions [22]. Polygon overlay is a generalization of line intersections [32]. Assume that one map shows diverse agricultural land use in a state, while another shows county boundaries. The task is to determine agricultural land use in each county. A special application of polygon intersection is polygon-to-grid transformation. This is useful, for instance, for transforming digitized soil maps into a cellular data structure for simpler query processing.

6.2 Proximity

Not all proximity problems are solved most efficiently by Voronoi diagrams. For instance, finding the closest, or farthest, pair of points among a set of points requires a specialized algorithm [40]. Shortest route problems, when the path is constrained to a surface, or to some graph imbedded in the surface, form a large class of their own [26].

6.3 Curves

Most GIS approximate curves by a sequence of straight-line segments because few geographic features can be described by simple mathematical curves. Irregular curves can, however, be approximated by *splines* of various families. Unfortunately, the economies of such representations are usually offset by the increased complexity of the processing algorithms. Currently, geometric curves are used only in applications where planar map projections are inadequate, as in long-distance navigation.

6.4 3-D GIS

While most GIS address essentially two-dimensional problems, in geology, mining, and some oceanographic problems the 3-D structure is essential. Rather than develop truly 3-D data

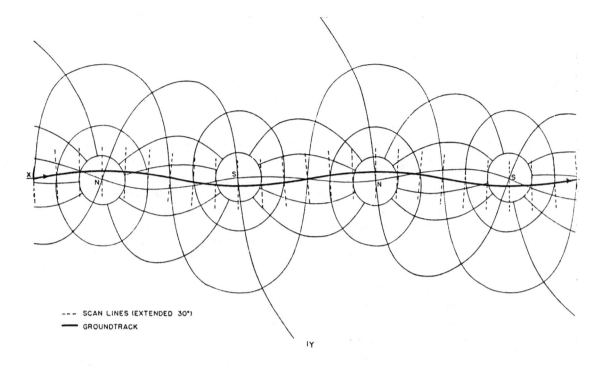

Figure 13: Two orbits of the Space Oblique Mercator projection, shown for Landsat. A narrow band along the groundtrack remains true to scale.

structures, a common approach is to consider parallel layers [41].

6.5 Interpolation

The simplest method to interpolate values defined on a grid is bilinear interpolation. Higher-order interpolation algorithms take into account more than the four immediate neighbors (eight in 3-D) of the query point. Triangulated surfaces allow linear interpolation. However, the resulting surfaces are not smooth, in the sense that they have discontinuities in their derivatives at the edges of the cells. When the data points are not uniformly distributed, Voronoi interpolation allows consideration of values at additional data points.

6.6 Physiographic features

The literature on landforms is remarkably short on algorithmic definitions. Most attempts to automate physiographic feature extraction have been based on discrete approximations to derivatives of the surface [12, 37]. These features are generally quite local and highly scale-sensitive, whereas the significance of terrain features depends on their size and location relative to similar features in the entire area. Metrics based on geometric visibility automatically take into account global relations.

Some examples of geometric visibility applied to topographic features are the following. The visibility region of significant peaks tends to be large, and includes most of the visibility regions of lesser peaks. Significant peaks also have many blocking segments and multiply-connected visible regions, which distinguishes them from points in broad valleys that also have high visibility. Ridges block the horizons of many observation points. Points that are intervisible are in the same valley, otherwise they are separated by ridges. In pits and valleys, the prospect is singly-connected and tends to change gradually. Even if landforms cannot be determined entirely according to visibility criteria, these may generate useful measures for ranking them.

6.7 Watersheds

The study of watersheds, originated more than a century ago by James Clerk Maxwell, illustrates a mathematically advanced aspect of the study of physiography. Critical points (*peaks*, *pits*, and *passes*), *elevation contours* and *slope lines*, *ridge lines* and *course lines*, and *dales* (a dale is a district whose slope lines run to the same pit) all have precise mathematical definitions in terms of partial derivatives of the terrain function [30]. Elaborate analyses and simulations are routinely carried out in planning dams and canals to mitigate flooding.

6.8   Navigation

Automated path planning (collision-free paths, shortest paths, cheapest paths, safest paths) is already an important aspect of robotics. Preliminary *gross planning* typically takes into account fixed aspects of the robot's environment, while *fine planning* copes with unpredicted obstacles, such as other vehicles, that may be sensed by the robot in motion. As the range of autonomous vehicles expands, path planning may become a routine component of GIS.

Horizons do provide an important clue for navigation in mountainous terrain. Discontinuities in visibility can be readily determined under poor conditions by a variety of sensors, and matched to stored or computed horizons to determine the location of the observer. Discernible terrain features guide airborne military vehicles [55]. The use of horizon lines for autonomous navigation by a Mars Rover has been considered.

6.9   Robust computation

There are two aspects to robust computation. One is the treatment of pathological cases. In geography, one cannot simply assume that points are in general positions and that special cases will not arise. The presence of collinear and co-circular points, exactly vertical or horizontal lines, duplicate entities, and so forth simply cannot be allowed to yield anomalous results or disrupt processing [20].

The second aspect is the accumulation of round-off error. Even if floating-point arithmetic is used, the finite word-length of computers allows only signed-integer computation, and overflow and underflow will result in arithmetic error. Consider, for instance, the intersection of two straight-line segments defined by endpoints placed on a grid. This grid corresponds to all the x and y coordinates that can be represented in a particular computer (if floating point arithmetic is used, then the size of the grid-cells changes in logarithmic increments). The intersection of the two line segments will not, in general, lie on the grid, and will thus have to be approximated by the coordinates of the nearest grid point. Therefore a straightforward arithmetic check for collinearity on the two endpoints of one of the segments and the computed point of intersection, will return an incorrect negative answer.

Arithmetic precision problems of this type can be circumvented by *rational arithmetic* [57]. However, on each successive intersection computation, the number of digits required in the numerator and the denominator tends to triple. An alternative, in the above problem, is to perturb the endpoints of the line segments in such a way that the intersection will fall on the grid. This can be accomplished with a method based on continued fractions, and results in a trade-off between positional accuracy and the preservation of collinearity relations [28]. A third method is demonstrated through the construction of a million-cell Voronoi diagram in [54].

Research on robust algorithms is booming because current methods often introduce slivered regions, region boundaries that double back on themselves or leak, and towns that are accidentally shifted from one bank of a river to the other.

References

[1] F. Aurenhammer. Voronoi diagrams—a survey of a fundamental geometric structure. *ACM Computing Surveys*, 23(2):354–406, September 1991.

[2] H. S. Baird. Fast algorithms for LSI artwork design. *Design Automation and Fault-tolerant Computing*, 2(2):165–178, May 1978.

[3] J. L. Bentley and T. A. Ottman. Algorithms for reporting and counting geometric intersections. *IEEE Trans. Computers*, 28:643–647, 1979.

[4] J. C. Cavendish. Automatic triangulation of arbitrary planar domains for the finite element method. *Int. Jour. Numer. Meth. Engrg.*, 8:679–696, 1974.

[5] J. C. Cavendish, D. A. Field, and W. H. Frey. Automating three-dimensional finite element mesh generation. In K. Baldwin, editor, *Modern Methods for Automating Finite Element Mesh Generation*, pages 61–72. American Society of Civil Engineers, New York, 1986.

[6] R. Cole and M. Sharir. Visibility problems for polyhedral terrains. *J. Symbolic Computation*, 7:11–30, 1989.

[7] B. Delaunay. Sur la sphere vide. *Bull. Acad. Sci. USSR(VII), Classe Sci. Mat. Nat.*, pages 793–800, 1934.

[8] M. B. Dillencourt. A non-Hamiltonian, nondegenerate Delaunay triangulation. *Information Processing Letters*, 25:149–151, 1987.

[9] H. Edelsbrunner. *Algorithms in Computational Geometry*. Springer-Verlag, Heidelberg, 1987.

[10] H. Edelsbrunner. An acyclicity theorem for cell complexes in *d*-dimensions. *Combinatorica*, 10:251–260, 1990.

[11] H. Edelsbrunner and R. Seidel. Voronoi diagrams and arrangements. *Discrete and Computational Geometry*, 1:25–44, 1986.

[12] B. Falcidieno and M. Spagnuolo. Polyhedral surface decomposition based on curvature analysis. In T. L. Kunii and Y. Shinagawa, editors, *Modern Geometric Computing*, pages 57–72. Springer-Verlag, Tokyo, 1992.

[13] J. P. Felleman. Landscape visibility. In R. C. Smardon, J. F. Palmer, and J. P. Felleman, editors, *Foundations of Visual Project Analysis*, pages 47–62. Wiley and Sons, NY, 1986.

[14] L. De Floriani. Surface representations based on triangular grids. *The Visual Computer*, 3:27–50, 1987.

[15] L. De Floriani, B. Falcidieno, G. Nagy, and C. Pienovi. On sorting triangles on a Delaunay tesselation. *Algorithmica*, 6:522–532, June 1991.

[16] L. De Floriani and P. Magillo. Visibility algorithms on triangulated terrain models. *Int. J. Geographic Information Systems*, 8(1):13–41, 1994.

[17] L. De Floriani, G. Nagy, and E. Puppo. Computing a line-of-sight network on a terrain model. In *Proc. Fifth Int. Symp. Spatial Data Handling*, pages 672–681, Charleston, SC, August 1993.

[18] S. Fotheringham and P. Rogerson. *Spatial Analysis and GIS*. Francis & Taylor, London, 1994.

[19] R. J. Fowler and J. J. Little. Automatic extraction of irregular network digital terrain models. *Computer Graphics*, 13:199–207, 1979.

[20] W. R. Franklin. Cartographic errors symptomatic of underlying algebra problems. *Proc. Int. Symp. Spatial Data Handling, Zurich*, pages 190–208, 1984.

[21] W. R. Franklin, V. Akman, and C. Verrilli. Voronoi diagram with barriers and on polyhedra for minimal path planning. *Visual Computing*, 1(2):133–150, October 1985.

[22] W. R. Franklin, M. Kankahalli, and C. Narayanaswari. Geometric computing and the uniform grid data technique. *Computer Aided Design*, 21(4):410–420, 1989.

[23] W. R. Franklin and C. Ray. Higher isn't necessarily better: visibility algorithms and experiments. In T. C. Waugh and R. G. Healey, editors, *Advances in GIS Research, Sixth International Symposium on Spatial Data Handling*, pages 751–770. Taylor & Francis, September 1994.

[24] A. Guttman. *R*-trees: a dynamic index structure for spatial searching. In *Proc. SIGMOD Conference*, pages 47–57, Boston, June 1984.

[25] K. Iwamura, Y. Nomoto, S. Kakumoto, and M. Ejiri. Geographical feature analysis using integrated information processing system. In *IAPR Workshop on Machine Vision Applications*. Tokyo, November 1990.

[26] H. Booth J. O'Rourke, S. Suri. Shortest paths on polyhedral surfaces. In *Proc. Second Symp. Theoretical Aspects of Computer Science*, volume 182 of *Lecture Notes in Computer Science*, pages 243–254, New York, 1985. Springer-Verlag.

[27] J. Lee. Analysis of visibility sites on topographic surfaces. *J. Geographical Information Systems*, 5(4):413–429, 1991.

[28] M. Mukherjee and G. Nagy. Collinearity constraints on geometric figures. In B. Falcidieno and T. Kunii, editors, *Modeling in Computer Graphics*, pages 115–125. Springer-Verlag, NY, 1993.

[29] K. Mulmuley. *Computational Geometry: An Introduction through Randomized Algorithms*. Prentice Hall, Englewood Cliffs, NJ, 1994.

[30] L. R. Nackman. Two-dimensional critical point configuration graphs. *IEEE Trans. Pattern Analysis and Machine Intelligence*, 6(4):442–450, 1984.

[31] G. Nagy. Terrain visibility. *Computers and Graphics*, 18(2):763–773, December 1994. Special issue on Modelling and Visualization of Spatial Data in GIS.

[32] J. Nievergelt and F. Preparata. Plane-sweep algorithms for intersecting geometric figures. *Comm. ACM*, 25(10):739–747, 1982.

[33] J. O'Rourke. *Art Gallery Theorems and Algorithms*. Oxford University Press, 1987.

[34] J. O'Rourke. Visibility graphs. *ACM SIGACT NEWS*, 24(1):20–25, 1993.

[35] J. O'Rourke. *Computational Geometry in C*. Cambridge University Press, 1994.

[36] M. H. Overmars and E. Welzl. A new method for computing visibility graphs. In *Proc. Fourth Annual ACM Symposium on Computational Geometry*, pages 164–171, Urbana-Champaign, IL, June 1988.

[37] T. K. Peucker and D. H. Douglas. Detection of surface specific points by local parallel processing of discrete terrain elevation data. *Computer Graphics and Image Processing*, 4:375–387, 1975.

[38] T. K. Peucker, R. J. Fowler, J. J. Little, and D. H. Mark. The triangulated irregular network. In *Proc. ASP-ACSM Symp. on DTMs*, St. Louis, 1978.

[39] D. J. Peuquet and D. F. Marble. *Introductory Readings in Geographic Information Systems*. Taylor & Francis, London, 1992.

[40] F. P. Preparata and M. I. Shamos. *Computational Geometry: An Introduction*. Springer-Verlag, New York, 1985.

[41] J. Raper. *Three Dimensional Applications in GIS*. Taylor & Francis, London, 1990.

[42] C. Ray. A new way to see terrain. *Military Review*, LXXIV(11), November 1994.

[43] D. Rhind. The information infrastructure of GIS. In *Proc. 5th Int. Symp. on Spatial Data Handling*, pages 1–20, Charleston, SC, 1992.

[44] D. Rhind, J. Raper, and H. Mounsey. *Understanding Geographical Information Systems*. Taylor & Francis, London, 1992.

[45] A. Saalfeld. Triangulated data structures for map merging and other applications in geographic information systems. *Proc. Int. Symp. on GIS*, 3:3–13, November 1987.

[46] H. Samet. *The Design and Analysis of Spatial Data Structures* and *Applications of Spatial Data Structures*. Addison Wesley, Reading, MA, 1989.

[47] M. Sharir. The shortest watchtower and related problems for polyhedral terrains. *Information Processing Letters*, 29:265–270, 1988.

[48] T. Shermer. Recent results in art galleries. *Proceedings of the IEEE*, 80(9):1384–1399, 1992.

[49] P. P. Silvester and R. L. Ferrari. *Finite Elements for Electrical Engineers*. Cambridge University Press, second edition, 1990.

[50] J. B. Snyder. The Space Oblique Mercator Projection. *Photogrammetric Engineering and Remote Sensing*, 44(5):585–596, 1978.

[51] J .P. Snyder. *Map Projections Used by the U.S. Geological Survey*. Number 1532 in Geological Survey Bulletin. US Gov. Printing Office, Washington, 1982.

[52] V. Srinivasan, L. R. Nackman, J. M. Tang, and S. N. Meshkat. Automatic mesh generation using the symmetric axis transformation of polygonal domains. *Proceedings of the IEEE*, 80(9):1485–1501, September 1992. Special Issue on Computational Geometry.

[53] J. Star and J. Estes. *Geographic Information Systems: An Introduction*. Prentice Hall, Englewood Cliffs, NJ, 1990.

[54] K. Sugihara and M. Iri. Construction of the Voronoi diagram for "one million" generators in single precision arithmetic. *Proceedings of the IEEE*, 80(9):1471–1484, September 1992. Special Issue on Computational Geometry.

[55] Y. Ansel Teng, Daniel De Menthon, and L. S. Davis. Stealth terrain navigation. *IEEE Trans. Systems, Man, and Cybernetics*, 23(1):96–110, Jan/Feb 1993.

[56] D. Tomlin. *Geographic Information Systems and Cartographic Modeling*. Prentice Hall, Englewood Cliffs, NJ, 1990.

[57] P. Y. Wu and W. R. Franklin. A logic programming approach to cartographic map overlay. *Canadian Computational Intelligence Journal*, 6(2):61–70, 1990.

[58] W. Zimmerman and S. Cunningham. *Visualization in Teaching and Learning Mathematics*. Number 19 in MAA Notes. MAA, Washington, DC, 1991.

On the Other Hands: Geometric Ideas in Robotics

Bud Mishra

Robotics Laboratory

Courant Institute of Mathematical Sciences

New York University, 719 Broadway

New York, NY, 10003

mishra@nyu.edu

1   Truth About Polydactility

I would like to start this essay with the following story based on an essay entitled "Eight Little Piggies" by the popular Harvard biologist Stephen Jay Gould [4].

The fossils of the oldest tetrapods were discovered in eastern Greenland by a Danish expedition in 1929. They date back to the very last phase of the Devonian period (the so-called "age-of-fish") and the two genera from that period that have been studied extensively are *Ichthyostega* and *Acanthostega*. The Swedish paleontologist, Gunnar Säve-Söderbergh, collected most of the material in 1931 and directed the project until his untimely death in 1948. One of the greatest Swedish paleontologists Erik Jarvik took over the project and published thorough anatomical studies of these two genera. See Figure 1 for a picture of the Ichthyostega taken from Erik Jarvik's tome *Basic Structure and Evolution of Vertebrates, Vol 1*.

Although no specimens preserved enough of the fingers or toes for an unambiguous count, Jarvik reconstructed Ichthyostega with *five* digits per limb. Why?

One needs to look at the history in order to understand such an unshakable faith in "pentadactylity" (five-fingered-ness). Richard Owen (one of England's greatest anatomists and a contemporary of Darwin) had actually hypothesized an archetypical tetrapod pentadactyl vertebrate. He was so pleased with it that he had a picture of this archetype engraved on his personal emblem. While Darwin held on to a more worldly view, he seemed to have been impressed by Owen's archetype which he described as a *"real representation as far as the most consummate skill and loftiest generalization*

can represent the parent form of the vertebrata."

In light of this history, it seems hardly surprising that Jarvik would accept the pentadactylity of vertebrates so easily, as many of my readers would. However, Jarvik did go a few steps ahead when he equated the "human culture" to the "basic pattern of our five-fingered hand."

However, the picture that Jarvik drew of an Ichthyostega came to some doubt beginning in 1984, when a Soviet paleontologist, O. A. Lebedev, discovered a fossil of another early tetrapod *Tulerpeton* and finally in 1990, a joint Copenhagen-Cambridge team found a hind limb of Ichthyostega and a fore limb of Acanthostega. They wrote, reporting their disagreement with Jarvik's construction:

> The proximal region of the hind limb of ichthyostega corresponds closely to the published description, but the tarsus [foot] and digits differ.

The back leg of Ichthyostega had *seven* toes.

Independently, quite a different line of investigation has emerged recently with our attempts to construct mechanized robot hands, capable of reproducing the same degree of dexterity as human hands. Notable among such hands are the Utah/MIT dextrous hand, the Stanford/JPL hand, the NYU Four Finger Manipulator (FFM), the Okada hand, and the Asada hand. Many early approaches had taken an anthropomorphic view and had justified the resulting kinematic structure with the belief that human hands represent some Platonic archetypal structure. However, some rather beautiful geometric analyses of manipulative and grasping tasks have begun to seed doubts about the necessity of pentadactylity even in this mechanical domain. This essay studies these *other hands* and

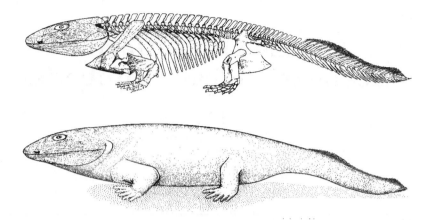

Figure 1: Erik Jarvik's rendition of Ichthyostega.

explores the underlying geometric and algorithmic ideas employed in this context.

To make matters somewhat concrete, we envisage an idealized dextrous hand, consisting of several independently movable force-sensing fingers. These fingers move as points in three-dimensional space. Here, we focus on the problem of *grip selection* for an object in the absence of static friction between the surface of the object and the fingers. This model is justified by the argument that presence of friction only improves the grasp and hence non-frictional grips represent in some sense the most pathological situation. Another argument favoring this grip model is that even when there is uncertainty about how much friction is available, the grasps synthesized under this restricted model remain immune to such uncertainties.

Such non-frictional grips have come to be known as *positive grips*. Since the fingers are assumed to be point fingers, a finger can only apply a force on the object along the surface-normal at the point of contact, directed inward.

If the shape of the object is precisely known, then the problem of *grip selection* reduces to that of choosing a set of GRIP POINTS and a set of associated FORCE TARGETS. We then ask:

- *Can an arbitrary object be gripped (positively) with a finite number of fingers?*

- *If so, what are the grip points and the magnitudes of the forces exerted by the fingers (force targets) for such a grip?*

It can be then shown that almost every smooth object allows a positive grip with only a bounded (relatively small—but not five) number of fingers.

2 A Little Physics

A good starting place for us would be to understand how an object in equilibrium can be characterized. There are two ways of doing this: Either we can assume that the forces and torques acting on an object are *monogenic* (i.e., force/torques are derived from the potential energy as would be the case if the fingers are assumed to be compliant) or that the forces are *polygenic* (the force/torques applied at the fingers are generated by some actuators whose mechanics need not concern us). In the first case, we may proceed by looking at the local minima (stable equilibrium points) of the scalar function characterizing the energy. In the second case, we need to understand the *resultant* force and torque equation, as in the classical Newtonian mechanics. Here, our focus will be on the models corresponding to polygenic forces, as these models demonstrate the close connection between robotics and combinatorial geometry.

Consider a rigid body subject to a set of external polygenic forces $f_1, \ldots, f_k$, applied respectively at the points $p_1, \ldots, p_k$, as in Figure 2. Then the necessary and sufficient condition for the rigid body to be in equilibrium is that *the resultant force and the resultant torque must be null vectors*. In mathematical notations, this condition can be stated as follows:

$$\sum_{i=1}^{k} f_i = 0 \quad \text{and} \quad \sum_{i=1}^{k} p_i \times f_i = 0,$$

where the cross product $\tau = p \times f$ is defined as

$$\begin{aligned}
\tau_x &= p_y f_z - p_z f_y, \\
\tau_y &= p_z f_x - p_x f_z, \quad \text{and} \\
\tau_z &= p_x f_y - p_y f_x.
\end{aligned}$$

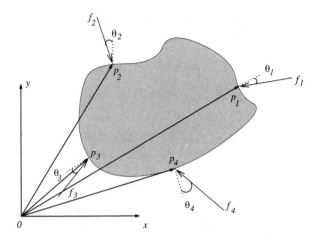

Figure 2: A planar object subject to four forces f_1, f_2, f_3 and f_4.

Thus, in order to hold an object in equilibrium with a multi-fingered hand (say, with k fingers), we need to place these fingers at points $p_1, \ldots, p_k$ on the boundary of the objects and apply forces $f_1, \ldots, f_k$ in such a manner that the equilibrium condition is satisfied. However, in this context, we need to satisfy two other conditions; namely,

- The normal components of the forces must be directed inward. Note that, otherwise, the fingers cannot maintain contact with the object.

- Furthermore, if the ith force f_i makes an angle θ_i with the surface normal at point p_i, then in order for the finger not to slip, $\tan \theta_i \leq \mu$, where the *coefficient of static friction* between the body and the finger is denoted by the constant μ.

Going back to the concept of *positive grip*,[1] we realize that in this special case, we have no friction; the coefficient of static friction $\mu = 0$ and each θ_i must be zero. Thus if we write n_i for the unit normal to the surface of the body at the point p_i and directed inward, then the finger force f_i must be a nonnegative multiple of n_i. Thus

$$f_i = \alpha_i n_i, \quad \text{and} \quad p_i \times f_i = \alpha_i (p_i \times n_i),$$
$$\text{where } \alpha_i \geq 0, \text{ scalar.}$$

In order to better understand the effect of the requirements imposed by positive grip, we may consider the following somewhat easier problem:

---

[1] A grip simply denotes any arbitrary placement of fingers. A grasp, on the other hand, is a special case of a grip that must satisfy an additional equilibrium constraint. Note, however, that in the robotics literature both of these terms have been used interchangeably.

Given: k grip points

$$\{p_1, p_2, \ldots, p_k\},$$

on the boundary of the body B.

Determine: If the object can be grasped (positively) by placing the fingers at the grip points.

For example, consider a planar rectangular object with four grip points at the midpoints of the edges (shown in Figure 3.) Let the grip points be denoted as p_1, p_2, p_3 and p_4 and the respective unit surface normals as n_1, n_2, n_3 and n_4. Then we wish to determine if there are four scalar quantities α_1, α_2, α_3 and α_4 such that

$$\alpha_1 n_1 + \alpha_2 n_2 + \alpha_3 n_3 + \alpha_4 n_4 = 0$$
$$\alpha_1 (p_1 \times n_1) + \alpha_2 (p_2 \times n_2) +$$
$$\alpha_3 (p_3 \times n_3) + \alpha_4 (p_4 \times n_4) = 0$$

$$\alpha_1 \geq 0, \alpha_2 \geq 0, \alpha_3 \geq 0, \alpha_4 \geq 0 \text{ and not all } 0.$$

Note that, for this example, any choice of $\alpha_1 = \alpha_3$ and $\alpha_2 = \alpha_4$ will satisfy the conditions (assuming that at least two of them are nonzero and all of them are nonnegative). In particular, we could have chosen all the α's to be $1/4$!

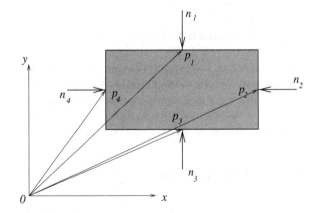

Figure 3: A planar rectangular object with designated grip points $\{p_1, p_2, p_3, p_4\}$.

To make matters a little more abstract, we should define a *wrench map*, Γ, taking a point on the boundary of the object B to a point in the d-dimensional wrench space $\mathbf{R}^d$. Note that the term wrench space is used to denote a vector space consisting of all the wrenches—vectors, each comprising the components of a force vector followed by the components of corresponding torque vector. Its dimension d is 1, 3 or 6, depending on whether the object belongs to 1-, 2-, or 3-dimensional space.

$$\Gamma : \partial B \rightarrow \mathbf{R}^d$$
$$: p_i \mapsto (n_i, p_i \times n_i).$$

Thus the wrench map Γ maps a point $p_i \in \partial B$ on the boundary of the body B to a wrench (a force/torque combination) that would be created if we apply a unit normal force directed inward at the point p_i. Then the feasibility of a positive grip can be expressed in terms of the existence of a solution of the following system of linear equations and inequalities:

$$\sum_{i=1}^{k} \alpha_i \Gamma(p_i) = 0$$
$$\alpha_i \geq 0, \quad i = 1, \ldots, k,$$
$$\sum_{i=1}^{k} \alpha_i = 1.$$

The last condition is added only for convenience, since if other α_i's were used, one could simply normalize their sum by dividing all the terms by a suitable denominator.

Geometrically, we are asking if some convex combination of the $\Gamma(p_i)$'s would yield the null vector. More compactly, we ask

$$0 \in \text{convex hull } (\Gamma(p_1), \ldots, \Gamma(p_k))?$$

If the answer to the preceding question is yes, then we can hold the object in equilibrium with the given grip points by applying forces whose magnitudes simply correspond to the coefficients used in the convex combination to express the null vector.

3   A Little Geometry

In this section, we shall provide some definitions, in order to discuss the geometry of wrench space in terms of the standard geometric vocabulary.

A d-dimensional space, $\mathbf{R}^d$, equipped with the standard linear operations, is said to be a *linear space*.

1. A *linear combination* of vectors $p_1, \ldots, p_n$ from $\mathbf{R}^d$ is a vector of the form

$$\alpha_1 p_1 + \cdots + \alpha_n p_n,$$

where $\alpha_1, \ldots, \alpha_n$ are in $\mathbf{R}$.

2. An *affine combination* of vectors $p_1, \ldots, p_n$ from $\mathbf{R}^d$ is a vector of the form

$$\alpha_1 p_1 + \cdots + \alpha_n p_n,$$

where $\alpha_1, \ldots, \alpha_n$ are in $\mathbf{R}$, with $\alpha_1 + \cdots + \alpha_n = 1$.

3. A *positive (linear) combination* of vectors $p_1, \ldots, p_n$ from $\mathbf{R}^d$ is a vector of the form

$$\alpha_1 p_1 + \cdots + \alpha_n p_n,$$

where $\alpha_1, \ldots, \alpha_n$ are in $\mathbf{R}_{\geq 0}$, the set of nonnegative real numbers.

4. A *convex combination* of vectors $p_1, \ldots, p_n$ from $\mathbf{R}^d$ is a vector of the form

$$\alpha_1 p_1 + \cdots + \alpha_n p_n,$$

where $\alpha_1, \ldots, \alpha_n$ are in $\mathbf{R}_{\geq 0}$ with $\alpha_1 + \cdots + \alpha_n = 1$.

By convention, we allow the empty linear combination (with $n = 0$) to take the value 0. We also assume that the empty linear combination is neither an affine combination nor a convex combination. Note that affine, positive and convex combinations are all linear combinations, and a convex combination is both affine and positive combinations.

A nonempty subset $L \subseteq \mathbf{R}^d$ is said to be a

1. *linear subspace*: if it is closed under linear combinations;

2. *affine subspace (or, flat)*: if it is closed under affine combinations;

3. *positive set (or, cone)*: if it is closed under positive combinations; and

4. *convex set*: if it is closed under convex combinations.

The intersection of any family of linear subspaces of $\mathbf{R}^d$ is again a linear subspace of $\mathbf{R}^d$. For any subset M of $\mathbf{R}^d$, the intersection of all linear subspaces containing M (i.e., the smallest linear subspace containing M) is called the *linear hull* of M (or, the linear subspace *spanned* by M), and is denoted by lin M.

Similarly, the intersection of any family of affine subspaces, or positive sets or convex sets of $\mathbf{R}^d$ is again, respectively, an affine subspace or positive set or convex set. Thus for any subset M of $\mathbf{R}^d$, we can define

1. the *affine hull* (denoted by aff M) to be the smallest affine subspace containing M,

2. the *positive hull* (denoted by pos M) to be the smallest positive set containing M, and

3. the *convex hull* (denoted by conv M) to be the smallest convex set containing M.

They are also called, respectively, the affine subspace, positive set and convex set *spanned* by M.

Equivalently, the linear hull lin M can be defined to be the set of all linear combinations of vectors from M. Similarly, the affine hull aff M (respectively, the positive hull pos M, the convex hull conv M) can be defined to be the set of all affine (respectively, positive, convex) combinations of vectors from M.

A set $p_1, \ldots, p_n$ of n vectors from $\mathbf{R}^d$ is said to be *linearly independent* if a linear combination

$$\alpha_1 p_1 + \cdots + \alpha_n p_n$$

can have the value 0, only when $\alpha_1 = \cdots = \alpha_n = 0$; otherwise, the set is said to be *linearly dependent*.

A set $p_1, \ldots, p_n$ of n vectors from $\mathbf{R}^d$ is said to be *affinely independent* if a linear combination

$$\alpha_1 p_1 + \cdots + \alpha_n p_n \quad \text{with} \quad \alpha_1 + \cdots + \alpha_n = 0$$

can have the value 0, only when $\alpha_1 = \cdots = \alpha_n = 0$; otherwise, the set is said to be *affinely dependent*.

A *linear basis* of a linear subspace L of $\mathbf{R}^d$ is a set M of linearly independent vectors from L such that $L = \text{lin } M$. The dimension dim L of a linear subspace L is the cardinality of any of its linear basis.

An *affine basis* of an affine subspace A of $\mathbf{R}^d$ is a set M of affinely independent vectors from L such that $A = \text{aff } M$. The dimension dim A of an affine subspace A is one less than the cardinality of any of its affine basis.

Let C be any convex set. Then by *d-interior* of C, denoted $\text{int}_d C$, we mean the set of points p such that, for some d-dimensional affine subspace, A, p is interior to $C \cap A$ relative to A. If c is the dim aff C, then by an abuse of notation, we write int C to mean $\text{int}_c C$.

Equivalently, we could have defined a convex set as follows:

A subset of a linear space is *convex* if it contains with any two of its points the line segment defined by them. (See Figure 4.)

Examples of convex sets include: a point, a line segment, a simplex, a cone, a half space, an affine

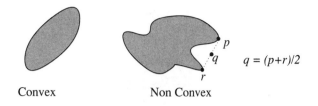

Figure 4: Convex and non-convex planar sets.

subspace or a linear subspace. Note that a closed half space can be defined to be the set of points

$$\{p = (x_1, \ldots, x_d) \in \mathbf{R}^d : a_1 x_1 + \cdots + a_d x_d \geq b\}.$$

Thus if p and p' are two points in the closed half space then clearly every point on the line segment pp' also belongs to the closed half space. Note that since the intersection of a family of convex sets is a convex set, we can also define the convex hull of a subset $M \subseteq \mathbf{R}^d$ to be the *intersection of the family of all closed half spaces containing M*.

Thus, given a finite set of points $M \subseteq \mathbf{R}^d$, we can enumerate the family of closed half spaces containing M and bounded by the affine hull of some subset of d points in M and then take their intersection to generate the convex hull of M. Such an algorithm would have a time complexity of $O(|M|^d)$. This naive algorithm can be improved significantly; techniques employed to construct a convex hull efficiently occupy a central place in the nascent field of Computational Geometry [3, 13].

3.1 Two Theorems from Convexity Theory

The following two theorems (Carathéodory's and Steintz's Theorems) from convexity theory have interesting implications for the theory of grasping. Subsequently, we will see how some of the most important results about grasping can be derived as simple consequences of these theorems. Proofs of these theorems can be found in any book on convexity theory [2, 17].

Carathéodory's Theorem: Let

$$X \subseteq \mathbf{R}^d \quad \text{and} \quad p \in \text{conv } X.$$

Then there exists some subset $Y \subseteq X$ such that

$$|Y| \leq d + 1 \quad \text{and} \quad p \in \text{conv } Y. \quad \square$$

For example, if $d = 1$, then Carathéodory's theorem implies that a point in the convex hull of a set of points on a line belongs to a line segment defined by some two points from the set. Similarly, for

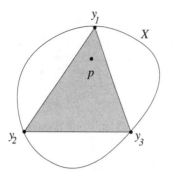

Figure 5: Example of Carathéodory's Theorem for $d = 2$.

$d = 2$, if $p \in \operatorname{conv} X$ then p belongs to the triangle formed by some three points of X (see Figure 5), i.e.,

$$\exists \{y_1, y_2, y_3\} \subseteq X, p \in \operatorname{conv} \{y_1, y_2, y_3\}.$$

Steinitz's Theorem: Let

$$X \subseteq \mathbf{R}^d \quad \text{and} \quad p \in \operatorname{int} \operatorname{conv} X.$$

Then there exists some subset $Y \subseteq X$ such that

$$|Y| \leq 2d \quad \text{and} \quad p \in \operatorname{int} \operatorname{conv} Y. \quad \square$$

Both Carathéodory's and Steinitz's theorems are examples of a general family of theorems, related to **Helly's Theorem.**

Helly's Theorem: Suppose $\mathcal{K}$ is a family of at least $d+1$ convex sets in $\mathbf{R}^d$, and $\mathcal{K}$ is finite or each member of $\mathcal{K}$ is compact. Then if all $d+1$ members have a common point, there is a point common to all members of $\mathcal{K}$. $\quad \square$

It is not hard to derive Carathéodory's theorem from Helly's theorem (using *polar duals*) and it can

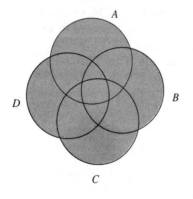

Figure 6: Example of Helly's Theorem for $d = 2$.

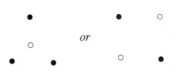

Figure 7: Example of Radon's Theorem for $d = 2$.

also be shown that Carathéodory's theorem implies Helly's theorem [2]. Other members of the family of so-called Helly-type theorems include Radon's theorem, Tverberg's theorem, theorems of Kirchberger and Krasnosd'skiĭ, etc. Radon's theorem states that each set of $d + 2$ or more points in $\mathbf{R}^d$ can be expressed as the union of two disjoint sets whose convex hulls have a common point.

The proof of Carathéodory and Steinitz's theorems are not that involved. In order just to get a flavor of the techniques employed in combinatorial discrete geometry, we shall describe the proofs for the cases where $X \subseteq \mathbf{R}^2$ is a finite set of points in the plane.

We begin with a proof of Carathéodory's theorem. Assume that $P = \operatorname{conv} X$ is a bounded polygon and p, a point of P. Let $v \ (\neq p)$ be a vertex of the polygon and let $\vec{vp}$ be the ray originating from the vertex v and passing through p. Let $\overline{vw} = P \cap \vec{vp}$ be the line segment with the end points v and $w \in \partial P$ on the boundary of P. There are two cases to consider: either w is a vertex of P or $w \in \overline{st}$ belongs to an edge $\overline{st}$ of P. In the first case, let $y_1 = v$ and $y_2 = w$ and in the latter case, let $y_1 = v$, $y_2 = s$ and $y_3 = t$. By construction, $p \in \operatorname{conv} \{v, w\}$ and in the second case $w \in \operatorname{conv} \{s, t\} \Rightarrow p \in \operatorname{conv} \{v, s, t\}$. Now let $Y = \{y_1, y_2\}$ or $= \{y_1, y_2, y_3\}$, as the case may be. In either case, $Y \subseteq X$ and $|Y| \leq d + 1 = 3$. Similar constructive proofs for Carathéodory's theorem can be provided for higher dimension.

Now the proof of Steinitz's theorem can be given with a slight modification to the preceding argument. Let $P = \operatorname{conv} X$ be as before and p, a point in the interior of P. Now construct a ray $\vec{vp}$ as before, except that v is now chosen to be an interior point on an edge $\overline{ab}$ of P. Let $\overline{vw} = P \cap \vec{vp}$ be the line segment as before with w on the boundary of P. Again there are two cases to consider: either w is a vertex of P or $w \in \overline{st}$ belongs to the interior of an edge $\overline{st}$ of P. In the first case, let $y_1 = a$, $y_2 = b$ and $y_3 = w$ and in the latter case, let $y_1 = a$, $y_2 = b$, $y_3 = s$ and $y_4 = t$. Now let $Y = \{y_1, y_2, y_3\}$ or $= \{y_1, y_2, y_3, y_4\}$, as the case may be. In either case, $Y \subseteq X$ and $|Y| \leq 2d = 4$

and $p \in \text{int conv } Y$. The extension to higher dimension is also fairly straightforward. The fact that the bound $2d$ is tight for Steinitz's theorem can be seen from the two-dimensional example shown in Figure 8.

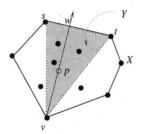

(a) Caratheodory's Theorem

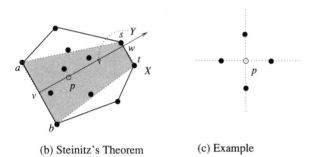

(b) Steinitz's Theorem   (c) Example

Figure 8: Proof sketches.

4 A Little Robotics

Equipped with our understanding of the geometric structures in convexity theory, we are now ready to tackle one of the simplest (but rather interesting) problems in grasping theory, the existence of positive grips.

Given: An arbitrary rigid 3-dimensional object B and some number k.

Determine: Whether one can choose k (finite) grip points, $\{p_1, p_2, \ldots, p_k\} \subseteq \partial B$ on the boundary of B such that the object can be grasped (positively) by placing fingers at those grip points.

$$\left(\exists ? \{p_1, \ldots, p_k\} \subseteq \partial B \right)$$
$$\left[0 \in \text{conv}\,(\Gamma(p_1), \ldots, \Gamma(p_k)) \right].$$

Surprisingly, the answer to the problem turns out to be "yes" and the necessary number of fingers is 7.

That is, there is a "universal hand" with seven fingers that can grasp *any* rigid object by judiciously choosing the grip points. Of course when we say *any*, we actually make some reasonable assumptions about the object. Namely, we assume that *B is a closed bounded connected object with piecewise smooth boundary ∂B*.

The proof proceeds in three simple steps:

STEP 1: Show that

$$0 \in \text{conv } \Gamma(\partial B),$$

where $\Gamma \colon \partial B \to \mathbf{R}^6 : p \mapsto (n, p \times n)$. This is a simple consequence of the fact that an object under uniform pressure remains in equilibrium. The proof of this claim can be given rigorously using the *Divergence theorem of Gauss*.

STEP 2: By Carathéodory's theorem

$$\left(\exists\, \{\Gamma(p_1), \ldots, \Gamma(p_k)\} \subseteq \Gamma(\partial B) \right)$$
$$\left[k \leq 7 \text{ and } 0 \in \text{conv}\,(\Gamma(p_1), \ldots, \Gamma(p_k)) \right].$$

Hence there are positive nonnegative scalar quantities $\alpha_1, \ldots, \alpha_k$ such that:

$$\alpha_1 n_1 + \cdots + \alpha_k n_k = 0,$$
$$\alpha_1 (p_1 \times n_1) + \cdots + \alpha_k (p_k \times n_k) = 0.$$

STEP 3: The positive grip is then selected by choosing

$$\text{Grip Points} = \{p_1, \ldots, p_k\} \subseteq \partial B,$$
$$\text{Force Magnitudes} = \alpha_1, \ldots, \alpha_k,$$

with k no larger than 7.

4.1 Problems with Equilibrium Grasps

Similar arguments in the plane imply that *any*[2] planar object can be grasped by at most *four* fingers. The number four is arrived at by taking the dimension of the wrench space and adding one to it, as implied by the Carathéodory's theorem. It is also instructive to examine a set of equilibrium grasps for three planar objects: a rectangle, a triangle and a disk. First consider the grasps for the rectangle. See Figure 9. Clearly, the grasps (a)

---

[2]Closed, connected, and bounded with piecewise smooth boundary.

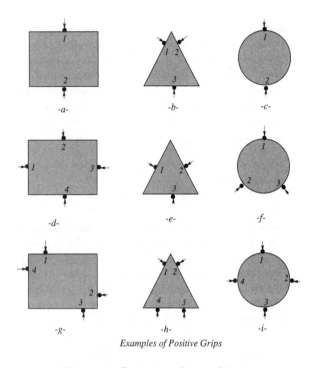

Examples of Positive Grips

Figure 9: Grasping planar objects.

and (d) are not as secure as (g)—a horizontal external force will break the grasp (a) [i.e., the equilibrium condition cannot be made to be satisfied, resulting in an infinitesimal movement] and an external torque about the center of the rectangle will break the grasp (d). In comparison, the grasp (g) is immune to such external disturbances, provided of course that such disturbances are relatively small in magnitude. Similar examination will show that the grasp (h) is the most secure for a triangle. However, in the case of the disk, while the grasps (f) and (i) are better than (c), there is simply no way to resist an external torque about the center irrespective of how many fingers are used.

The kinds of secure grasps described in the preceding paragraph have been characterized as *closure grasps*. Furthermore, exactly those objects that do not allow closure grasps can also be characterized in purely geometric terms, and are referred to as *exceptional objects*. While we shall not go into a detailed description of the exceptional objects (see [11]), it should suffice for the present purpose to say that the only planar bounded exceptional object is a disk and the only spatial bounded exceptional object is an object bounded by a surface of revolution.[3]

Closure Grasps: A set of grip points on an object B is said to constitute a *closure grasp* if and only if any arbitrary external force/torque combination acting on the object can be balanced by simply *pressing the fingertips against this object at the selected fixed grip points.*

Thus our job is to solve the following.

Given: An arbitrary non-exceptional rigid 3-dimensional object B and some number k.

Determine: If one can choose k (finite) grip points, $\{p_1, p_2, \ldots, p_k\} \subseteq \partial B$ on the boundary of B such that the object can be grasped (with closure) by placing fingers at those grip points.

In other words, we must have for every $g \in \mathbf{R}^6$ ($-g$ is the external wrench), a set of nonnegative force magnitudes

$$\{\alpha_1, \ldots, \alpha_k\} \qquad \alpha_i \geq 0$$
$$\alpha_1 \Gamma(p_1) + \cdots + \alpha_k \Gamma(p_k) = g$$

Equivalently,

$$\mathrm{pos}\,(\Gamma(p_1), \ldots, \Gamma(p_k)) = \mathbf{R}^6;$$

$\Gamma(p_1)$, ..., $\Gamma(p_k)$ positively span the entire wrench space. Since this is also equivalent to the following condition

$$\left(\forall g \in \mathbf{R}^6 \; \exists \epsilon > 0\right)$$
$$\left[\epsilon g = \sum \alpha_i \Gamma(p_i), \sum \alpha_i = 1, \alpha_i \geq 0\right],$$

we can also express this problem as asking

$$\left(\exists?\{p_1, \ldots, p_k\} \subseteq \partial B\right)$$
$$\left[0 \in \mathrm{int}\,\mathrm{conv}\,(\Gamma(p_1), \ldots, \Gamma(p_k))\right].$$

Note that the first condition implies that a sufficiently small 6-dimensional ball (of radius ϵ) can fit in the convex hull of the wrenches and hence leads to the second condition.

The answer to the problem turns out to be "yes" (for all non-exceptional objects) and the necessary number of fingers is 12 (twice the dimension of the wrench space).[4]

The proof proceeds again in three simple steps:

---

[3]If one allows unbounded objects then in 3-D, we have to include unbounded prisms and helical objects and in 2-D an unbounded strip of constant width. These objects in 3-D describe the so-called Reuleux pairs, studied almost a

century ago.

[4]For different approaches and an improved bound of 7 fingers, you may consult the results of Markenscoff et al. [8] and Meyer [10].

	2D Objects	3D Objects
Equilibrium Grasps		
Piecewise Smooth	4	7
Smooth	3	5
Convex, Smooth	2	2
Closure Grasps		
Piecewise Smooth	6 (excluding disks)	12 (excluding objects with a surface of revolution)

Table 1: Summary of Results. The numbers in the table are upper bounds on the required number of fingers.

STEP 1: Show that, if B is non-exceptional

$$0 \in \text{int conv } \Gamma(\partial B),$$

where $\Gamma \colon \partial B \to \mathbf{R}^6 : p \mapsto (n, p \times n)$. This is only true if B is non-exceptional, as otherwise $\Gamma(\partial B)$ spans only a low-dimensional subspace.

STEP 2: By Steinitz's theorem

$$\left(\exists \; \{\Gamma(p_1), \dots, \Gamma(p_k)\} \subseteq \Gamma(\partial B) \right)$$

$$\left[k \leq 12 \text{ and } 0 \in \text{int conv } (\Gamma(p_1), \dots, \Gamma(p_k)) \right].$$

STEP 3: The closure grasp is then selected by choosing as grip points

$$\text{Grip Points } = \{p_1, \dots, p_k\} \subseteq \partial B,$$

with k no larger than 12.

Some of the results about grasping (including the ones discussed earlier) have been summarized in Table 1.

5 A Simple Algorithm

At this point, it is natural for a roboticist to ask how one (a robot) can construct a grasp for a specific object and what sorts of computation this may entail. The answer turns out to be very interesting and shows a close connection of this problem to a classical algorithm, the simplex method, used for solving linear programming problems.

Thus, suppose we have a polyhedral object with n faces. Since this object is "non-exceptional," in principle, we should be able to grasp it by a closure grasp using no more than twelve fingers. In our terminology, we wish to simply identify no more than

twelve grip points on the faces of the polyhedron—but we wish to do so *constructively*, and furthermore, as quickly as possible.

We proceed in a manner not very dissimilar from the ways we proved the existences of such a grasp. We first create a closure grasp with an extremely large number of fingers: about $15n$ grip points, where n is the number of faces of the polyhedron. Of course, this is all done by our imagination (or by a mathematical model in the computer memory); we don't need to physically construct a hand with $15n$ fingers! Next, step by step, we can eliminate one finger in each step while maintaining the closure grasp as long as the number of grip points at the beginning of that step is strictly larger than twelve. The algorithm terminates when we are left with no more than twelve grip points.

5.1 Algorithmic Preliminaries

In order to understand the process by which the fingers are eliminated, we shall digress to consider an algorithmic approach to *algebraic manipulation with positive linear combinations*.

Given: A set of vectors $\{V_1, V_2, \dots, V_l\} \subseteq \mathbf{R}^d$ and $V \in \mathbf{R}^d$ such that

$$\alpha_1 V_1 + \cdots + \alpha_l V_l = \alpha V$$
$$\alpha_i \geq 0, \alpha > 0, \; V \neq 0.$$

Find: A subset $m \leq d$ vectors

$$\{V_{i_1}, V_{i_2}, \dots, V_{i_m}\} \subseteq \{V_1, \dots, V_l\} \text{ and } \alpha' > 0$$

such that

$$\alpha'_1 V_{i_1} + \cdots + \alpha'_m V_{i_m} = \alpha' V$$
$$\alpha'_i \geq 0, \; (\alpha' > 0, \; V \neq 0).$$

The problem can be solved by the algorithm described below. If you are already familiar with the simplex algorithm for linear programming problems, then you should realize that the basic step of the algorithm shown resembles the "pivot step" of the simplex algorithm.

Reduction Algorithm

if $l \leq d$ then HALT;

else repeat

> Choose d vectors from $\{V_1, \ldots, V_l\}$
> (Say, the first d): $\{V_1, \ldots, V_d\}$.
> There are two cases to consider, depending on whether the vectors $V_1, \ldots, V_d$ are *linearly dependent* or not.

Case 1: $V_1, \ldots, V_d$ **are linearly dependent.**

We can write

$$\beta_1 V_1 + \cdots + \beta_d V_d = 0,$$

not all $\beta_i = 0$.

Assume that at least one $\beta_i < 0$ (otherwise, replace each β_i by $-\beta_i$ in the equation to satisfy the condition).
Let

$$\gamma = \min_{\beta_i < 0}(\alpha_i/\beta_i) < 0.$$

(For specificity, we may assume $\gamma = \alpha_1/\beta_1$.)
Put $\alpha_i' = \alpha_i - \gamma\beta_i$ for $1 \leq i \leq d$.
Hence by adding the equation $(\sum_{i=1}^{l} \alpha_i V_i = \alpha V)$ to $(-\gamma \sum_{i=1}^{d} \beta_i V_i = 0)$, we get

$$\alpha_2' V_2 + \cdots + \alpha_d' V_d + \alpha_{d+1} V_{d+1} + \cdots + \alpha_l V_l = \alpha V,$$

and by construction $\alpha_2', \ldots, \alpha_d' \geq 0$.

Case 2: $V_1, \ldots, V_d$ **are linearly independent.**

We can write

$$\beta_1 V_1 + \cdots + \beta_d V_d = V.$$

Assume that at least one $\beta_i < 0$ (otherwise, we have nothing more to do!).
Let

$$\gamma = \min_{\beta_i < 0}(\alpha_i/\beta_i) < 0.$$

(For specificity, we may assume $\gamma = \alpha_1/\beta_1$.)
Put $\alpha_i' = \alpha_i - \gamma\beta_i$ for $1 \leq i \leq d$, and $\alpha' = \alpha - \gamma > 0$.

Hence by adding the equation $(\sum_{i=1}^{l} \alpha_i V_i = \alpha V)$ to $(-\gamma \sum_{i=1}^{d} \beta_i V_i = -\gamma V)$, we get

$$\alpha_2' V_2 + \cdots + \alpha_d' V_d + \alpha_{d+1} V_{d+1} + \cdots + \alpha_l V_l = \alpha' V,$$

and by construction $\alpha_2', \ldots, \alpha_d' \geq 0$.

Note that this process terminates after at most $(l - d)$ repetitions of the basic step and each basic step involves some matrix operations involving $d \times d$ matrices, thus using in each basic step an amount of computer time that is cubic in d. [In algorithmic terminology, we would write that "the reduction algorithm has a time complexity of $O(ld^3)$."] In our grasping application, d will turn out to be a constant $(= 6)$ and l no more than $15n$. Thus we will see that this algorithm will give us a grasping algorithm whose time complexity will be proportional to n, the number of faces of the polyhedron it is trying to grasp.

—*The end of digression.*

5.2 Grasping Algorithms

Let us get back to our original question about grasping a polyhedron B with n faces. As hinted earlier, we shall start with a closure grasp of B using no more than $15n$ grip points. Assume that B is provided with a triangulation of each face, and

$$t_1, t_2, \ldots, t_N$$

is the set of triangles partitioning ∂B. For each triangle t_i, choose three non-collinear grip points p_{i_1}, p_{i_2} and $p_{i_3} \in t_i$ such that $(p_{i_1} + p_{i_2} + p_{i_3})/3$ is the centroid of t_i. In totality they will give us the initial $3N$ grip points. Using Euler's formula and some simple combinatorics, one can show that $N \leq 5n - 12$ and the total number of grip points is no more than $15n - 36$ [11].

Now, it can be shown that if one chooses p_{i_j}'s, $1 \leq i \leq N$, $j = 1, 2, 3$, as the grip points then they give rise to a closure grasp. In particular, we can see (by using linear algebraic manipulations [11]) that

$$\frac{\text{Area}(t_1)}{3}\Gamma(p_{1_1}) + \frac{\text{Area}(t_1)}{3}\Gamma(p_{1_2})$$
$$+ \frac{\text{Area}(t_1)}{3}\Gamma(p_{1_3}) + \cdots + \frac{\text{Area}(t_N)}{3}\Gamma(p_{N_1})$$
$$+ \frac{\text{Area}(t_N)}{3}\Gamma(p_{N_2})$$
$$+ \frac{\text{Area}(t_N)}{3}\Gamma(p_{N_3}) = 0,$$

and that

$$\text{pos}\left(\Gamma(p_{1_1}), \Gamma(p_{1_2}), \ldots, \Gamma(p_{N_3})\right) = \mathbf{R}^6.$$

Henceforth, rewriting these grip points as $\{p_1, p_2, \ldots, p_l\}$, and the "area terms" as the magnitudes of the coefficients, $\alpha_1, \alpha_2, \ldots, \alpha_l$, we have

$$\alpha_1\Gamma(p_1) + \alpha_2\Gamma(p_2) + \cdots + \alpha_l\Gamma(p_l) = 0, \quad (1)$$

where $\alpha_i > 0$. Furthermore, since

$$\text{lin}\left(\Gamma(p_1), \Gamma(p_2), \ldots, \Gamma(p_l)\right) = \mathbf{R}^6,$$

without loss of generality, assume that the first six wrenches are linearly independent, thus spanning the entire wrench space, i.e.,

$$\text{lin}\left(\Gamma(p_1), \ldots, \Gamma(p_6)\right) = \mathbf{R}^6.$$

Synthesizing an Equilibrium Grasp with Seven Fingers: Let us now see how we can go from here to get a simple equilibrium grasp with no more than seven fingers. Note first that we can rewrite our equation 1 (for l-fingered grip) as

$$\frac{\alpha_1}{\alpha_l}\Gamma(p_1) + \cdots + \frac{\alpha_{l-1}}{\alpha_l}\Gamma(p_{l-1}) = -\Gamma(p_l),$$

where $\alpha_i > 0$ and $\Gamma(p_i) \in \mathbf{R}^6$. Now, we can use the "Reduction Algorithm" to find

$$\{p_{i_1}, p_{i_2}, \ldots, p_{i_m}\} \subseteq \{p_1, \ldots, p_{l-1}\}$$

satisfying the conditions below:

$$\alpha_1'\Gamma(p_{i_1}) + \cdots + \alpha_m'\Gamma(p_{i_m}) = -\alpha'\Gamma(p_l),$$

and $m \leq 6$. Thus we have

$$\alpha_1'\Gamma(p_{i_1}) + \cdots + \alpha_m'\Gamma(p_{i_m}) + \alpha'\Gamma(p_l) = 0,$$

with $\alpha_1' \geq 0, \ldots, \alpha_m' \geq 0$ and $\alpha' > 0$. Of course, this is our equilibrium grasp using no more than $m + 1 \leq 7$ fingers, placed at grip points $p_{i_1}, \ldots,$

p_{i_m}, p_l with associated force magnitudes $\alpha_1', \ldots, \alpha_m', \alpha'$.

Synthesizing a Closure Grasp with Twelve Fingers: Recall that the initial l grip points are so chosen that

$$\text{lin}\left(\Gamma(p_1), \ldots, \Gamma(p_6)\right) = \mathbf{R}^6.$$

Let

$$V = -(\Gamma(p_1) + \ldots + \Gamma(p_6)).$$

Express V using all the wrenches as follows:

$$\alpha_1\Gamma(p_1) + \alpha_2\Gamma(p_2) + \cdots + \alpha_l\Gamma(p_l) = V,$$

which exploits the fact that the original set of l grip points form a closure grasp (i.e., $\Gamma(p_i)$'s positively span the entire wrench space).

Now, we can again use the "Reduction Algorithm" to find

$$\{p_{i_1}, p_{i_2}, \ldots, p_{i_m}\} \subseteq \{p_1, \ldots, p_{l-1}\}$$

satisfying the conditions below:

$$\alpha_1'\Gamma(p_{i_1}) + \cdots + \alpha_m'\Gamma(p_{i_m}) = V,$$

and $m \leq 6$. We now choose as the desired grip points

$$\{p_{i_1}, p_{i_2}, \ldots, p_{i_m}\} \cup \{p_1, \ldots, p_6\},$$

numbering no more than $m + 6 \leq 12$. We claim that these give rise to a closure grasp.

To see why, consider some arbitrary external wrench $f \in \mathbf{R}^6$. We wish to show that this f can be expressed as a positive linear combination of

$$\{\Gamma(p_{i_1}), \ldots, \Gamma(p_{i_m})\} \cup \{\Gamma(p_1), \ldots, \Gamma(p_6)\}.$$

First, note that we can write f as a linear combination of $\Gamma(p_i)$, $i = 1, \ldots, 6$:

$$f = \sum_{i=1}^{6} \beta_i\Gamma(p_i),$$

and suppose that not all $\beta_i \geq 0$, since otherwise we have nothing more to prove. Now let

$$\gamma = \min_{1 \leq i \leq 6} \beta_i < 0.$$

Thus

$$
\begin{aligned}
f &= \sum_{i=1}^{6} \beta_i \Gamma(p_i) \\
&= \sum_{i=1}^{6} (\beta_i - \gamma) \Gamma(p_i) + (-\gamma) \sum_{i=1}^{6} -\Gamma(p_i) \\
&= \sum_{i=1}^{6} (\beta_i - \gamma) \Gamma(p_i) + (-\gamma) V \\
&= \sum_{i=1}^{6} (\beta_i - \gamma) \Gamma(p_i) + \sum_{j=1}^{m} (-\gamma \alpha_j') \Gamma(p_{i_j}).
\end{aligned}
$$

Since $-\gamma$ is positive and $\gamma < \beta_i$ and since α_j''s are positive, all the coefficients in the above equation are nonnegative.

Thus, we have

$$
\mathrm{pos}\left(\{\Gamma(p_{i_1}), \ldots, \Gamma(p_{i_m})\} \cup \{\Gamma(p_1), \ldots, \Gamma(p_6)\}\right) \\
= \mathbf{R}^6,
$$

which means we have shown that the chosen grip points $\{p_{i_1}, \ldots, p_{i_m}\} \cup \{p_1, \ldots, p_6\}$ indeed form a closure grasp.

6   Final Remarks

Most of the questions dealt with here come from one of the first papers I wrote in this area about fifteen years ago with Jack Schwartz and Micha Sharir [11]. Since then the area has grown substantially and researchers have addressed many more interesting questions dealing with different finger models, different concepts of closure, various measures of goodness of a grasp, regrasping (also called finger-gaiting), fixturing and workholding. Of course, it is not possible to go into all these topics here. The readers wishing to learn more about these topics must consult the references given at the end of the paper. Also, a website designed by Ken Goldberg (FixtureNet, URL http://teamster.usc.edu/fixture/) at University of Southern California can automatically find for you how a polygonal (2D) object can be fixtured. You may want to check it out to sharpen your intuition.

Acknowledgements

I am grateful to many of my colleagues for their help, advice and comments: Hans Moravec and Rod Brooks, who got me interested in robotics; Jack Schwartz, Micha Sharir, Chee Yap, David Kirkpatrick, S. Rao Kosaraju, Fred Hansen, Jia-Wei Hong, Xiao-Nan Tan, Gerardo Lafferriere, Zexiang Li, Naomi Silver, Marek Teichmann and Richard Wallace, who have collaborated with me in my research on robot hands; Dayton Clark, Lou Salkind, Chris Fernandes and Marco Antoniotti, who worked with me in examining a varied class of problems in robotics; and finally, Randy Brost, Joe Burdick, John Canny, Bruce Donald, Mike Erdmann, Ken Goldberg, Pradeep Khosla, Dan Koditschek, Vladimir Lumelsky, Matt Mason, Christos Papadimitrou, Elon Rimone, Jeff Trinkle and the rest of the robotics community, for making it fun.

References

[1] L. Danzer, B. Grünbaum, and V. Klee. Helly's theorem and its relatives. In *Convexity*, volume 7, pages 101–180, Providence, RI, 1963. Proc. of Symposia in Pure Math, AMS.

[2] J. Eckhoff. Helly, radon, carathéodory type theorems. In P.M. Gruber and J.M. Willis, editors, *Handbook of Convex Geometry*, volume A, pages 389–448. North-Holland, New York, 1993.

[3] H. Edelsbrunner. *Algorithms in Combinatorial Geometry*. Springer-Verlag, New York, 1987.

[4] S.J. Gould. *Eight Little Piggies*. W.W. Norton & Company, New York, 1993. Chapter 4.

[5] J. Hong, G. Lafferriere, B. Mishra, and X. Tan. Fine manipulation with multifinger hands. In *IEEE International Conference on Robotics and Automation*, pages 1568–1573, May 1990.

[6] D. Kirkpatrick, B. Mishra, and C. Yap. Quantitative steinitz's theorem with applications to multifingered grasping. *Discrete & Computational Geometry*, 7(3):295–318, 1992.

[7] K. Lakshminarayana. The mechanics of form closure. *The International Journal of Robotics Research*, 1978.

[8] X. Markenscoff, L. Ni, and C. H. Papadimitriou. The geometry of grasping. *The International Journal of Robotics Research*, 9(1), 1990.

[9] M. T. Mason and J. K. Salisbury, Jr. *Robot Hands and the Mechanics of Manipulation*. Cambridge University Press, MIT Press, 1985.

[10] W. Meyer. Seven fingers allow force-torque closure grasps on any convex polyhedron. *Algorithmica*, 9(3):278–292, 1993.

[11] B. Mishra, J.T. Schwartz, and M. Sharir. On the existence and synthesis of multifinger positive grips. *Algorithmica*, 2:541–558, 1987.

[12] B. Mishra and N. Silver. Some discussion of static gripping and its stability. *IEEE Transactions on Systems, Man and Cybernetics*, 19:783–796, 1989.

[13] J. O'Rourke. *Computational Geometry in C*. Cambridge University Press, Cambridge, 1994.

[14] J.T. Schwartz and C.-K. Yap. *Advances in Robotics, Vol. I: Algorithmic and Geometric Aspect of Robotics*. Lawrence Erlbaum Associates, Hillsdale, New Jersey, 1987.

[15] E. Steinitz. Bedingt konvergente reihen und konvexe systeme. *J. reine angew. Math*, 1913. (I) **143**:128–175, 1913; (II) **144**:1–48, 1914; and (III) **146**:1–52, 1916.

[16] M. Teichmann. *Grasping and Fixturing: a Geometric Study and an Implementation*. PhD thesis, New York University, New York, 1995.

[17] F. A. Valentine. *Convex Sets*. McGraw-Hill, New York, 1964.

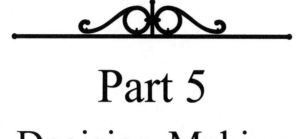

Part 5

Decision-Making Process

Decisions through Triangles

Donald G. Saari

Department of Mathematics

Northwestern University

Evanston, Illinois 60208–2730

What makes problems from the social sciences appealing to some mathematicians is that the answers matter to large numbers of people and the mathematics differs from the physical sciences. The social sciences worry about how we decide, how society should choose, how resources are allocated, and how we should be governed. So, mathematical contributions can provide important insights into how we, as a society, should function. Intellectual attraction derives from the mathematical approaches which differ from those used in the physical, engineering, and biological sciences. In the social sciences, the mathematics needs to capture the much higher dimensional nature of the problems. This is manifested by the deep insight into complex issues already provided by elementary geometry. I offer a flavor of this by describing "voting." The reader interested in a more complete discussion should consult [5].

Voting is a standard part of our life. We vote to decide within social groups, families, organizations, and political units. We vote to choose which kind of pizza to order, what to name a pet turtle, or where to go on vacation. We check the voting results on the sports page to learn who is the MVP of basketball, who won in figure skating or gymnastics, which city will host the next Olympics, and who made the weekly ranking of collegiate football teams. We vote to choose who should be hired, who should be Chair of a department or group, who should represent a district, and who should be President of our country. Our elected representatives vote to determine which laws will govern our lives. Without argument, voting is a critical tool of society. So, before proceeding, I invite the reader to reflect about what can go right and what can go wrong with voting.

It is easy to describe what is "right"; answers are readily available from any high school civics course. As for what can go wrong, responses tend to emphasize social issues such as voter apathy, fraud, and so forth. My concern is more fundamental; it involves the hidden mathematical properties of standard voting procedures. In particular, I argue that our commonly used method—the plurality vote where each voter votes for one candidate—is seriously flawed.

To explain this serious charge, start with the acknowledged fact that bad election decisions can and have been made. When this occurs, we may blame the "other voters" by prominently displaying a tacky bumper sticker announcing *"Don't blame me; I voted for—."* A more appropriate message, however, might be *"Don't blame us; we used the plurality vote"* because, as shown here, often the election of an inferior choice can be completely blamed on the pernicious, hidden mathematical properties of our lousy method. Inherent in this assertion is a host of related issues: What can go wrong? What are the mathematical problems? Does the choice of a voting method matter? These questions are addressed with geometry.

1 A Simple Example

To demonstrate my concern, suppose 25 voters are to rank the three candidates $\{A, B, C\}$ where 12 have the preferences $A \succ C \succ B$, 11 have $B \succ C \succ A$ and the last two have $C \succ B \succ A$. There is nothing controversial here; the plurality election outcome of $A \succ B \succ C$ is supported by the tally $12 : 11 : 2$. While we might worry about the $\{A, B\}$ relative ranking, C received so few votes that, clearly, these voters express no interest in her candidacy.

Before becoming comfortable with this conclusion, compare these candidates by using pairwise elections. Here C, a candidate who barely registers in the plurality tally, beats all comers! (The $C \succ A$ and $C \succ B$ outcomes are supported by the respective tallies of $13 : 12$ and $14 : 11$.) Adding to the surprise is that the pairwise election between top-ranked A and middle-ranked B also is reversed; the $B \succ A$ outcome is given by the $13 : 12$ tally.

What is going on? This example proves that the plurality top-ranked candidate (A) can be beaten in all head-to-head competitions while a bottom-ranked candidate (C)—even if she hardly receives any votes—can beat everyone! In other words, instead of the $A \succ B \succ C$ plurality outcome, the pairwise votes strongly suggest that the reversed $C \succ B \succ A$ more accurately reflects the wishes of the voters of this example. Why? What else can happen? Can simple conditions be found to indicate when such a topsy-turvy event can occur?

These conflicting election conclusions suggest checking the outcome by tallying the ballots in still another way. To select a procedure, notice from the *profile*—the list of voters' preferences—that by registering only each voter's top-ranked candidate, the plurality vote misses the critical fact that only C never is bottom-ranked by the voters. To recover this lost information, we might use the "antiplurality" method where a voter votes for everyone except his bottom-ranked candidate. (In essence, a voter votes against his bottom-ranked candidate—thus the name "antiplurality.") So, for three candidates, a voter votes for two of them. As this method acknowledges lower-ranked candidates, we might expect the antiplurality outcome to be $C \succ B \succ A$. Is it? To motivate later constructions, I ask the reader to compute the outcome and to determine whether still other rankings emerge by assigning a different weight to the second-ranked candidate.

2 Some Geometry

To address these issues via the geometry of the three-dimensional space R^3, let values on the x, y, z axes denote, respectively, the tallies for A, B, and C. In this manner, election outcomes are converted into points in R^3; e.g., the example election defines the point $(12, 11, 2)$. Nothing changes in the election ranking if the actual tally is replaced with each candidate's fraction of the total number of points. Thus the *normalized plurality election tally* for the example is $(\frac{12}{25}, \frac{11}{25}, \frac{2}{25})$ while the *normalized antiplurality tally* is the point $(\frac{12}{50}, \frac{13}{50}, \frac{25}{50})$. An advantage of using normalized election tallies is that they allow us to analyze and compare all possible election outcomes through the geometry of the simplex

$$Si(3) = \{(x, y, z) \in R^3 \mid x + y + z = 1,\ x, y, z \geq 0\}$$

which connects the three positive unit vectors of R^3 as represented in Figure 1. The reader might wish to plot the two normalized election points on the simplex because they are used in our later discussion.

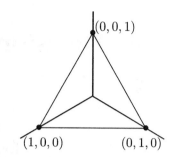

Figure 1: The simplex of normalized election outcomes.

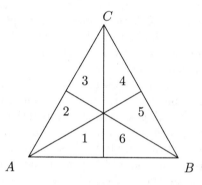

Figure 2: The representation triangle and ranking regions.

This identification of R^3 coordinates requires the $x = y$ plane to represent a tie vote between A and B. Geometrically, the plane divides the simplex into $A \succ B$ and $B \succ A$ regions separated by the $A \sim B$ indifference line (the intersection of $x = y$ with $Si(3)$). Similarly, the $x = z$ and the $y = z$ planes (which capture, respectively, $\{A, C\}$ and $\{B, C\}$ rankings), divide the simplex into thirteen regions; I call them *ranking regions* because each region defines a unique ordinal ranking of the candidates. To see the geometry and to verify the count, notice that the simplex is an equilateral triangle where its intersection with a dividing plane defines a line from a vertex to the midpoint of the opposite side. Therefore, the three lines defined by the dividing planes intersect at the barycenter $(\frac{1}{3}, \frac{1}{3}, \frac{1}{3})$, leading to Figure 2 (where I suppress the coordinate axes).

The identification of the *ranking regions* is immediate. As a point has the $A \succ B$ ranking if and only if it is on the $x > y$ side of the $x = y$ plane, it is on the side closest to the A-vertex (the vertex $(1, 0, 0)$). Because a similar argument holds for the other two dividing planes, the ranking defined by a point is based on its distances to the different vertices where "closer is better." For instance, our election example of $(\frac{12}{25}, \frac{11}{25}, \frac{2}{25})$ is closest to the A vertex, next closest to the B vertex, and farthest from the C vertex, so it is in the $A \succ B \succ C$ ranking region. Each normalized election tally is in a ranking region; the name of the region is the election ranking. In Figure 2, the strict rankings are designated by the following "type" numbers.

So, the six small open triangles correspond to the six rankings without ties, the six line segments separating triangles correspond to rankings with one tie, and the barycenter $(\frac{1}{3}, \frac{1}{3}, \frac{1}{3})$ represents the complete tie outcome $A \sim B \sim C$. The resulting figure—the simplex (equilateral triangle) divided into ranking regions—is called a *representation triangle* because I use it to *represent* both election outcomes and voter profiles. (Recall, a "profile" lists each of the six strict rankings of the alternatives. Ties are not included for voter's preferences primarily because they complicate the exposition without adding insight about the problems of election procedures.)

3 A Dimensional Digression Into Complexity

I claim that the principal obstacle in understanding elections with three or more candidates is manifested by the large dimensions required by the implicitly defined geometry. To understand my assertion, notice that two candidates $\{A, B\}$ admit only two types of voters—those who like A (with the preference $A \succ B$) and those who prefer B (with $B \succ A$). If u represents the fraction of all voters that are of the first type, the fraction of voters of both types is $(u, 1 - u)$. Thus, an analysis of two-candidate elections is a one-dimensional problem captured by the single independent variable $0 \le u \le 1$. There is nothing difficult here; e.g., if $u > \frac{1}{2}$, then A wins.

Everything changes with three candidates. With an election outcome for each candidate, the range of the election mapping is R^3. As described earlier, we can reduce the election outcomes to the two-dimensional simplex $Si(3)$; this reduction means we only need to consider two dependent variables (the normalized election tallies for two candidates) as the third value can be recovered from the equation $x + y + z = 1$. The troubles come from the domain of the election mapping which captures all the ways voters can vote. Three candidates define the $3! = 6$ *different voter types specified in Table 1—these are the independent variables*. Listing the number of voters of each type in an array defines a vector in

Type	Ranking	Type	Ranking
1	$A \succ B \succ C$	4	$C \succ B \succ A$
2	$A \succ C \succ B$	5	$B \succ C \succ A$
3	$C \succ A \succ B$	6	$B \succ A \succ C$

Table 1: Votes and types.

the six-dimensional space R^6. To reduce the dimension, consider the *fraction* of all voters with a particular preference type. Even here the independent variables reside in the five-dimensional simplex

$$Si(3!) =$$

$$\{\mathbf{p} = (p_1, p_2, \dots, p_6) \in R^6 \mid \sum_{j=1}^{6} p_j = 1, \, p_j \ge 0\}$$

which, geometrically, is the hyperplane passing through the six positive unit vectors in R^6.

To illustrate, the 30 voter integer profile $(6, 3, 2, 8, 9, 2)$ lists the number of voters of each type, and its *normalized profile* is $(\frac{6}{30}, \frac{3}{30}, \frac{2}{30}, \frac{8}{30}, \frac{9}{30}, \frac{2}{30})$. Conversely, the common denominator of the fractions in a normalized profile identifies associated integer profiles. For instance, this normalized profile represents integer profiles with multiples of 30 voters. (By using $Si(6)$ as a domain, some profiles have irrational components. But, because the rational points are dense, this causes no difficulty.) The dependent and independent variables are connected through the election procedures; different procedures define different mappings.

So, to understand the problems of voting, we need to understand the geometric relationship of the larger dimension domain and the smaller image space as connected by an election mapping. A plurality election, for instance, just counts what fraction of voters have each candidate top-ranked. Thus, the three-candidate plurality election mapping can be identified with a system of two linear equations in five unknowns.

Insight about the source of the election problems already comes from those first algebra lessons about m equations in n unknowns, $n > m$. They warn us to expect a surprising number of ways to connect domain points with image values. Namely, by solving these two equations in five unknowns, we learn that, in general, *each normalized election outcome is supported by a three-dimensional space of normalized profiles*.

Observe the radical difference between a two and a three candidate election. In a two candidate election, the one-to-one nature of the election mapping requires a normalized outcome to come from a unique normalized profile. If Bobbi wins with two-thirds of the vote, then two-thirds of all voters support her. But, the "three-dimensional subspace to one" nature of three-candidate elections allows many surprising domain-image relationships. The unexpected connections correspond to "voting paradoxes" (such as the one given by the example profile) which create worry that, inadvertently, our choice of a voting procedure might cause us to select badly.

This geometry warns us to expect the complexities and problems to significantly escalate with more candidates. To select a realistic number of candidates to investigate, recall that the 1995 Presidential election in France started with nine candidates and the early days of the 1996 Republican Presidential Primaries had 11 candidates. Each fall there are at least 20 candidates (teams) in the weekly "elections" for the collegiate football standings. Thus, it is reasonable to consider elections involving, say, ten candidates. Here, the normalized election tallies are in the nine-dimensional simplex $Si(10)$; this hyperplane connects the positive unit vectors of the ten-dimensional space R^{10}. The complexities of ten-candidate elections are caused by the domain of normalized profiles; it is the unit simplex $Si(10!)$ that connects the positive unit vectors in a $10! = 3,628,800$-dimensional space! When expressed in terms of algebra, we "only" need to analyze nine equations in 3,628,799 unknowns. No wonder we need geometry!

This huge dimensional differential between the domain and range promises all sorts of new, strange paradoxes and serious electoral problems—and they occur. For instance, ten candidates define $2^{10} - 11 = 1013$ subsets with two or more candidates. For each subset, use the weekly lotto to rank the candidates. Even though this assignment ensures that any relationship among the rankings of the different subsets is purely coincidental, a profile of voters can be constructed so that when these voters sincerely vote on the candidates for each of the 1013

subsets, the plurality outcome agrees with the randomly generated outcomes. This assertion does not add confidence to our standard tool of democracy. For a sample of other troubling outcomes, see [6]. I return to the ten-candidate setting later.

4 Profile Representations and Computing Votes

In a two-candidate election, if $\frac{1}{4}$ of the voters prefer $A \succ B$, then B is the winner with $\frac{3}{4}$ of the vote. Usually, such quick answers are not possible with three-candidate elections. This is manifested, for instance, by my request to compute the antiplurality outcome for the example profile. While the arithmetic is trivial, the answer required at least partial computations. This, of course, reflects the dimensional differential between the domain and range for three-candidate elections which allows a three-dimensional subspace of independent variables (profiles) to determine each outcome.

To illustrate with another example, try to quickly interpret the following profile which, as I explain later, is of historical importance. Who should win? What are the plurality, antiplurality, and pairwise outcomes?

Part of the problem in analyzing a profile is that this traditional listing is not overly useful. To improve upon it, consider what an ideal profile representation should offer. First, it should help us extract and discover election relationships. Second, it should avoid the five-dimensional profile domain $Si(3!)$. Yet (third) it should capture the relational geometry of this five-dimensional space.

A convenient way to satisfy this wish-list is to place the entries of a profile in a representation triangle. Namely, in each ranking region, list the number of people with that ranking; this is illustrated in Figure 3a with the entries from Table 2. (I leave it to the reader to extract why this representation captures enough of the $Si(3!)$ geometry to derive the relationships which follow.) As another example, can the reader find anything about the three-voter profile $A \succ B \succ C$, $B \succ C \succ A$, $C \succ A \succ B$ that catches your attention? (This profile was discovered by the mathematician Condorcet in the 1780s.) By listing this profile in a representation triangle (Figure 3b), its two-dimensional (120°) symmetry becomes apparent.

To develop election relationships from this profile representation, start with the fact that A's plurality vote is the number of voters who have her top-ranked. But if a ranking region has A top-ranked,

Number	Type	Number	Type
30	$A \succ B \succ C$	1	$C \succ B \succ A$
1	$A \succ C \succ B$	10	$B \succ C \succ A$
10	$C \succ A \succ B$	29	$B \succ A \succ C$

Table 2: An election profile.

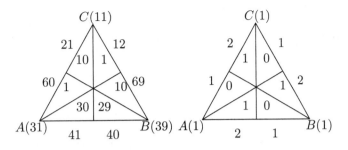

Figure 3: Representing profiles and elections.

A is a vertex. Therefore, A's plurality vote is the sum of the numbers in the two regions with A as a vertex. In this manner, each candidate's election tally in Figures 3a-b is determined; they are given by the numbers near the vertices. Notice how the symmetry of the Figure 3b profile is manifested by the tie plurality outcome. This tie outcome should be expected from the geometry because the profile's symmetrical arrangement favors no candidate over another; each candidate is in first, second, and last place once.

To find a geometric representation for the $\{A, B\}$ pairwise tallies, recall that all voters with the preference $A \succ B$ are on the $x > y$ side of the $x = y$, or $A \sim B$, line. Therefore, a candidate's pairwise outcome is the sum of voters on the appropriate side of the $A \sim B$ line. These outcomes are listed on the bottom of the triangles. The outcomes for the other two pairwise elections, obtained in a similar manner, are given by the entries along the respective edges. Notice how the pairwise outcomes suggest that the correct ranking for Figure 3a is $A \succ B \succ C$ while the plurality ranking is $B \succ A \succ C$.

A disturbing feature of the pairwise outcomes for the Condorcet triplet (Figure 3b) is that they define a nontransitive outcome—the *cycle* $A \succ B, B \succ C, C \succ A$ where each tally is 2 : 1. Actually, this cyclic outcome can be explained by the symmetry displayed by this profile. After all, the pairwise vote can be viewed as projecting the entries of the Condorcet profile onto each edge of the triangle to be added, so the profile's two-dimensional symmetry must be manifested in some manner. It is—with a cycle. This raises the interesting question whether a voting cycle always requires the profile to exhibit some of this two-dimensional symmetry. (An algebraic answer comes from solving the three pairwise election equations with cyclic outcomes. A geometric answer is harder to find; if you have difficulty, see [5].)

A reason for the difference between the pairwise and plurality rankings now is apparent from the geometry; these procedures use different information about the voters' preferences. By exploiting this difference, we discover the following *triangle inequality rule* which identifies when a plurality tally imposes no constraints on the accompanying pairwise election rankings.

Theorem 1 *Suppose the plurality tally for each of the three candidates $\{A, B, C\}$ satisfies the (plurality) triangle inequality rule whereby the sum of the tallies for any two candidates is strictly greater than the tally for the remaining candidate. Choose any strict ranking for each of the three pairs of candidates. There exists a profile supporting both the specified plurality tallies and pairwise election rankings. The pairwise rankings can include a tie vote if and only if, in addition, the total number of voters is even.*

If the plurality tallies do not satisfy the triangle inequality rule, then certain pairwise rankings never can accompany the plurality tallies.

To illustrate this, since the plurality tallies (1000000, 999900, 200) satisfy the triangle inequality condition, the theorem ensures there is a profile defining both this plurality outcome and the pairwise election rankings which correspond to the reversed $C \succ B \succ A$ ranking. Other profiles yield this same plurality tally and the pairwise *cycle* $C \succ A, A \succ B, B \succ C$.

What makes this rule particularly easy to use is that it is satisfied if and only if no candidate receives a strict majority of the votes! Thus examples using actual three-candidate elections can be constructed if there is no majority winner. Immediate choices are the Bush, Clinton, Perot 1992 US Presidential election and the Edwards, Duke, Roemer 1991 race for Governor of Louisiana. Did the voters elect who they really wanted?

Proof. Suppose given plurality tallies satisfy the triangle inequality. Let $\mathcal{B}(c_i, c_j)$ be c_i's tally in a $\{c_i, c_j\}$ binary election and let $\mathcal{T}_P(c_j)$ be c_j's plurality tally. As $\mathcal{T}_P(A) = p_1 + p_2$, $\mathcal{T}_P(B) = p_5 + p_6$, $\mathcal{T}_P(C) = p_3 + p_4$, the $\{A, B\}$ pairwise election tallies can be expressed as

$$
\begin{aligned}
(\mathcal{B}(A,B), & \mathcal{B}(B,A)) \\
&= (p_1 + p_2 + p_3, p_4 + p_5 + p_6) \\
&= (\mathcal{T}_P(A) + p_3, \mathcal{T}_P(B) + p_4) \\
&= (\mathcal{T}_P(A) + p_3, \mathcal{T}_P(B) + \mathcal{T}_P(C) - p_3).
\end{aligned}
$$

The pairwise outcome $B \succ A$ requires selecting p_3 (which is bounded above by $\mathcal{T}_P(C)$) so that $\mathcal{B}(B,A) > \mathcal{B}(A,B)$. This defines the inequality

$$0 \leq 2p_3 < \mathcal{T}_P(B) + \mathcal{T}_P(C) - \mathcal{T}_P(A). \quad (1)$$

The triangle inequality assumption ensures that an appropriate p_3 can be selected. If, instead, we wish to have the pairwise outcome $A \succ B$, add $2p_4$ to both sides of inequality 1, use the definition of $\mathcal{T}_P(C)$, and collect terms to obtain the relevant condition that $p_4 \leq \mathcal{T}_P(C)$ needs to be selected so that

$$0 \leq 2p_4 < \mathcal{T}_P(A) + \mathcal{T}_P(C) - \mathcal{T}_P(B). \quad (2)$$

Again, the triangle inequality assumption ensures that such a p_4 exists. To obtain an $A \sim B$ pairwise outcome, change the strict inequalities of Equations 1 and 2 to an equality. Since $2p_j$ is an even number, the triangle equality expression also must be even. As $N = \mathcal{T}_P(A) + \mathcal{T}_P(B) + \mathcal{T}_P(C)$ is the total number of voters, it is easy to show that the triangle inequality expression is even if and only if N is an even number.

To complete the proof, notice from symmetry that a similar expression holds for all three pairs and that the independent variable involves the voter types where the top-ranked candidate is the candidate missing from the pair.

It remains to show that if the triangle inequality rule is not satisfied, then not all pairwise election outcomes can occur. If this rule is not satisfied, then some combination, say the right-hand side of Equation 1, is not positive. As an appropriate *non-negative* p_j cannot be found, the pairwise ranking cannot occur. $\square$

An interesting project is to assume that the triangle inequality condition does not hold, and then characterize all possible pairwise rankings. An instructive exercise is to extend this theorem to $n = 4$ (or to $n \geq 4$) candidates.

To illustrate with the example of $\mathcal{T}_P(A) = 1,000,000$, $\mathcal{T}_P(B) = 999,900$, $\mathcal{T}_P(C) = 200$, the inequalities to create the pairwise rankings defining $C \succ B \succ A$ are

$$
\begin{aligned}
0 \leq 2p_3 &< \mathcal{T}_P(B) + \mathcal{T}_P(C) - \mathcal{T}_P(A) = 100 \\
0 \leq 2p_2 &< \mathcal{T}_P(C) + \mathcal{T}_P(A) - \mathcal{T}_P(B) = 300 \\
0 \leq 2p_6 &< \mathcal{T}_P(C) + \mathcal{T}_P(B) - \mathcal{T}_P(A) = 100
\end{aligned}
$$

where the extreme solution $p_2 = p_3 = p_6 = 0$ defines the supporting profile $(1000000, 0, 0, 200, 999900, 0)$. Similarly, a profile where the pairwise cycle $C \succ A$, $A \succ B$, $B \succ C$ accompanies the same plurality outcome is $p_1 = 1,000,000$, $p_3 = 200$, $p_5 = 999,900$, $p_2 = p_4 = p_6 = 0$. It is easy to create other examples for this plurality tally where, say, the pairwise outcomes define the remaining cycle $A \succ C$, $B \succ A$, $C \succ B$.

For comparison, because the triangle inequality is not satisfied for $\mathcal{T}_P(A) = 100$, $\mathcal{T}_P(B) = 80$, $\mathcal{T}_P(C) = 10$, we know that certain pairwise rankings cannot accompany these plurality tallies. A quick computation shows that we must have $A \succ B$, $A \succ C$, but any $\{B, C\}$ ranking is admissible. An interesting exercise is to show that even if $\mathcal{T}_P(A) = 1000$, $\mathcal{T}_P(B) = 980$, $\mathcal{T}_P(C) = 0$—so C receives *no votes*—profiles can be constructed where C beats B in a pairwise election.

One might worry whether this theorem addresses a common or a rare situation. To develop a sense about which setting prevails, plot the normalized plurality outcomes satisfying the conditions of this theorem. Denoting the normalized coordinates by (q_A^P, q_B^P, q_C^P), the specified region (Figure 4) identifies plurality tallies where no normalized coordinate has a value greater than or equal to $\frac{1}{2}$. (This assertion follows immediately by using the $q_A^P + q_B^P + q_C^P = 1$ constraint with the triangle inequality condition.) This region, where the plurality tallies alone permit' no constraints on the pairwise election outcomes, captures most of the truly contested elections and it meets all thirteen ranking

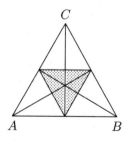

Figure 4: Plurality outcomes allowing any pairwise vote.

regions. We find from the geometry, then, that *any plurality ranking can be accompanied by any set of pairwise rankings*. In particular, the introductory example only hints at the true magnitude of problems experienced by the plurality vote! The next assertion captures the severity of the situation.

Corollary 1 *For three candidates, choose a ranking of the three candidates and choose rankings for each of the three pairs of candidates. These selections can be made in any desired manner, even by use of dice or other random devices. There exists a profile with these pairwise election rankings where the plurality outcome is the specified one.*

This corollary, which emphasizes the chaotic nature of plurality election outcomes, continues to erode our confidence in our commonly used election procedure.

5   Other Important Procedures

The geometry of two other methods will be discussed. The first is the Borda Count (BC) discovered in 1770 by the French mathematician J.C. Borda [1]. Borda, who can be considered as the father of the mathematics of voting procedures, proposed assigning the weights 2, 1, 0, respectively, to a voter's top-, middle-, and bottom-ranked candidate. It turns out that c_j's BC tally, denoted by $\mathcal{T}_{BC}(c_j)$, is the sum of points she receives in all binary elections [1, 4, 5]. Thus $\mathcal{T}_{BC}(A) = \mathcal{B}(A,B) + \mathcal{B}(A,C)$. For the Figure 3a example, add the numbers under the edges to obtain $\mathcal{T}_{BC}(A) = 41 + 60 = 101$, $\mathcal{T}_{BC}(B) = 40 + 69 = 109$, $\mathcal{T}_{BC}(C) = 21 + 12 = 33$ or the BC ranking $B \succ A \succ C$. For the Figure 3b example, the BC tally for each candidate is $2 + 1 = 3$, so the BC ranking for the Condorcet triplet is $A \sim B \sim C$.

The antiplurality tally, $\mathcal{T}_A(c_j)$, is the number of voters who have c_j at least second-ranked. On the representation triangle, A's votes come from the four ranking regions defining a U shape where two of the regions have A as a vertex and the other two share an edge with one of these first two regions. In Figure 3a, the tallies are $\mathcal{T}_A(A) = 10 + 1 + 30 + 29 = 70$, $\mathcal{T}_A(B) = 30 + 29 + 10 + 1 = 70$, $\mathcal{T}_A(C) = 1 + 10 + 1 + 10 = 22$ defining the antiplurality ranking $A \sim B \succ C$. The reader should determine the antiplurality outcome for the Condorcet triplet.

The introductory example suggests that the antiplurality method is better than the plurality procedure; it is not. To show this, I indicate (leaving the proof to the reader) that if the antiplurality

tallies satisfy a particular "triangle inequality" relationship, then this tally can be accompanied by any selection of pairwise outcomes. Consequently the "randomness" assertion of Corollary 1 also applies to the antiplurality vote! In stating this assertion, notice that with N voters, $N - \mathcal{T}_A(c_i)$ of the voters have c_i bottom-ranked; this is the number of voters who would vote against her.

Theorem 2 *a. For N voters, suppose the antiplurality tallies satisfy the (antiplurality) triangle inequality rule where the sum of any two of the terms*

$$(N - \mathcal{T}_A(A)), \ (N - \mathcal{T}_A(B)), \ (N - \mathcal{T}_A(C))$$

is strictly greater than the remaining term. (Equivalently, N plus the antiplurality tally for any candidate is strictly greater than the sum of the remaining two antiplurality tallies.) For any choice of strict rankings of the three pairs, there exists a profile supporting both the specified antiplurality tallies and the rankings of the pairs. (The pairwise rankings also can be a tie vote if and only if, in addition, N is an even integer.) If, however, the antiplurality tallies do not satisfy the triangle inequality, then not all pairwise rankings are admitted.

b.   If (q_A^P, q_B^P, q_C^P) and (q_A^A, q_B^A, q_C^A) are, respectively, the normalized plurality and antiplurality tallies for a profile $\mathbf{p}$, then

$$\frac{1}{2}q_j^P \le q_j^A \le \frac{1}{2}, \quad j = A, B, C. \tag{3}$$

c.   A candidate who beats all others in pairwise elections cannot be BC bottom-ranked. Similarly, a candidate who loses to all others in pairwise elections cannot be BC top-ranked.

I leave the proof of (a), along with an extension of Corollary 1, to the reader. Once the proof is understood, it becomes easy to generate disturbing examples.

Part (a) raises an interesting issue. The plurality vote has troubles because it ignores lower-ranked candidates; the reader should find a similar explanation for the difficulties of the antiplurality outcomes. Also, Figure 4 depicts the normalized plurality outcomes which admit such difficulties; find a similar geometric description for the antiplurality vote.

Part (b) is a technical assertion used later. To prove it, notice that the voters who vote for A in the plurality election also vote for A in an antiplurality election. In computing the normalized antiplurality vote, the total number of points cast is $2N$—twice that of the plurality vote. This leads to the q_j^A lower bound. A proof for the upper bound is given in the next section.

Part (c) proves that not all possible pairwise rankings can accompany a BC tally. This important assertion means that instead of allowing total randomness in election outcomes, the BC imposes order on electoral conclusions. To prove this assertion, notice that a candidate who wins both pairwise elections must (by the above way to compute the BC tallies) receive more than the average number of points. (See [4] or [5].)

6   All Possible Positional Methods

Already in the early days of the mathematics of voting, Laplace and other prominent mathematicians worried whether Borda's assignment of $(2, 1, 0)$ points could be justified. Why not assign $(10, 3, 0)$ points? In general, a positional method is where w_j points are assigned to a voter's jth ranked candidate where these values satisfy the obvious conditions $w_1 \geq w_2 \geq w_3 = 0$ and $w_1 > 0$. I call (w_1, w_2, w_3) a *voting vector*. So, the plurality vote is defined by voting vector $(1, 0, 0)$, the antiplurality vote by $(1, 1, 0)$, and the BC by $(2, 1, 0)$. (Because these weights can be normalized to require $w_1 = 1$, any method where $w_1 - w_2 = w_2 - w_3$ is the BC.)

Theorems 1 and 2 establish that the plurality and antiplurality methods admit chaotic settings where their rankings of the candidates may have absolutely nothing to do with the candidate's pairwise rankings. This troubling assertion has to cause doubt and concern about the meaning of these elections. On the other hand, the BC—the "average" of the two extreme methods—admits consistency. Are there any other positional methods which provide the needed consistency in election outcomes? The surprising answer is no.

Indeed, it is an interesting exercise to show that if a positional method is not the BC, then it admits a conclusion similar to that of Theorems 1 and 2 along with an interpretation such as Corollary 1. (One proof is to show that the system of five equations in five unknowns defined by the three pairwise and two components for the positional elections has full rank. By having full rank, all possible outcomes can occur.) In other words, *only the BC offers relief from all possible election paradoxes; only the BC provides consistency in the election rankings of the three candidates and the pairs.* The flaws of all other methods cast serious doubt on the meaning and validity of their election outcomes.

Instead of proving this assertion (I recommend that the reader proves the statement; a geometric proof is in [5]), I show how different positional election rankings can result from a fixed profile and how they are related. To do this via the geometry of the representation triangle, we need to normalize the voting vector so that the sum of its components equals unity. This is done by multiplying the voting vector $(w_1, w_2, 0)$ by $1/(w_1 + w_2)$. This scalar multiplication does not affect the outcome, only the value of the tally. This normalization, for instance, converts the plurality, BC, and antiplurality voting vectors into, respectively, the equivalent voting vectors $(1, 0, 0)$, $(\frac{2}{3}, \frac{1}{3}, 0)$, $(\frac{1}{2}, \frac{1}{2}, 0)$. As another example, the method $(6, 1, 0)$ (where 6 and 1 points are assigned, respectively, to a voter's top and second ranked candidates) has the normalized form $\frac{1}{6+1}(6, 1, 0) = (\frac{6}{7}, \frac{1}{7}, 0)$. Obviously, all voting vectors have a $\mathbf{w}_s = (1 - s, s, 0)$ form for some $s \in [0, \frac{1}{2}]$.

So, I have referred to an election as a linear mapping. At this point, it is appropriate to write it down in vector form. Using our identification for coordinates, the vote cast by a type-4 voter (with preferences $C \succ B \succ A$) using $\mathbf{w}_s$ is $(0, s, 1 - s)$ to reflect that the most points are given to C, the second most to B, and no points to A. Similarly, the *vector ballot* for a type-five voter ($B \succ C \succ A$) is $(0, 1 - s, s)$. To simplify the notation, let $[\mathbf{w}_s]_j$ be the vector ballot cast by a type-j voter, $j = 1, \ldots, 6$.

The vote is determined by the number of voters of each type. As there are p_j type-j voters, the total type-j vote is $p_j[\mathbf{w}_s]_j$. Adding all partial tallies defines the expression

$$f(\mathbf{p}, \mathbf{w}_s) = \sum_{j=1}^{6} p_j[\mathbf{w}_s]_j \qquad (4)$$

To illustrate with the introductory example, the $\mathbf{w}_s$ outcome is

$$\frac{12}{25}[\mathbf{w}_s]_2 + \frac{11}{25}[\mathbf{w}_s]_5 + \frac{2}{25}[\mathbf{w}_s]_4$$
$$= \frac{12}{25}(1 - s, 0, s) + \frac{11}{25}(0, 1 - s, s) + \frac{2}{25}(0, s, 1 - s)$$
$$= \frac{1}{25}(12 - 12s, 11 - 9s, 2 + 21s) \qquad (5)$$

By using normalized voting vectors and profiles, the sum of all components in the outcome equals unity. As intended, the normalizations force the election tally on the representation triangle.

Incidentally, to find a geometric interpretation of $\mathbf{w}_s$-elections, plot the six points $\{[\mathbf{w}_s]_j\}_{j=1}^{6}$ on the representation triangle. (Each point is on an edge of the triangle.) Next, connect these points with straight lines to create their *convex hull*. Equation 4 ensures that the election outcome is in this hull; conversely, any point in this hull is the election outcome of some profile. In particular, a complex coordinate of a point corresponds to a normalized profile. Using this construction with $\mathbf{w}_{\frac{1}{2}}$, we obtain a geometric proof of the Equation 3 assertion that $q_i^A \leq \frac{1}{2}$.

Notice how $\mathbf{w}_s$ can be expressed as the linear expression

$$\mathbf{w}_s = (1 - 2s)(1, 0, 0) + 2s(\frac{1}{2}, \frac{1}{2}, 0)$$
$$= (1 - 2s)\mathbf{w}_0 + 2s\mathbf{w}_{\frac{1}{2}} \quad (6)$$

of the plurality ($\mathbf{w}_0$) and the antiplurality ($\mathbf{w}_{\frac{1}{2}}$) voting vectors. In words, the normalized election vectors live on the line segment defined by the $\mathbf{w}_0$ and $\mathbf{w}_{\frac{1}{2}}$ endpoints. Before introducing details, let me describe how to exploit this relationship.

According to Equation 4, once $\mathbf{w}_s$ is specified, the election is represented by a linear mapping in the six variables defining the profile. Similarly, holding a profile $\mathbf{p}$ fixed converts the election mapping into a linear expression in the $\mathbf{w}_s$ variables. Remember, a linear mapping transfers a linear object, such as a line, into another linear object, such as a line or a point. Therefore, for a fixed profile $\mathbf{p}$, the election mapping picks up the line defined by Equation 6 and places it in the representation triangle. Because the image line describes all (normalized) election outcomes for all $\mathbf{w}_s$-procedures (for the selected profile), I call it the *procedure line*. The properties of the procedure line provide answers to all sorts of deep questions about election procedures [4, 5].

To provide supporting details, notice that if Equation 6 is used with Equation 7, we have

$$f(\mathbf{p}, \mathbf{w}_s) = \sum_{j=1}^{6} p_j [\mathbf{w}_s]_j$$
$$= \sum_{j=1}^{6} p_j [(1 - 2s)\mathbf{w}_0 + 2s\mathbf{w}_{\frac{1}{2}}]$$
$$= (1 - 2s) \sum_{j=1}^{6} p_j [\mathbf{w}_0]_j + 2s \sum_{j=1}^{6} p_j [\mathbf{w}_{\frac{1}{2}}]_j$$
$$= (1 - 2s)f(\mathbf{p}, \mathbf{w}_0) + 2sf(\mathbf{p}, \mathbf{w}_{\frac{1}{2}}). \quad (7)$$

In other words, Equation 7 establishes my assertion that $\mathbf{p}$'s $\mathbf{w}_s$ outcome is on the line connecting $\mathbf{p}$'s plurality and antiplurality outcomes.

The *procedure line* is found by plotting and connecting the normalized plurality and antiplurality outcomes. Each point on the line is a particular $\mathbf{w}_s$ outcome for this profile. For instance, according to Equation 7, the BC outcome ($s = \frac{1}{3}$) is two-thirds of the way from the plurality to the antiplurality endpoint, while the $(6, 1, 0)$ outcome ($s = \frac{1}{7}$) is two-sevenths of the way. To find all possible election outcomes for the profile, just check which ranking regions are crossed by the procedure line. For instance, the Figure 3a profile has the plurality ranking $B \succ A \succ C$ while the antiplurality ranking, $A \sim B \succ C$, is on the boundary of this region. As the ranking region is convex, the procedure line cannot stray outside of the $B \succ A \succ C$ region. Consequently, the election ranking for all remaining positional methods is $B \succ A \succ C$. (Condorcet concocted the Figure 3a profile in the 1780s to show that the pairwise outcomes can disagree with all positional outcomes! (See [2].) Condorcet's purpose was to argue that all positional methods have a serious flaw. His argument has been generally accepted until recently; there now are strong arguments [5] to indicate that he is wrong.)

To further illustrate the procedure line with the introductory example, once the normalized plurality and antiplurality outcomes, $(\frac{12}{25}, \frac{11}{25}, \frac{2}{25})$ and $(\frac{11}{50}, \frac{14}{50}, \frac{25}{50})$, are plotted (Fig. 5), we find that the connecting line passes through the seven ranking regions $A \succ B \succ C$, $A \succ B \sim C$, $A \succ C \succ B$, $A \sim C \succ B$, $C \succ A \succ B$, $C \succ A \sim B$, $C \succ B \succ A$. The s values which support the different rankings can be determined with elementary algebra. For instance, to find all procedures with the ranking $A \succ B \succ C$, first find the boundary s value defined by $A \succ B \sim C$. This s-value is the solution of the equation (obtained by setting the y and z coordinates of Equation 5 equal) $11 - 9s = 2 + 21s$ or $s = \frac{3}{10}$. Thus, for this profile, the $\mathbf{w}_s$ ranking for $0 \leq s < \frac{3}{10}$ is $A \succ B \succ C$ and it is $A \succ B \sim C$ for $s = \frac{3}{10}$. A similar computation shows that the $\mathbf{w}_s$ ranking is $C \succ B \succ A$ for $\frac{1}{3} < s \leq \frac{1}{2}$. The reader can determine which $\mathbf{w}_s$ define the remaining rankings for this profile.

The geometry of the procedure line can be used to answer all sorts of interesting questions. For instance, we now know that the introductory example admits seven different election rankings just by varying the choice of $\mathbf{w}_s$. Can an example be found with, say, eight different rankings? What is the maximum number of rankings that can supported by a single profile? This deep problem is easy to resolve with the procedure line. All that is required is to determine the maximum number of ranking regions a straight line can cross. This number, of course, is seven.

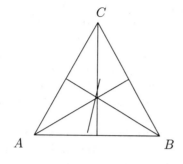

Figure 5: The procedure line.

To discover other properties, we need to determine which lines in the representation triangle can be procedure lines. This can be a difficult task when the emphasis is placed on specific profiles. Therefore, to turn the issue around, we seek properties of lines which ensure that *some profile* defines the line.

Proposition 1 [4] *Suppose the endpoints of a line in the representation triangle satisfy Equation 3. There is a profile* **p** *where the* **p** *procedure line is the specified line.*

The proof of this assertion involves setting up the four equations in five unknowns corresponding to the plurality and antiplurality elections. It is not overly difficult to show that these equations have maximal rank. This rank condition allows all choices of plurality and antiplurality tallies to admit solutions. Unfortunately, some solutions require inadmissible negative p_j values. So, one has to fiddle with the equations to determine which tallies admit solutions where all $p_j \geq 0$. The surprising fact is that it is any choice admitted by Equation 3.

The proposition imposes minimal restrictions on the procedure line. For instance, it allows any line with endpoints not too far from the barycenter to be the procedure line for some profile. So, just by drawing line segments in the representation triangle, we obtain important new conclusions. Some are listed below; others are left for the reader to discover by experimentation. To find a supporting profile for a procedure line with interesting properties, use the plurality and antiplurality tallies with Equation 7 and solve for the profile.

1. If the plurality ranking for a profile is $A \sim B \sim C$, then the ranking of any other $\mathbf{w}_s$ defines the ranking for all remaining positional procedures. (As one endpoint of the procedure

line is on the barycenter, the rest of the line can be in at most one other ranking region.)

2. Choose a ranking of three candidates. There exists a profile where not only do all positional methods have this ranking, but *all procedures have the same normalized tally*. (Just choose the plurality and antiplurality endpoints to agree. This forces the procedure line to collapse into the point.)

3. Choose an integer between one and seven. There exists a profile which admits precisely this number of different election rankings by varying the choice of $\mathbf{w}_s$. (Draw a line crossing exactly this number of regions. If the line is sufficiently close to the barycenter, it is the procedure line for some profile.)

4. If profile $\mathbf{p}$ is such that for a particular s_1, $0 < s_1 < \frac{1}{2}$, the $\mathbf{w}_{s_1}$ outcome is $A \sim B \sim C$, then either all positional outcomes are a tie vote, or there are two other rankings where the $\mathbf{w}_s$ ranking for $0 \leq s < s_1$ is the opposite of the ranking for $s_1 < s \leq \frac{1}{2}$. (Any line where the point $(1 - 2s_1)$ of the distance from one endpoint to the other is on the barycenter must have two segments in reversed ranking regions.)

5. For the positional methods $(6, 1, 0)$ and $(7, 2, 0)$, there is a profile $\mathbf{p}$ where its $(6, 1, 0)$ outcome is $A \succ C \succ B$ while its $(7, 2, 0)$ outcome is $B \succ A \succ C$. (These positional methods are, respectively, $\mathbf{w}_{\frac{1}{7}}$ and $\mathbf{w}_{\frac{2}{9}}$, so their outcomes are distinct points on the procedure line. Draw such a line so that first point is in the $A \succ C \succ B$ ranking region while the second one is in the $B \succ A \succ C$ region. Your choice of a line determines all other rankings. What rankings never can occur no matter how this line is chosen?)

6. Suppose a profile allows two different positional methods to have the same normalized election tally. This is the election tally for all positional methods. (Equation 7 requires the outcome for each method to be a particular distance along the procedure line from the plurality endpoint. The only way both outcomes could be at the same point is if the procedure line is a point—thus all normalized outcomes agree. This result is illustrated by the Condorcet profile (Figure 3b) where both the plurality and BC outcomes are the $A \sim B \sim C$ barycenter. This requires the $A \sim B \sim C$ outcome to hold for

all positional methods—including the antiplurality vote.)

7 More Candidates

The same geometric analysis extends to any number of candidates. The main difference is that the procedure line is replaced with a *procedure hull*. For instance, with four candidates, a normalized voting vector can be described as $\mathbf{w}_{s,t} = (1 - (s + t), s, t, 0)$ where $1 - (s + t) \geq s \geq t \geq 0$. All such voting vectors can be expressed as a linear combination of the three normalized voting vectors $(1, 0, 0, 0)$, $(\frac{1}{2}, \frac{1}{2}, 0, 0)$, $(\frac{1}{3}, \frac{1}{3}, \frac{1}{3}, 0)$. (The first is the plurality vote, the second and third are, respectively, normalized forms of the methods where you are instructed to vote for two or for three candidates.) Here, the voting vectors are in a triangle. Again, for a fixed $\mathbf{p}$, the election mapping is linear in the choice of the positional method. Therefore, the mapping lifts this triangle and places it in the representation simplex $Si(4)$. Points on this transported triangle correspond to the different election outcomes for this profile.

With geometric reasoning, surprising results about these hulls can be uncovered. (To learn about these arguments, see [3].) For instance, with $n = 10$, this approach circumvents the serious dimensional problems described earlier and allows new results to be derived. To illustrate, we now know that with ten candidates, profiles can be found where over 84 different election rankings occur just by varying the choice of a positional method. Ah, that should read over 84 *million* different rankings! So, which one of these millions upon millions of different outcomes best reflects the true wishes of the voters? Remember, the voters already voted by marking their ballots, so they are not changing their opinions; these highly contradictory outcomes are direct consequences of the geometric properties of how voting vectors are related.

This assertion underscores the need for a mathematical study of voting. What method or methods best capture the wishes of the voters? The importance of this issue is obvious; we must avoid choosing badly. Recent research (see [5, 6] and their references) indicates that the "best" method is the BC (where, for n candidates, $n-j$ points are assigned to a voter's jth ranked candidate; $j = 1, \ldots, n$.) However, more work and careful thought are needed. I invite you—and your students—to join this investigation! The conclusions can be important and the research is enjoyable.

References

[1] J. C. Borda. Memoire sur les elections au Scrutin. *Histoire de l'Academie Royale des Sciences*, 1781.

[2] I. McLean and F. Hewitt. *Condorcet*. Elgar, 1994.

[3] Donald G. Saari. Millions of outcomes from a single profile. *Social Choice & Welfare*, 9:277–306, 1991.

[4] Donald G. Saari. *Geometry of Voting*. Springer-Verlag, 1994.

[5] Donald G. Saari. *Basic Geometry of Voting*. Springer-Verlag, 1995.

[6] Donald G. Saari. A chaotic exploration of aggregation paradoxes. *SIAM Review*, pages 37–52, 1995.

Geometry in Learning

Kristin P. Bennett
Department of Mathematical Sciences
Rensselaer Polytechnic Institute
Troy, NY 12180
bennek@rpi.edu

Erin J. Bredensteiner
Department of Mathematics
University of Evansville
Evansville, IN 47722
eb6@evansville.edu

Abstract

One of the fundamental problems in learning is identifying members of two different classes. For example, to diagnose cancer, one must learn to discriminate between benign and malignant tumors. Through examination of tumors with previously determined diagnosis, one learns some function for distinguishing the benign and malignant tumors. Then the acquired knowledge is used to diagnose new tumors. The perceptron is a simple biologically inspired model for this two-class learning problem. The perceptron is trained or constructed using examples from the two classes. Then the perceptron is used to classify new examples. We describe geometrically what a perceptron is capable of learning. Using duality, we develop a framework for investigating different methods of training a perceptron. Depending on how we define the "best" perceptron, different minimization problems are developed for training the perceptron. The effectiveness of these methods is evaluated empirically on four practical applications: breast cancer diagnosis, detection of heart disease, political voting habits, and sonar recognition. This paper does not assume prior knowledge of machine learning or pattern recognition.

1 Introduction

Imagine that your job is to determine whether breast tumors are benign or malignant. A surgeon inserts a needle into the breast tumor and aspirates a small amount of tissue. A microscope slide of the fine needle aspirate is prepared. Your job is to examine the cells on the slide, assess important attributes of the cells such as the uniformity of the cell shape and variability in the cell size, and then determine a diagnosis of benign or malignant. You would learn to do this by examining many tumors that were previously determined to be benign or malignant by an expert pathologist using surgical biopsies. Probably somebody would help you by pointing out which attributes were important. You would generalize the knowledge you learned by applying it to diagnosing new tumors.

At the University of Wisconsin-Madison, a computer system has been developed that has "learned" to diagnose breast cancer [32, 34, 35]. The prepared slide of the fine needle aspirate is inserted into a computer imaging system that measures and determines low-level features of the nuclei of the cells within the tumor. The tumor is then described as a vector of real numbers. Each number represents one attribute of the cells. The vector is input into a computer program that produces a suggested diagnosis. The computer program was "trained" by giving it hundreds of examples of tumors that are known to be benign or malignant. During training a mathematical function was developed to classify the given examples as benign or malignant. This function is subsequently used to diagnose new cases. The computer learned in the sense that it generated a classification function based on observing examples with known classification.

In this paper, we examine the underlying problem of constructing a function to discriminate between examples from two classes. Our goal is to create a geometrical investigation of the problem from initial conception to evaluation of the computational results. We do not assume prior knowledge of machine learning or pattern recognition. Using the biology of the brain as an inspiration, we examine a simple mathematical model of learning called the perceptron. Geometrically, we describe what concepts a perceptron can learn. Then geometrical arguments are used to motivate algorithms for training the perceptron. Duality is used to provide different mathematical models of perceptron training. Depending on how we characterize the "best" perceptron, different optimization methods are constructed to train the perceptron. The tradeoffs of the different methods are discussed. An empirical comparison of the methods is performed.

For each classification problem, we are given examples or points from two classes. Each example x is represented by an n-dimensional real vector. Each of the dimensions represents an attribute of the example. In the heart disease problem, each example represents a patient. The attributes of each example include the patient's age, the patient's sex, and the patient's cholesterol level. In the training phase, we are also given the class to which each example belongs. For example, the set $\mathcal{A}$ could correspond to patients with heart disease and set $\mathcal{B}$ could correspond to patients without heart disease. Our problem is to construct, using the two sets of examples, a function $f(x)$ that returns 1 if x belongs to $\mathcal{A}$ and 0 if x belongs to $\mathcal{B}$. In the training phase, the function is constructed using the two sets of sample points, one set from each class. In the testing phase, the function is used to classify future points whose classification is unknown. Currently there is no mathematical definition of the "best" function for any given problem. The goal is to make this function as accurate as possible on future points; i.e., the function should generalize well. But we can only guess what the future points will be while the function is being constructed. Many types of classification functions are possible, but in this paper we will restrict ourselves to a linear function called the perceptron [27, 21].

While the perceptron model is quite simple, it works very well on many practical problems including the Wisconsin Breast Cancer Diagnosis problem described above. We present computational results for several methods of training a perceptron for real-world classification problems. Specifically, we will examine the performance of the perceptron on the following problems: breast cancer diagnosis, detection of heart disease, determination of the party affiliation of United States Representatives based on their voting habits, and sonar recognition of mines.

The following notational conventions will be used. For a column vector x in the n-dimensional real space R^n, x_i denotes the ith component of x. The notation $A \in R^{m \times n}$ will signify a real $m \times n$ matrix. For such a matrix, A_i will denote the ith row. The transpose of x and A are denoted x' and A' respectively. The dot product of two vectors x and w will be denoted by $x'w$. A vector of ones in a space of arbitrary dimension will be denoted by e. The scalar 0 and a vector of zeros are both

represented by 0. Thus, for $x \in R^m$, $x > 0$ implies that $x_i > 0$ for $i = 1, \ldots, m$. In general, for $x, y \in R^m$, $x > y$ implies that $x_i > y_i$ for $i = 1, \ldots, m$. Similarly, $x \geq y$ implies that $x_i \geq y_i$ for $i = 1, \ldots, m$. Several norms are used. The 1-norm of x, $\sum_{i=1}^{n} |x_i|$, is denoted by $\|x\|_1$. The 2-norm or Euclidean norm of x, $\sqrt{\sum_{i=1}^{n} x_i^2} = \sqrt{x'x}$, is denoted by $\|x\|$. Frequently, the 2-norm is squared to make it differentiable: $\|x\|^2 = x'x$. The infinity norm of x, $\max_{i=1\ldots n}(|x_i|)$, is denoted by $\|x\|_\infty$.

2   A Simple Learning Model

The perceptron model can be motivated biologically. The brain consists of an interconnecting network of about 10^{11} neurons or nerve cells [13]. Each neuron receives stimuli from other cells. A stimulus may be excitatory or inhibitory. If the combined stimuli exceed some threshold then the cell "fires". In the perceptron, the stimuli are modeled with an n-dimensional real input vector x. Each stimulus x_i has an associated real weight w_i. If the w_i is positive, the stimulus is excitatory. If w_i is negative, the stimulus is inhibitory. If the weighted sum of the stimuli $\sum_{i=1}^{n} w_i x_i = x'w$ is greater than some threshold γ then the perceptron fires and we say that x is in class $\mathcal{A}$. Otherwise we say that x is in class $\mathcal{B}$. Mathematically we define the perceptron as follows:

Definition 2.1 (Perceptron) *Let $x \in R^n$ be a point to be classified. A perceptron with weights $w \in R^n$ and threshold $\gamma \in R$ is defined as*

$$\begin{aligned} x'w - \gamma > 0 &\Rightarrow x \in \mathcal{A} \\ x'w - \gamma < 0 &\Rightarrow x \in \mathcal{B} \end{aligned} \tag{1}$$

In theory, if $x'w = \gamma$ then the class of the point x is undefined. In practice, we use the convention that if $x'w = \gamma$, then x is in $\mathcal{B}$.

Two questions immediately arise: What exactly can a perceptron learn and how does one train the perceptron, i.e., determine the weights w and threshold γ? The classic book by Minsky and Papert [21] may be consulted for an extensive discussion of both these questions. In our geometric approach, we will answer the first question using basic geometric arguments and then show how the geometry naturally motivates optimization methods for training a perceptron.

3   Geometry of a Perceptron

Geometrically training a perceptron corresponds to finding a plane that separates the two given example sets $\mathcal{A}$ and $\mathcal{B}$. The perceptron determines the separating plane $x'w = \gamma$, where w is the normal of the plane and $\frac{|\gamma|}{\|w\|}$ is the Euclidean distance of the plane from the origin. Let the coordinates of the points in $\mathcal{A}$ be given by the m rows of the $m \times n$ matrix A. Let the coordinates of the points in $\mathcal{B}$ be given by the k rows of the $k \times n$ matrix B. By definition of the perceptron model (2.1) we know that a perceptron exists that correctly classifies the two finite sets if and only if there exist w and γ satisfying the inequalities

$$\begin{aligned} Aw &> e\gamma \\ Bw &< e\gamma \end{aligned} \qquad (2)$$

where e is a vector of ones of appropriate dimension. If such an w and γ exist, we say that the sets are **linearly separable**. Of course, w and γ are the weights and threshold, respectively, of the separating perceptron. An example of two linearly separable sets in R^2 is given in Figure 1. Notice all of the points in $\mathcal{A}$ are in the open half space $\{x \in R^2 | x'w > \gamma\}$ and all the points in $\mathcal{B}$ are in the open half space $\{x \in R^2 | x'w < \gamma\}$.

In Figure 1, the convex hulls of $\mathcal{A}$ and $\mathcal{B}$ are shown as the areas enclosed by the dashed lines. Note that when the sets are separable by a plane, the convex hulls of the two sets do not intersect. Recall that the convex hull of $\mathcal{A}$ is the set of all points that can be written as convex combinations of the points in $\mathcal{A}$. A convex combination c of $\mathcal{A}$ is defined by

$$c' \;=\; u_1 A_1 + u_2 A_2 + \cdots + u_m A_m \;=\; u'A \quad (3)$$

where $u \in R^m$, $u \geq 0$, and $\sum_{i=1}^m u_i = e'u = 1$. Similarly, the convex hull of $\mathcal{B}$ is the set of all points that can be written as convex combinations of the points in $\mathcal{B}$. A convex combination d of $\mathcal{B}$ is defined by

$$d' \;=\; v_1 B_1 + v_2 B_2 + \cdots + v_k B_k \;=\; v'B \quad (4)$$

where $v \in R^k$, $v \geq 0$, and $\sum_{i=1}^k v_i = e'v = 1$.

Thus, we know a perceptron that correctly classifies $\mathcal{A}$ and $\mathcal{B}$ exists if and only if no point can be written as a convex combination of both $\mathcal{A}$ and $\mathcal{B}$, i.e.,

$$\left.\begin{aligned} A'u &= B'v \\ e'u &= e'v = 1 \\ u &\geq 0 \;\; v \geq 0 \end{aligned}\right\} \quad \text{has no solution.} \quad (5)$$

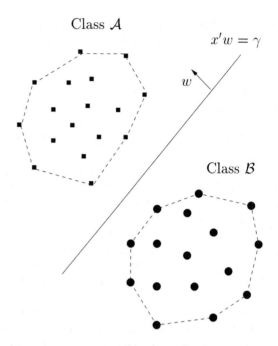

Figure 1: Separating plane for two linearly separable sets

At first glance, these two characterizations of what a perceptron can learn, equations (2) and (5), look unrelated. But in fact, using Gordan's Theorem of the Alternative [16], it can be shown that equations (2) have a solution if and only if equations (5) have no solution (see Theorem A.1 in Appendix). The two formulations are in different spaces but they solve the same underlying problem. Either problem can be used as a definition of linear separability of two sets. This is one of many examples of how duality is an important tool in geometry. In our case, the dual of the problem of finding the separating plane is the problem of determining whether the two convex hulls intersect. In the next section, we will use duality to develop methods for training the perceptron.

4   Training: Linearly Separable Case

In this section we will use an intuitive geometric argument to motivate several different optimization approaches to constructing a perceptron for linearly separable problems.

A geometric procedure to construct the "best" plane to separate two linearly separable sets is the following. Find the two closest points in the convex hulls of $\mathcal{A}$ and $\mathcal{B}$. Construct a line segment between

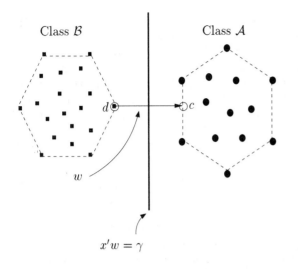

Figure 2: The two closest points of the convex hulls determine the separating plane.

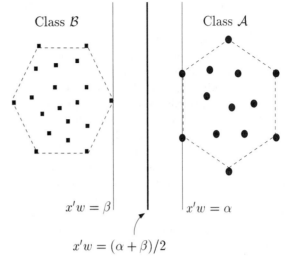

Figure 3: The dual problem maximizes the distance between two parallel supporting planes.

the two points. The plane that is orthogonal to the line segment and bisects the line segment is the separating plane. An example of such a separating plane is given in Figure 2. Intuitively this plane is "best" because the two sets are as far away from the separating plane as possible, thus improving the likelihood of correctly classifying future points. Any definition of "best", based on the training sets, is only an approximation of the actual goal of finding the plane that generalizes best.

The problem of finding two closest points in the convex hulls can be written as an optimization problem:

$$\min_{u,v} \quad \tfrac{1}{2} \left\| A'u - B'v \right\|^2$$
$$\text{s.t.} \quad e'u = 1 \quad e'v = 1 \tag{6}$$
$$u \geq 0 \quad v \geq 0.$$

The perceptron, (w, γ), is constructed from the results of Problem (6). The weights w are the normal of the separating plane. The normal w is exactly the vector between the two closest points in the convex hulls. Let $\bar{u}$ and $\bar{v}$ be an optimal solution of (6). The two closest points, c and d, are $c = A'\bar{u}$ and $d = B'\bar{v}$. The weights of the perceptron are the difference of these two points: $w = c - d = A'\bar{u} - B'\bar{v}$. The threshold, γ, is the distance from the origin to the point halfway between the two closest points along the normal w:

$$\gamma = \left(\frac{c+d}{2} \right)' w = \frac{(\bar{u}'Aw + \bar{v}'Bw)}{2}.$$

Using duality we can transform this problem to an optimization problem in the space of the percep-

tron. The dual of the following problem is Problem (6). See Theorem A.2 in the Appendix for the derivation.

$$\min_{w,\alpha,\beta} \quad \tfrac{1}{2} \left\| w \right\|^2 - (\alpha - \beta)$$
$$\text{s.t.} \quad Aw - \alpha e \geq 0 \tag{7}$$
$$-Bw + \beta e \geq 0.$$

There is a simple geometric interpretation of Problem (7). Examine Figure 3. All the points in $\mathcal{A}$ are constrained to be on the "greater than" side of the α-plane $x'w = \alpha$. All the points in $\mathcal{B}$ are constrained to be on the "less than" side of the β-plane $x'w = \beta$. The α-plane is a support plane of $\mathcal{A}$ and the β-plane is a support plane of $\mathcal{B}$. The distance between these two parallel support planes and thus between the two convex hulls is

$$\frac{\alpha - \beta}{\left\| w \right\|}. \tag{8}$$

The objective function of (7) minimizes $\left\| w \right\|^2$ and maximizes $\alpha - \beta$. Thus Problem (7) maximizes the distance between the two parallel planes. Let $\hat{w}$, $\hat{\alpha}$, and $\hat{\beta}$ be a solution of (7). For separable problems, $\hat{\alpha} - \hat{\beta} > 0$. The two parallel supporting planes are shown in Figure 3. The final separating plane is the plane halfway between the two parallel planes: $x'\hat{w} = (\hat{\alpha} + \hat{\beta}/2$.

Both Problems (6) and (7) are quadratic programming problems with linear constraints. They can be solved using standard mathematical programming packages [25, 22]. The choice of which problem to use in practice depends on the characteristics of the underlying problem. In Problem (6), the constraints are very simple, and the number of variables depends only on the total number of points. Thus, when the number of attributes is very large, Problem (6) would be preferable.

Problem (7) provides a unifying framework for explaining other prior optimization approaches to constructing the perceptron. By transforming (7) into mathematically equivalent optimization problems, different algorithms result. There are two simplifications that can be performed to transform the problem. Either $\|w\|^2 = 1$ is added as a constraint and $\alpha - \beta$ is maximized, or $\alpha - \beta$ is fixed at some positive value and $\|w\|^2$ is minimized.

The former approach results in the following problem:

$$\begin{aligned} \max_{w,\alpha,\beta} \quad & \alpha - \beta \\ \text{s.t.} \quad & Aw - \alpha e \geq 0 \\ & -Bw + \beta e \geq 0 \\ & \|w\|^2 = 1. \end{aligned} \tag{9}$$

The difficulty with this approach is that the resulting Problem (9) is much harder to solve than Problem (7) since the constraints are now nonlinear and nonconvex. However, by substituting the constraint $\|w\|_\infty = 1$, the problem can be solved in polynomial time using $2n$ linear programs. Polynomial algorithms exist for linear programming problems. In practice, existing general purpose linear program solvers are very fast and numerically stable [22]. By changing the norm used in the constraints, the quality of the solution is not degraded and the solution of the problem becomes much easier.

Specifically, the first n linear programs are defined for $j = 1, \ldots, n$ as

$$\begin{aligned} \max_{w,\alpha,\beta} \quad & \alpha - \beta \\ \text{s.t.} \quad & Aw - \alpha e \geq 0 \\ & -Bw + \beta e \geq 0 \\ & -e \leq w \leq e \\ & w_j = 1. \end{aligned} \tag{10}$$

The second n linear programs are defined for $j = 1, \ldots, n$ as

$$\begin{aligned} \max_{w,\alpha,\beta} \quad & \alpha - \beta \\ \text{s.t.} \quad & Aw - \alpha e \geq 0 \\ & -Bw + \beta e \geq 0 \\ & -e \leq w \leq e \\ & w_j = -1. \end{aligned} \tag{11}$$

A solution of one the $2n$ problems with the great-est value of $\alpha - \beta$ is the optimal answer.[1] This approach, called the Multisurface Method of Pattern Recognition (MSM) [15], was used in the initial implementation of the automated breast cancer diagnosis system described in the introduction [19, 34].

The second general method is to fix $\alpha - \beta > 0$. If we set $\alpha - \beta = 2$ by defining $\alpha = \gamma + 1$ and $\beta = \gamma - 1$, then Problem (7) becomes

$$\begin{aligned} \min_{w,\gamma} \quad & \tfrac{1}{2}\|w\|^2 \\ \text{s.t.} \quad & Aw - (\gamma + 1)e \geq 0 \\ & -Bw + (\gamma - 1)e \geq 0. \end{aligned} \tag{12}$$

Problem (12) is exactly the "Optimal Plane" proposed by Vapnik [33]. By using optimality conditions it can be shown that Problem (12) and Problem (6) are equivalent on separable problems. The proof is provided in the Appendix.

The above are only a few of the many existing methods for training a perceptron. We provide a few pointers to other approaches. This is not a comprehensive list. A notable class of algorithms is comprised of Rosenblatt's Perceptron algorithm [4, 10, 27] and the Motzkin-Schoenberg algorithm for finding the solution of linear inequalities [23]. A perceptron is also known as a linear discriminant. So any linear discriminant algorithms such as in [7] may be used. A single linear program can be used to construct a separating plane in polynomial time [11, 14]. Edelsbrunner proposed an algorithm with $O(\log m + \log k)$ complexity [8]. Statistical methods such as Fisher's Linear Discriminant may also be applied [9].

The problems formulated in this section are only for the linearly separable case. Care must be taken when applying any method for the linearly separable case to the linearly inseparable case. One pitfall is that the meaningless solution $w = 0$ is feasible and optimal for Problems (6) and (7) since the convex hulls of $\mathcal{A}$ and $\mathcal{B}$ intersect. Problem (12) is not even feasible in the inseparable case. Thus solving these problems yields no meaningful solution. In the next section we will discuss approaches for the linearly inseparable case.

5   Training: Linearly Inseparable Case

Frequently the sets $\mathcal{A}$ and $\mathcal{B}$ cannot be separated by any plane. For the linearly inseparable case, the separation problem becomes more difficult. There

---

[1]When applied to inseparable problems, the solution that misclassified the least number of points is selected.

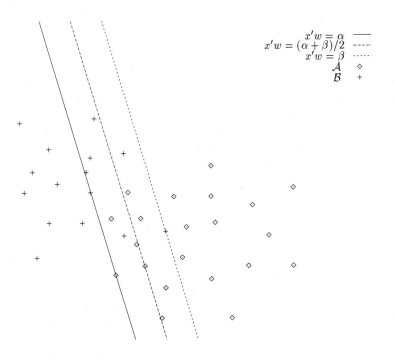

$$x'w = \alpha \quad \text{———}$$
$$x'w = (\alpha + \beta)/2 \quad \text{– – –}$$
$$x'w = \beta \quad \text{· · · ·}$$
$$\mathcal{A} \quad \diamond$$
$$\mathcal{B} \quad +$$

Figure 4: Minimizing the maximum error in the inseparable case.

is no clear definition of what constitutes the "best" separating plane. The general approach is to develop some measure of the misclassification error and then minimize that error. Depending on the choice of error measurement, different optimization problems arise. We will begin by generalizing the above approach to the case when the two sets are not strictly linearly separable.

Figure 4 illustrates how Problem (9) is applied to a linearly inseparable problem. Recall Problem (9) was

$$\begin{aligned} \max_{w,\alpha,\beta} \quad & \alpha - \beta \\ \text{s.t.} \quad & Aw - \alpha e \geq 0 \\ & -Bw + \beta e \geq 0 \\ & \|w\|^2 = 1. \end{aligned}$$

Geometrically, all the points in $\mathcal{A}$ are on the greater-than side of the plane $x'w = \alpha$ and all the points in $\mathcal{B}$ are on the less-than side of the plane $x'w = \beta$. Since the convex hulls of the sets $\mathcal{A}$ and $\mathcal{B}$ do intersect, we know that only α and β satisfying $\alpha - \beta \leq 0$ are feasible. Thus by maximizing $\alpha - \beta$ the two planes are being moved as close together as possible. At optimality, the point with the largest misclassification error is at a distance of exactly $|\alpha - \beta|/2$ from the separating plane. Thus Problem (9) minimizes the maximum distance of any misclassified point to the separating plane. However this property is also a limitation of the method, since adding or subtracting one "noisy" or hard-to-

classify point can dramatically change the solution. Nor is it clear that minimizing the maximum error results in the best separating plane.

An alternative method is to minimize the sum of the misclassification errors. We will measure the misclassification error as the distance of a misclassified point from the appropriate supporting plane. We say a point A_i in $\mathcal{A}$ is misclassified if $-A_iw + \alpha \geq 0$. Define $y_i = -A_iw + \alpha$ if $-A_iw + \alpha > 0$, i.e., if A_i is misclassified. Otherwise let $y_i = 0$. Similarly, let $z_j = B_jw - \beta$ if $B_jw - \beta > 0$, i.e., if B_j is misclassified. Otherwise let $z_j = 0$. We can construct a linear program to minimize the sum of the y_i, $i = 1, \ldots, m$, and the sum of the z_j, $j = 1, \ldots, k$ where m and k are the number of points in $\mathcal{A}$ and $\mathcal{B}$ respectively.

Consider the following variant of the Robust Linear Programming (RLP) approach of Bennett and Mangasarian [2]:

$$\begin{aligned} \min_{w,\alpha,\beta,y,z} \quad & \tfrac{1}{m}e'y + \tfrac{1}{k}e'z \\ \text{s.t.} \quad & Aw - \alpha e + y \geq 0 \\ & -Bw + \beta e + z \geq 0 \\ & \alpha - \beta = 2 \\ & y \geq 0 \quad z \geq 0. \end{aligned} \quad (13)$$

The nonnegative slack variable y relaxes the constraints that all points in $\mathcal{A}$ be on the greater-than side of the plane $x'w = \alpha$. If a point A_i is misclassified it will be on the less-than side of the plane

$x'w = \alpha$. If the point A_i is classified correctly, then $y_i = 0$. If the point A_i is misclassified then $y_i = -A_iw + \alpha > 0$ is proportional to the distance of that point from the plane $x'w = \alpha$. Similarly, variable z relaxes the constraints that all points in $\mathcal{B}$ be on the less-than side of the plane $x'w = \beta$. The same arguments hold for z_j, the point B_j, and the plane $x'w = \beta$. The coefficients $\frac{1}{m}$ and $\frac{1}{k}$ were chosen to guarantee that there always exists a meaningful optimal solution $w \neq 0$. This is proved using optimality conditions in [2]. If $w = 0$ then the perceptron puts all the points in one class and the solution is useless. Problem (13) is used in the most recent version of the breast cancer diagnosis system described in the introduction [32, 35].

Both MSM (10 and 11) and RLP (13) have been successfully used in practice [19, 34, 35]. They do, however, have some limitations. MSM works well for the separable case. As noted above on problems with noisy data, the method performs poorly since it minimizes the maximum error [2]. RLP achieves very strong results for inseparable problems [3, 35]. For separable problems, however, any separating plane, if scaled appropriately, is optimal for RLP since $e'y = e'z = 0$. So RLP is not very well defined in the separable case. The next two approaches preserve the benefits of both RLP and MSM.

We present two ways of combining MSM with RLP (13). The first method minimizes the 2-norm of w as follows:

$$\min_{w,y,z,\alpha,\beta} \quad (1-\lambda)(\tfrac{1}{m}e'y + \tfrac{1}{k}e'z) + \tfrac{\lambda}{2}w'w$$
$$\text{s.t.} \quad Aw - \alpha e + y \geq 0$$
$$\qquad\quad -Bw + \beta e + z \geq 0 \qquad (14)$$
$$\qquad\quad \alpha - \beta = 2$$
$$\qquad\quad y \geq 0 \quad z \geq 0.$$

where $0 < \lambda < 1$ is a fixed constant. This problem minimizes the misclassification error in two ways. The term $(1-\lambda)(\tfrac{1}{m}e'y + \tfrac{1}{k}e'z)$ reduces the average distance of the misclassified points from the relaxed supporting planes. The term $\tfrac{\lambda}{2}w'w$ decreases the maximum classification error which corresponds to the distance between the relaxed supporting planes. Since this problem is a very minor variation of the Generalized Optimal Plane of Cortes and Vapnik [5, 33], we refer to it as GOP. The problem is a nonlinear perturbation of the RLP. When λ is close to 0 the RLP objective is emphasized. Using theorems on nonlinear perturbation of linear programming in [18], we know that there exists some positive number $\bar{\lambda}$ such that if $\lambda \in (0, \bar{\lambda}]$, there exists a solution of GOP (14) that also solves RLP (13). This means that for λ sufficiently small, GOP will choose one

of the solutions of RLP that minimizes the distance between the supporting planes.

The dual of GOP (14) is closely related to the problem of finding the two closest points in the convex hulls. The dual of GOP (14) (see Appendix for derivation) is

$$\min_{u,v,\delta} \quad \tfrac{1}{2\lambda}\|A'u - B'v\|^2 - 2\delta$$
$$\text{s.t.} \quad e'u = e'v = \delta \qquad\qquad (15)$$
$$\qquad\quad \tfrac{1-\lambda}{m}e \geq u \geq 0 \quad \tfrac{1-\lambda}{k}e \geq v \geq 0.$$

To better understand the problem, consider the case when $\delta = 1$. The points $A'u$ and $B'v$ are in the convex hulls of $\mathcal{A}$ and $\mathcal{B}$ respectively. The objective still minimizes the distance between these points. Each $u_i > 0$ corresponds to an example or vector that determines the convex combination c of $\mathcal{A}$. Vapnik refers to these examples with positive multipliers as support vectors [33]. Recall that the convex combination c of $\mathcal{A}$ is $c = A'u$ with $e \geq u \geq 0$ and $e'u = 1$. In this problem, the u and v are further constrained: $e > \tfrac{1-\lambda}{m}e \geq u \geq 0$ and $e > \tfrac{1-\lambda}{k}e \geq v \geq 0$. This forces more u_i and v_j to be positive, and therefore more points in $\mathcal{A}$ and $\mathcal{B}$ are used in the convex combinations. Thus the solution is dependent on more points from each of the classes. Roughly speaking, when the convex combination c is forced to contain many points, then c is forced away from the edge of the convex hull. Thus the two points selected must be more centrally located in the convex hulls of the two sets. As λ increases, more and more points in $\mathcal{A}$ and $\mathcal{B}$ must be used. Increasing δ has a similar effect.

A linear programming version of GOP can be constructed by replacing the norm used to minimize the weights w. Recall that the GOP objective minimizes the square of the 2-norm of w, $\|w\|^2 = w'w$. The 1-norm of w, $\|w\|_1 = e'|w|$, is used instead. The absolute value function can be removed by introducing the variable s and the constraints $-s \leq w \leq s$. The GOP objective is then modified by substituting $e's$ for $\tfrac{w'w}{2}$. At optimality, $s = |w|$. The resulting Perturbed Robust Linear Program (RLP-P) is

$$\min_{w,\alpha,\beta,y,z,s} \quad (1-\lambda)(\tfrac{1}{m}e'y + \tfrac{1}{k}e'z) + \lambda e's$$
$$\text{s.t.} \quad Aw - \alpha e + y \geq 0$$
$$\qquad\quad -Bw + \beta e + z \geq 0$$
$$\qquad\quad \alpha - \beta = 2 \qquad\qquad (16)$$
$$\qquad\quad -s \leq w \leq s$$
$$\qquad\quad y \geq 0 \quad z \geq 0 \quad s \geq 0.$$

As in the GOP method, the RLP-P method minimizes both the average distance of the misclassified

Figure 5: Results of MSM, RLP, RLP-P, and GOP on a linearly inseparable problem.

points from the relaxed supporting planes and the maximum classification error. The main advantage of the RLP-P method over the GOP problem is that RLP-P is a linear program solvable in polynomial time while GOP is a quadratic program.

We have discussed several methods for training the perceptron. Figure 5 illustrates the results of the four methods, MSM, RLP, RLP-P, and GOP, on a sample problem. On this simple dataset, GOP appears to choose a more intuitive plane. Recall the best plane is the plane that generalizes best. In the next sections, the generalization of these methods is examined through an empirical comparison on practical classification problems.

6   Practical Applications

The perceptron model is simple yet it can be very successful on many real-world applications. These applications are very diverse and some address challenging scientific questions. We will perform experiments on the following problems: diagnosis of heart disease, the diagnosis of breast cancer, using voting records of congressmen to determine party affiliation, and using sonar signals to differentiate mines from rocks. These are just a few of the many ap-

plications possible. In practice most applications involve linearly inseparable datasets. However, the dataset used in the sonar example is linearly separable. A brief description of each of these applications follows.

The first application is determining whether or not a patient has heart disease. Several relevant attributes, such as age, cholesterol level, and resting blood pressure, are collected for each patient. The perceptron model is applied to a dataset containing these attributes for a number of patients whose heart disease status is known. By evaluating a new patient's attributes with respect to the separating plane a diagnosis is made. The Cleveland Heart Disease Database (Heart) is a publicly available dataset that contains information on 297 patients using 13 attributes [6].

A second application, as discussed previously, is the diagnosis of breast cancer. To evaluate whether a tumor is benign or malignant, a fine needle aspiration is performed collecting a small amount of tissue from the tumor for examination. Several measurements such as clump thickness, uniformity of cell size, and uniformity of cell shape are collected. A mathematical programming approach incorporating the RLP has been employed in clinical practice [20, 34]. The data were collected before the computer imaging techniques were used for measuring attributes as discussed in the introduction. Researchers report 100% correctness on 131 new cases that have been diagnosed. Additionally, many studies have been made on the Wisconsin Breast Cancer Database (Cancer) [34]. This publicly available dataset contains information on 682 patients using nine integer attributes.

The voting patterns of congressmen can be used to determine party affiliation. A specific example of this application is the 1984 United States Congressional Voting Records Database (House Votes) [30]. This is a publicly available dataset. Each instance of the dataset represents a U.S. House of Representatives Congressman. Information on 435 congressmen is given. Each congressman is classified as either Democrat or Republican. The attributes consist of 16 key votes. These attributes have values of y (yea), n (nay), and "?". A value of "?" indicates that a position was not made known. To solve the problem numerically, we let y, n, and "?" be 2, -2, and 0 respectively. An interesting use for this dataset is to determine whether a congressman is supporting his party's views. A congressman who is difficult to classify may represent someone who tends to vote outside party lines.

The final application presented is the use of sonar signals to distinguish mines from rocks. Mine attributes are obtained by bouncing sonar signals off a metal cylinder. The sonar signal is transmitted at various angles with rises in frequency. A similar procedure is performed to obtain the rock attributes. The publicly available Sonar dataset represents 208 mines and rocks [12]. Sixty real-valued attributes between 0.0 and 1.0 are collected for each mine or rock. The value of the attribute represents the amount of energy within a particular frequency band, integrated over a certain period of time. As was stated previously, this dataset is an example of a practical application with linearly separable points. Most datasets for practical applications are linearly inseparable.

We have limited this investigation to datasets that are publicly available via the World Wide Web. All of the above datasets are available via anonymous file transfer protocol (ftp) from the UCI Repository of Machine Learning Databases and Domain Theories [24] at ftp://ftp.ics.uci.edu/pub/machine-learning-databases. The following section contains computational results for the MSM, RLP, RLP-P, and GOP methods on these datasets.

7   Computational Results

This section contains a computational comparison of the four methods, MSM, RLP, RLP-P and GOP. We report results on four datasets: Heart, Cancer, Sonar, and House Votes. The previous section contains a description of each dataset and its application.

All four methods were implemented using the MINOS [25] mathematical programming software package. Other optimization packages could easily be substituted. The results of GOP and RLP-P are dependent on the parameter λ. We report results for GOP using $\lambda = .1$ and $\lambda = .05$ and for RLP-P using $\lambda = .05$ and $\lambda = .02$.

The best algorithm is the one that generalizes best, i.e., the algorithm most accurate on future points. Since the future points are unknown, we use an experimental technique called cross validation [31] to estimate the accuracy on future points. In 10-fold cross validation a dataset is divided randomly into ten disjoint parts of equal size. The method is then trained on nine of these parts. These nine parts are called the training set. The tenth part, the testing set, is reserved to test the accuracy of the plane obtained during training. This procedure is repeated ten times allowing each part to be used for testing. The accuracies on the training and testing sets are averaged over the 10 trials. Both training and testing set accuracies are reported. The testing set accuracy is an estimate of generalization. While training set accuracy is important too, it is possible for the training set accuracy to be high while the testing set accuracy is low. This phenomenon, called overfitting, results when the perceptron learns the wrong concept. Ideally, the training set accuracy should be just slightly greater than the testing set accuracy. The training and testing percent accuracies are given in Table 1. The best results are shown in bold.

Table 1 indicates that the GOP and RLP-P methods have larger testing set accuracies than both the RLP and MSM methods on three of the four datasets. However, many of the differences reported are not statistically significant. On the Heart dataset, GOP ($\lambda = .05$) performed as well as RLP and better than GOP ($\lambda = .1$), MSM and RLP-P. This illustrates one drawback to the GOP and RLP-P methods: The quality of the solution is extremely dependent on the choice of λ. With other choices of the parameter λ, the testing set accuracies for the GOP and RLP-P methods could continue to improve.

The Sonar dataset is linearly separable, i.e., a

Dataset		MSM	RLP	GOP (.1)	GOP (.05)	RLP-P (.05)	RLP-P (.02)
Heart	Training	75.01	85.07	85.11	**85.22**	84.92	84.89
	Testing	72.44	**83.50**	82.83	**83.50**	82.15	83.16
Cancer	Training	94.93	**97.72**	97.57	97.54	97.46	97.56
	Testing	95.01	97.21	97.51	97.36	**97.65**	97.07
House Votes	Training	93.54	**97.52**	96.14	97.34	95.35	95.91
	Testing	93.79	95.17	95.63	**95.86**	94.48	95.40
Sonar	Training	**100.0**	**100.0**	79.43	81.30	79.38	79.91
	Testing	74.05	69.71	75.00	76.44	75.48	**76.92**

Table 1: Average Training and Testing Accuracy (%)

single hyperplane can successfully separate the two classes of points. For linearly separable datasets, RLP and MSM will always choose planes that completely separate the two classes. Recall that MSM attempts to push apart two parallel supporting planes of the two convex hulls of the classes. RLP can pick any linearly separable plane. Thus we anticipated MSM would test better than RLP on these datasets. The testing set accuracies reported for the Sonar dataset indicate that MSM did perform better than RLP. Roughly speaking GOP and RLP-P try to push apart two relaxed supporting planes of the convex hulls. If λ is sufficiently large, GOP and RLP-P will not necessarily result in a separating plane even when the sets are linearly separable. This was the case on the Sonar Data. By selecting a plane with errors, the GOP and RLP-P methods found better solutions than both MSM and RLP. This is an illustration of overfitting: the best training set accuracy did not result in the best testing set accuracies.

As indicated in the previous section, these datasets along with many others are publicly available. Thus, with the aid of a mathematical programming software package, the reader can reconstruct results reported in this section. Further computational studies may also be of interest. For example, an investigation of the affects of varying choices of λ have on the GOP and RLP-P solutions could be performed.

8   Extension to Nonlinear Separation

As we discovered in Section 3, perceptrons are limited to problems that are totally or almost linearly separable. When the underlying problem is highly nonlinear, the perceptron will fail. We provide a few references to three major ways to extend the per-

ceptron model to nonlinear separators. The most popular is the multilayer perceptron or neural network. A neural network is created from an interconnecting network of threshold/perceptron type units. We invite the reader to consult [13, 17] or one of the many books on this subject. Another approach is decision trees. In decision trees, a linear separation is constructed that divides the attribute space into two parts. If the parts contain points all or largely all of one class the algorithm stops. If not, the process is repeated in each of the half spaces until the desired accuracy is achieved. Perceptrons are one such way to construct the decision [1, 4, 17, 26]. The final method is to construct nonlinear mappings of the attributes and then construct a linear discriminant in the augmented attribute space. The resulting problem is linear in its parameters but a nonlinear decision surface is created. For example to create a quadratic separator with two attributes, x_1 and x_2, you would add three more attributes, x_1^2, x_2^2, and $x_1 x_2$. The separating surface would now be $x_1 w_1 + x_2 w_2 + x_1^2 w_3 + x_2^2 w_4 + x_1 x_2 w_5 = \gamma$. Support Vector Networks [33] and Polynomial Networks [28, 29] are examples of this type of approach.

9   Conclusions

We have studied the problem of training a perceptron to classify points from two sets. We showed that a perceptron can only correctly classify two sets if the sets are either linearly separable or, equivalently, their convex hulls do not intersect. In the separable case, we argued geometrically that the perceptron can be constructed by finding the two closest points in the convex hulls of the two sets. This problem was formulated as a minimization problem. The dual problem for the separable case maximized the distance between the two parallel supporting planes of the two sets. From these two formulations we derived several existing optimization problems for training a perceptron. We explored different ways of extending these methods to the linearly inseparable case. Different approaches result depending on how we measure the misclassification error. Four different algorithms were explored both theoretically and computationally on practical problems. Our computational results indicated that the Generalized Optimal Plane and the Perturbed Robust Linear Programming method maintained the benefits of the other two approaches while avoiding some of their limitations. The datasets we investigated are publicly available, so interested readers can conduct their own experiments.

Acknowledgements

Thanks to Olvi Mangasarian, Harry McLaughlin and Anne McEntee for their valuable suggestions. This material is based on research supported by the National Science Foundation Grant IRI-9409427.

Appendix

A   Supporting Proofs

The details of proofs and arguments of the main text are supplied in this appendix.

In this theorem we show that the existence of a separating plane and the non-intersection of the convex hulls of two sets are equivalent definitions of linear separability.

Theorem A.1 (Characterization of linear separability) *Let $\mathcal{A}$ and $\mathcal{B}$ be nonempty point sets in R^n with m and k points respectively. The following are equivalent:*

(a) *$\mathcal{A}$ and $\mathcal{B}$ are linearly separable; that is, there exist $w \in R^n$ and $\gamma \in R$ such that*

$$Aw - e\gamma > 0 \\ -Bw + e\gamma > 0. \tag{17}$$

(b) *The convex hulls of $\mathcal{A}$ and $\mathcal{B}$ do not intersect; that is, there does not exist $u \in R^m$ and $v \in R^k$ such that*

$$A'u = B'v \\ e'u = e'v = 1 \tag{18} \\ u \geq 0 \quad v \geq 0.$$

Proof. Gordan's Theorem of the Alternative [16] states:

For each given matrix C, either

I $Cx > 0$ has a solution x, or

II $C'y = 0$, $y \geq 0$ has a nonzero solution y

but never both.

Theorem A.1 follows directly with

$$C = \begin{bmatrix} A & -e \\ -B & e \end{bmatrix}, x = \begin{bmatrix} w \\ \gamma \end{bmatrix}, y = \begin{bmatrix} u \\ v \end{bmatrix}.$$

$\square$

The problem of finding the two closest points in the convex hulls of two sets is the dual of the problem of finding the best separating plane.

Theorem A.2 (Dual of separating problem) *Let $\mathcal{A}$ and $\mathcal{B}$ be nonempty point sets in R^n with m and k points respectively. The primal problem,*

$$\min_{w,\alpha,\beta} \quad \frac{1}{2}\|w\|^2 - (\alpha - \beta) \\ s.t. \quad Aw - \alpha e \geq 0 \tag{19} \\ -Bw + \beta e \geq 0,$$

has the following dual problem:

$$\min_{u,v} \quad \frac{1}{2}\|A'u - B'v\|^2 \\ s.t. \quad e'u = 1 \quad e'v = 1 \tag{20} \\ u \geq 0 \quad v \geq 0.$$

Proof. The dual problem maximizes the Lagrangian function of (19), $L(w, \alpha, \beta, u, v)$, subject to the constraints that the partial derivatives of the Lagrangian with respect to the primal variables are equal to zero [15]. Specifically, the dual of (20) is

$$\max_{w,\alpha,\beta,u,v} \quad L(w, \alpha, \beta, u, v) \\ = \frac{1}{2}\|w\|^2 - (\alpha - \beta) \\ -u'(Aw - e\alpha) - v'(-Bw + e\beta)$$

$$s.t. \quad \frac{\partial L}{\partial w} = w - A'u + B'v = 0 \tag{21}$$

$$\frac{\partial L}{\partial \alpha} = -1 + e'u = 0$$

$$\frac{\partial L}{\partial \beta} = 1 - e'v = 0$$

$$u \geq 0, \quad v \geq 0$$

where $\alpha, \beta \in R$, $w \in R^n$, $u \in R^k$, and $v \in R^m$. To simplify the problem, substitute in $w = (A'u - B'v)$:

$$\max_{\alpha,\beta,u,v} \quad \frac{1}{2}\|A'u - B'v\|^2 - (\alpha - \beta) \\ -u'A(A'u - B'v) + e'u\alpha \\ +v'B(A'u - B'v) - e'v\beta \tag{22} \\ s.t. \quad e'u = e'v = 1 \\ u, v \geq 0.$$

Using the fact that

$$\|A'u - B'v\|^2 = (u'A - v'B)(A'u - B'v),$$

this simplifies to

$$\min_{u,v} \quad \tfrac{1}{2}\|A'u - B'v\|^2$$
$$\text{s.t.} \quad e'u = e'v = 1 \tag{23}$$
$$u, v \geq 0.$$

□

The dual Problem (19) of finding the closest two points is equivalent to the Optimal Separating Plane problem (24) proposed by Vapnik [33]. Specifically, every solution of one problem can be used to construct a corresponding solution of the other.

Theorem A.3 (Equivalence of two separating problems) *For separable sets, Problem (19) and the following problem are equivalent*

$$\min_{w,\gamma} \quad \tfrac{1}{2}\|w\|^2$$
$$\text{s.t.} \quad Aw - (\gamma + 1)e \geq 0 \tag{24}$$
$$-Bw + (\gamma - 1)e \geq 0.$$

Proof. Since both problems are convex minimization problems with linear constraints, all we must show is that if the optimality conditions of one problem are satisfied then the optimality conditions of the other are satisfied.

First we will show that any solution of Problem (24) can be used to construct a solution of Problem (19). Let $\bar{u}, \bar{v}, \bar{w}, \bar{\gamma}$ be an optimal solution of Problem (24), then the following Karush-Kuhn-Tucker optimality conditions are satisfied [15]:

$$Aw - (\bar{\gamma} + 1)e \geq 0$$
$$-B\bar{w} + (\bar{\gamma} - 1)e \geq 0$$
$$\bar{u}'(A\bar{w} - (\bar{\gamma} + 1)e) = 0$$
$$\bar{v}'(-B\bar{w} + (\bar{\gamma} - 1)e) = 0 \tag{25}$$
$$\bar{w} = A'\bar{u} - B'\bar{v}$$
$$e'\bar{u} = e'\bar{v}$$
$$\bar{u} \geq 0, \bar{v} \geq 0.$$

Let $\delta = e'\bar{u} = e'\bar{v}$. Define $\hat{u} = \frac{\bar{u}}{\delta}$, $\hat{v} = \frac{\bar{v}}{\delta}$, $\hat{w} = \frac{\bar{w}}{\delta}$, $\hat{\alpha} = \frac{\bar{\gamma}+1}{\delta}$, and $\hat{\beta} = \frac{\bar{\gamma}-1}{\delta}$ Then $(\hat{u}, \hat{v}, \hat{w}, \hat{\alpha}, \hat{\beta})$ satisfy

$$\hat{w} = A'\hat{u} - B'\hat{v}$$
$$A\hat{w} - \hat{\alpha}e = \tfrac{1}{\delta}(A\bar{w} - (\bar{\gamma}+1)e) \geq 0$$
$$-B\hat{w} + \hat{\beta}e = \tfrac{1}{\delta}(-B\bar{w} + (\bar{\gamma}-1)e) \geq 0$$
$$\hat{u}'(A\hat{w} - \hat{\alpha}e) = \tfrac{\bar{u}'}{\delta^2}(A\bar{w} - (\bar{\gamma}+1)e) = 0 \tag{26}$$
$$\hat{v}'(-B\hat{w} + \hat{\beta}e) = \tfrac{\bar{v}'}{\delta^2}(-B\bar{w} + (\bar{\gamma}-1)e) = 0$$
$$e'\hat{u} = e'\hat{v} = 1$$
$$\hat{u} \geq 0, \hat{v} \geq 0.$$

which are the optimality conditions of Problem (19).

Now we will show that any solution of Problem (19) can be used to construct a solution of Problem (24). Let $(\hat{u}, \hat{v}, \hat{w}, \hat{\alpha}, \hat{\beta})$ satisfy the optimality conditions (26) of Problem (19). For the separable case, $\hat{\alpha} - \hat{\beta} > 0$. Define $\bar{u} = \frac{2\hat{u}}{\hat{\alpha}-\hat{\beta}}$, $\bar{v} = \frac{2\hat{v}}{\hat{\alpha}-\hat{\beta}}$, $\bar{w} = \frac{2\hat{w}}{\hat{\alpha}-\hat{\beta}}$, and $\bar{\gamma} = \frac{\hat{\alpha}+\hat{\beta}}{\hat{\alpha}-\hat{\beta}}$. Then we know

$$\bar{w} = A'\bar{u} - B'\bar{v}$$

$$A\bar{w} - (\bar{\gamma} + 1)e$$
$$= \frac{1}{\hat{\alpha}-\hat{\beta}}(2A\hat{w} - ((\hat{\alpha} + \hat{\beta}) + (\hat{\alpha} - \hat{\beta}))e)$$
$$= \frac{2}{\hat{\alpha}-\hat{\beta}}(A\hat{w} - \hat{\alpha}e) \geq 0$$

$$-B\bar{w} - (\bar{\gamma} - 1)e$$
$$= \frac{1}{\hat{\alpha}-\hat{\beta}}(-2B\hat{w} + ((\hat{\alpha} + \hat{\beta}) - (\hat{\alpha} - \hat{\beta}))e)$$
$$= \frac{2}{\hat{\alpha}-\hat{\beta}}(-B\hat{w} + \hat{\beta}e) \geq 0$$

$$\bar{u}'(A\bar{w} - (\bar{\gamma} + 1)e) = \frac{4\hat{u}'}{(\hat{\alpha}-\hat{\beta})^2}(A\hat{w} - \hat{\alpha}e) = 0$$

$$\bar{v}'(-B\bar{w} + (\bar{\gamma} - 1)e) = \frac{4\hat{v}'}{(\hat{\alpha}-\hat{\beta})^2}(-B\hat{w} + \hat{\beta}e) = 0$$

$$e'\bar{u} = e'\bar{v}$$
$$\bar{u} \geq 0, \bar{v} \geq 0.$$

Thus the optimality conditions of Problem (24) are also satisfied. □

The dual problem of the Generalized Optimal Plane is constructed as follows [33, 5]:

Theorem A.4 (Dual of generalized optimal plane) *The dual of problem*

$$\min_{w,y,z,\alpha,\beta} \quad (1-\lambda)(\tfrac{1}{m}e'y + \tfrac{1}{k}e'z) + \tfrac{\lambda}{2}w'w$$
$$\text{s.t.} \quad Aw - \alpha e + y \geq 0$$
$$-Bw + \beta e + z \geq 0$$
$$\alpha - \beta = 2$$
$$y \geq 0 \quad z \geq 0$$

is

$$\min_{u,v,\delta} \quad \tfrac{1}{2\lambda}\|A'u - B'v\|^2 - 2\delta$$
$$\text{s.t.}$$
$$e'u = e'v = \delta$$
$$\tfrac{1-\lambda}{m}e \geq u \geq 0 \quad \tfrac{1-\lambda}{k}e \geq v \geq 0.$$

Proof. Using multipliers u, v, s, t, and δ, the dual problem is

$$\max_{w,y,z,\alpha,\beta,u,v,s,t,\delta} \quad (1-\lambda)(\tfrac{1}{m}e'y + \tfrac{1}{k}e'z) + \tfrac{\lambda}{2}w'w$$
$$-u'(Aw - \alpha e + y)$$
$$-v'(-Bw + \beta e + z)$$
$$-\delta(\alpha - \beta - 2) - s'y - t'z$$

s.t.
$$\lambda w - A'u + B'v = 0$$
$$e'u = e'v = \delta$$
$$\tfrac{1-\lambda}{m}e - u = s \geq 0$$
$$\tfrac{1-\lambda}{k}e - v = t \geq 0$$
$$u \geq 0 \quad v \geq 0.$$

Substituting in $w = \frac{A'u - B'v}{\lambda}$ yields

$$\max_{y,z,\alpha,\beta,u,v,\delta} \quad (1-\lambda)(\tfrac{1}{m}e'y + \tfrac{1}{k}e'z)$$
$$+\tfrac{1}{2\lambda}(u'A - v'B)(A'u - B'v)$$
$$-\tfrac{1}{\lambda}u'A(A'u - B'v) + \alpha e'u$$
$$-u'y + \tfrac{1}{\lambda}v'B(A'u - B'v) + \beta e'v$$
$$-z'v - \delta(\alpha - \beta - 2) - \tfrac{1-\lambda}{m}e'y$$
$$+u'y - \tfrac{1-\lambda}{k}e'z + v'z$$

s.t.
$$e'u = e'v = \delta$$
$$\tfrac{1-\lambda}{m}e \geq u \geq 0$$
$$\tfrac{1-\lambda}{k}e \geq v \geq 0.$$

Finally we use the fact that $(u'A - v'B)(A'u - B'v) = \|A'u - B'v\|^2$ to get

$$\max_{u,v,\delta} \quad -\tfrac{1}{2\lambda}\|A'u - B'v\|^2 + 2\delta$$

s.t.
$$e'u = e'v = \delta$$
$$\tfrac{1-\lambda}{m}e \geq u \geq 0$$
$$\tfrac{1-\lambda}{k}e \geq v \geq 0.$$

□

References

[1] K. P. Bennett. Decision tree construction via linear programming. In M. Evans, editor, *Proceedings of the 4th Midwest Artificial Intelligence and Cognitive Science Society Conference*, pages 97–101, Utica, Illinois, 1992.

[2] K. P. Bennett and O. L. Mangasarian. Robust linear programming discrimination of two linearly inseparable sets. *Optimization Methods and Software*, 1:23–34, 1992.

[3] K. P. Bennett and O. L. Mangasarian. Bilinear separation of two sets in n-space. *Computational Optimization and Applications*, 2:207–227, 1993.

[4] C. E. Brodley and P. E. Utgoff. Multivariate decision trees. *Machine Learning*, 19(1):45–77, 1995.

[5] C. Cortes and V. N. Vapnik. Support vector networks. *Machine Learning*, 20:273–297, 1995.

[6] R. Detrano, A. Janosi, W. Steinbrunn, M. Pfisterer, J. Schmid, S. Sandhu, K. Guppy, S. Lee, and V. Froelicher. International application of a new probability algorithm for the diagnosis of coronary artery disease. *American Journal of Cardiology*, 64:304–310, 1989.

[7] R. O. Duda and P. E. Hart. *Pattern Classification and Scene Analysis*. John Wiley & Sons, New York, 1973.

[8] H. Edelsbrunner. Computing the extreme distances between two convex polytopes. *Journal of Algorithms*, 6:213–224, 1985.

[9] K. Fukunaga. *Statistical Pattern Recognition*. Academic Press, New York, 1990.

[10] S. Gallant. Optimal linear discriminants. In *Proceedings of the International Conference on Pattern Recognition*, pages 849–852. IEEE Computer Society Press, 1986.

[11] F. Glover. Improved linear programming models for discriminant analysis. *Decision Sciences*, 21:771–785, 1990.

[12] R.P. Gorman and T.J. Sejnowski. Analysis of hidden units in a layered network trained to classify sonar targets. *Neural Networks*, 1:75–89, 1988.

[13] J. Hertz, A. Krogh, and R. G. Palmer. *Introduction to the Theory of Neural Computation*. Addison-Wesley, Redwood City, California, 1991.

[14] O. L. Mangasarian. Linear and nonlinear separation of patterns by linear programming. *Operations Research*, 13:444–452, 1965.

[15] O. L. Mangasarian. Multi-surface method of pattern separation. *IEEE Transactions on Information Theory*, IT-14:801–807, 1968.

[16] O. L. Mangasarian. *Nonlinear Programming*. McGraw–Hill, New York, 1969.

[17] O.L. Mangasarian. Mathematical programming in neural networks. *ORSA Journal on Computing*, 5(4):349–360, 1993.

[18] O.L. Mangasarian and R.R. Meyer. Nonlinear perturbation of linear programs. *SIAM Journal on Control and Optimization*, 17(6):745–752, November 1979.

[19] O.L. Mangasarian, R. Setiono, and W.H. Wolberg. Pattern recognition via linear programming: Theory and application to medical diagnosis. In *Proceedings of the Workshop on Large-Scale Numerical Optimization, 1989*, pages 22–31, SIAM, Philadelphia, PA, 1990.

[20] O.L. Mangasarian, W. N. Street, and W.H. Wolberg. Breast cancer diagnosis and prognosis via linear programming. *Operations Research*, 43(4):570–577, 1995.

[21] M. Minsky and S. Papert. *Perceptrons: An Introduction to Computational Geometry*. MIT Press, Cambridge, Massachusetts, 1969.

[22] J. J. Moré and S. J. Wright. *Optimization Software Guide*. SIAM, Philadelphia, 1993.

[23] T. S. Motzkin and I. J. Schoenberg. The relaxation method for linear inequalities. *Canadian Journal of Mathematics*, 6:393–404, 1954.

[24] P.M. Murphy and D.W. Aha. UCI repository of machine learning databases. Department of Information and Computer Science, University of California, Irvine, California, 1992.

[25] B.A. Murtagh and M.A. Saunders. MINOS 5.4 user's guide. Technical Report SOL 83.20, Stanford University, 1993.

[26] S. Murthy, S. Kasif, and S. Salzberg. A system for induction of oblique decision trees. *Journal of Artificial Intelligence Research*, 2:1–32, 1994.

[27] F. Rosenblatt. The perceptron—a perceiving and recognizing automaton. Technical Report 85-460-1, Cornell Aeronautical Laboratory, Itahca, New York, January 1957.

[28] A. Roy, S. Govil, and R. Miranda. An algorithm to generate radial basis function (rbf)-like nets for classification problems. *Neural Networks*, 8(2):179–202, 1995.

[29] A. Roy, L. S. Kim, and S. Mukhopadhyay. A polynomial time algorithm for the construction and training of a class of multilayer perceptrons. *Neural Networks*, 6:535–545, 1993.

[30] J.C. Schlimmer. *Concept acquisition through representational adjustment*. PhD thesis, Department of Information and Computer Science, University of California, Irvine, CA, 1987.

[31] M. Stone. Cross-validatory choice and assessment of statistical predictions. *Journal of the Royal Statistical Society*, 36:111–147, 1974.

[32] W.N. Street, W.H. Wolberg, and O.L. Mangasarian. Nuclear feature extraction for breast tumor diagnosis. In *IS&T/SPIE 1993 International Symposium on Electronic Imaging: Science and Technology*, volume 1905, pages 861–870, San Jose, California, 1993.

[33] V. N. Vapnik. *The Nature of Statistical Learning Theory*. John Wiley & Sons, New York, 1996.

[34] W. H. Wolberg and O.L. Mangasarian. Multisurface method of pattern separation for medical diagnosis applied to breast cytology. *Proceedings of the National Academy of Sciences, U.S.A.*, 87:9193–9196, 1990.

[35] W.H. Wolberg, W. N. Street, and O.L. Mangasarian. Image analysis and machine learning applied to breast cancer diagnosis and prognosis. *Quantitative Cytology and Histology*, 17(2):77–87, 1995.

Part 6

Mathematics
and Science

The Geometry of Numbers

Antonie Boerkoel

Department of Mathematics

Emporia State University

Emporia, KS 60801

boerkoet@esumail.emporia.edu

1 Introduction

In the 17th century Pierre de Fermat and René Descartes, independently, introduced the idea of coordinate geometry. Very few people realized the tremendous impact this concept would have on the development of mathematics; in fact, it revolutionized mathematics. The entire machinery of algebra could be used to analyze problems in geometry, and geometry in return shed new light on algebra. The merger of these two fields turned out to be so useful that today the elements of analytic geometry are part of the high school curriculum. These days it is hard to imagine the Euclidean plane without the use of coordinates. We routinely identify real numbers with points on the 'number' line. Almost automatically, we link the geometric objects of circles and lines to the algebraic equations $(x-a)^2 + (y-b)^2 = r^2$ and $ax + by = c$.

With the tools of algebra, geometry became computable, and mathematics soared to new heights. The development of analytic geometry had a great impact on many areas of mathematics. It was a major player in the development of calculus. To quote Lagrange: "As long as algebra and geometry travelled separate paths their advance was slow and their applications limited. But when these two sciences joined company, they drew from each other fresh vitality and henceforward marched on at a rapid pace towards perfection." New fields emerged and because of the quantitative and descriptive power of algebra rejuvenating geometry, mathematics became even more important in the natural sciences. The use of algebra has enriched the field of geometry tremendously, and in return geometry has been used successfully to uncover deep results in algebra as well. In this chapter we will look at some applications of geometry to a beautiful branch of mathematics: number theory. In particular we will introduce the area of mathematics called the geometry of numbers, which was created single-handedly by Hermann Minkowski in the late 19th century.

2 Minkowski

Minkowski was born in Russia in 1864, finished the Gymnasium when he was 15 and went on to study mathematics in Königsberg and Berlin. He received his Ph.D. in 1885. At the early age of seventeen, Minkowski submitted a solution to a problem posed by the French Academy in Paris, in which he gave a treatment of the theory of quadratic forms, reaching far beyond the question posed originally. He was awarded the grand prize for this in 1883. Minkowski was a very gifted mathematician and a number theorist at heart, with a tremendous grasp of geometrical concepts and deeply interested in applications of mathematics, in particular to physical problems. In the 1890s Minkowski published his geometry of numbers centered around such theorems as the Convex Body Theorem, which was referred to by David Hilbert, a close friend of Minkowski, as 'a pearl' amongst Minkowski's inventions. The geometry of numbers is based on simple geometric principles, yet furnishes the tools for proofs of many central results in algebraic number theory. For example, one can prove Dirichlet's unit theorem, establish results on quadratic forms, show that the discriminant of an algebraic number field ($\neq \mathbf{Q}$) is greater than 1, and show that the class number is finite. Some of these results require more algebraic number theory than we can develop in this chapter and thus we will limit ourselves to three applications of the geometry of numbers of a more elementary nature: a theorem of Dirichlet about the approximation of real numbers by rational numbers, a theorem of Fermat about a sum of two squares, and Lagrange's famous theorem that every natural number can be written as the sum of four squares.

The main result we will use is Minkowski's Convex Body Theorem, one form of which states: If $\mathcal{B}$ is an open, bounded, convex, symmetric subset of $\mathbf{R}^2$ with area greater than 4, then $\mathcal{B}$ must contain a point with integral coordinates, not all equal to zero.

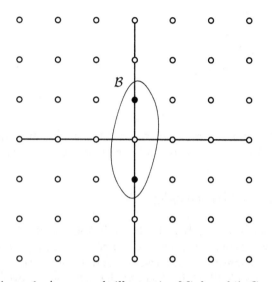

Figure 1: An example illustrating Minkowski's Convex Body Theorem.

We will state and prove a more general form of this theorem although the main content is still the same. The intuitive idea is amazingly simple: if the convex, symmetric body $\mathcal{B}$ is big enough it has to contain points with integral coordinates, not all zero.

3   Notation and Definitions

The geometry of numbers can be described most easily in the language of vectors. We will abuse notation slightly by identifying vectors and points in $\mathbf{R}^N$. We may talk about the 'point' $\vec{x} \in \mathcal{B}$ while actually referring to the endpoint of the vector $\vec{x}$. This should give no problems as the context will indicate the correct interpretation. We will use the following (vector) notations for the point $Q = (a, b, c, d)$ in $\mathbf{R}^4$:

$$\overrightarrow{OQ} = \vec{q} = \,<a, b, c, d> \, = \begin{pmatrix} a \\ b \\ c \\ d \end{pmatrix}.$$

The volume $v(\mathcal{B})$ of a subset $\mathcal{B}$ of $\mathbf{R}^N$ will be

calculated in the usual way,

$$v(\mathcal{B}) = \int \cdots \int_{\mathcal{B}} 1 \, dV \,.$$

At times we will lightly skip over some minor details, since our main goal is to give an introduction to the geometry of numbers. For example, we will develop the theory in two-dimensional space, nevertheless in our last example we will use the four-dimensional version without any hesitation. It should be noted though that most theorems can be extended to the N-dimensional case effortlessly.

We will begin by introducing some essential ingredients of the geometry of numbers: a lattice, a fundamental domain, convexity, and symmetry.

Let $\vec{e}_1$ and $\vec{e}_2$ be two linearly independent vectors in $\mathbf{R}^2$. By a **lattice** $\mathcal{L}$ in $\mathbf{R}^2$ we will mean the set of all linear combinations of $\vec{e}_1$ and $\vec{e}_2$ with integral coefficients

$$\mathcal{L} = \{\, a\vec{e}_1 + b\vec{e}_2 \,:\, a, b \in \mathbf{Z} \,\}.$$

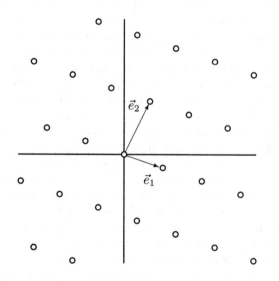

Figure 2: The lattice generated by the vectors $\vec{e}_1$ and $\vec{e}_2$.

Hence the lattice $\mathcal{L}$ in $\mathbf{R}^2$ is a $\mathbf{Z}$-module generated by the two vectors $\vec{e}_1$ and $\vec{e}_2$.

The set $\mathcal{F}$ of points $s\vec{e}_1 + t\vec{e}_2$ with $0 \le s, t < 1$ is called a **fundamental domain** of the lattice $\mathcal{L}$:

$$\mathcal{F} = \{s\vec{e}_1 + t\vec{e}_2 : 0 \le s, t < 1\}.$$

By $v(\mathcal{L})$ we will denote the area of this fundamental domain. Hence we have

$$v(\mathcal{L}) = v(\mathcal{F}) = |\det(\vec{e}_1 \ \vec{e}_2)|\,.$$

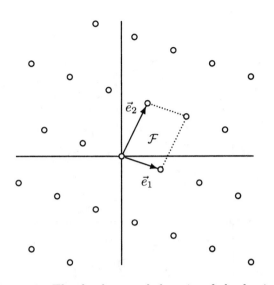

Figure 3: The fundamental domain of the lattice generated by the vectors $\vec{e}_1$ and $\vec{e}_2$.

It should be noted that a lattice can be spanned by different bases and therefore can have different fundamental domains. Fortunately $v(\mathcal{L})$ is the same for all fundamental domains of a lattice.

A subset $\mathcal{B}$ of $\mathbf{R}^2$ is called **convex** if

$$\lambda\,\vec{x} + (1-\lambda)\,\vec{y} \subseteq \mathcal{B}$$

for all $\vec{x}$ and $\vec{y}$ in $\mathcal{B}$ and $0 \leq \lambda \leq 1$.

And finally a subset $\mathcal{B}$ of $\mathbf{R}^2$ is called **symmetric** if

$$\vec{x} \in \mathcal{B} \implies -\vec{x} \in \mathcal{B}.$$

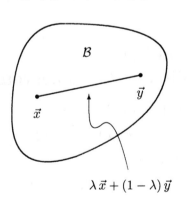

Figure 4: A convex region $\mathcal{B}$.

4   The Convex Body Theorem

We are now ready to state Minkowski's Convex Body Theorem. First we will give the two-dimensional version, which we will prove later.

Theorem 1 (Minkowski, 1896) *Let $\mathcal{B}$ be an open, bounded, symmetric, convex subset of $\mathbf{R}^2$, and let $\mathcal{L}$ be a lattice in $\mathbf{R}^2$ with $v(\mathcal{B}) > 4 \cdot v(\mathcal{L})$; then $\mathcal{B}$ contains a non-zero lattice point.*

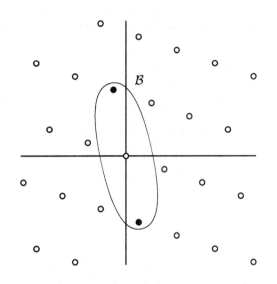

Figure 5: A convex, bounded, symmetric subset of $\mathbf{R}^2$.

Using basically the same ideas, we can prove the more general version.

Theorem 2 (Minkowski, 1896) *Let $\mathcal{B}$ be an open, bounded, symmetric, convex subset of $\mathbf{R}^N$, and let $\mathcal{L}$ be a lattice in $\mathbf{R}^N$ with $v(\mathcal{B}) > 2^N \cdot v(\mathcal{L})$; then $\mathcal{B}$ contains a non-zero lattice point.*

When we add the condition that $\mathcal{B}$ be closed, we don't need strict inequality.

Theorem 3 (Minkowski, 1896) *Let $\mathcal{B}$ be a compact, symmetric, convex subset of $\mathbf{R}^N$, and let $\mathcal{L}$ be a lattice in $\mathbf{R}^N$ with $v(\mathcal{B}) \geq 2^N \cdot v(\mathcal{L})$; then $\mathcal{B}$ contains a non-zero lattice point.*

To prove Theorem 1 we will need the lemma discussed next.

5   A Lemma

Minkowski's Convex Body Theorem is based on a geometric version of the pigeonhole principle. The pigeonhole, or box, principle states that when, for example, six marbles are distributed over five boxes there will be a box with at least two marbles. The geometric version that we will use says that when we have bits and pieces of tiles totaling for example

6 ft^2, and with those we tile a floor of only 5 ft^2, there will be some tiles that overlap. With this obvious idea we can prove the following lemma.

Lemma 4 *Let $\mathcal{L}$ be a lattice in $\mathbf{R}^2$ and let $\mathcal{B}$ be a bounded subset of $\mathbf{R}^2$ with $v(\mathcal{B}) > v(\mathcal{L})$. Then there exist two points $\vec{b}_1$ and $\vec{b}_2$ in $\mathcal{B}$ such that $\vec{b}_1 - \vec{b}_2$ is in $\mathcal{L}$.*

Proof Let $\mathcal{F}$ be a fundamental domain of the lattice $\mathcal{L}$:

$$\mathcal{F} = \{\, s\vec{e}_1 + t\vec{e}_2 \,:\, 0 \le s,\, t < 1 \,\}.$$

Note that we can tile all of $\mathbf{R}^2$, without any overlap, by translating $\mathcal{F}$ by lattice vectors:

$$\mathbf{R}^2 = \cup_{\vec{l} \in \mathcal{L}}(\vec{l} + \mathcal{F}).$$

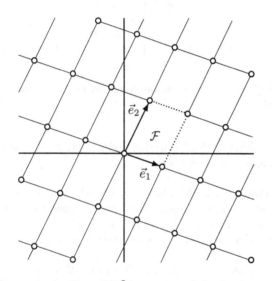

Figure 6: A tiling of $\mathbf{R}^2$ by copies of the fundamental domain $\mathcal{F}$.

Since $\mathcal{B}$ is bounded it is contained in only a finite number of these 'tiles' of the form $\vec{l}_i + \mathcal{F}$. Let $\mathcal{B}_i$ be the part of $\mathcal{B}$ in $\vec{l}_i + \mathcal{F}$. Each piece $\mathcal{B}_i$ can be translated into $\mathcal{F}$ by $-\vec{l}_i$:

$$\mathcal{B}_i - \vec{l}_i \subseteq \mathcal{F}.$$

But clearly

$$\begin{aligned}
\sum_i v(\mathcal{B}_i - \vec{l}_i) &= \sum_i v(\mathcal{B}_i) \\
&= v(\mathcal{B}) \\
&> v(\mathcal{L}) = v(\mathcal{F}).
\end{aligned}$$

Hence by the geometric version of the pigeonhole principle mentioned earlier, some pieces $\mathcal{B}_i - \vec{l}_i$ must

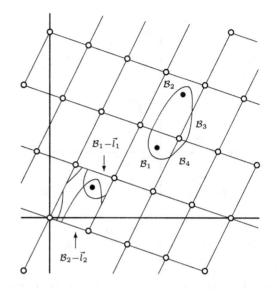

Figure 7: Translation of parts of the subset $\mathcal{B}$ to the fundamental region $\mathcal{F}$.

overlap, say $\mathcal{B}_1 - \vec{l}_1$ and $\mathcal{B}_2 - \vec{l}_2$. This means there are vectors $\vec{b}_1$ and $\vec{b}_2$ in $\mathcal{B}$ such that

$$\vec{b}_1 - \vec{l}_1 = \vec{b}_2 - \vec{l}_2,$$

and thus

$$\vec{b}_1 - \vec{b}_2 = \vec{l}_1 - \vec{l}_2 \in \mathcal{L}. \qquad \square$$

It is remarkable that with this elegant result, based on a simple geometric principle, we can prove such a marvelous theorem as Minkowski's Convex Body Theorem.

6 Proof of Minkowski's Theorem

Let $2\mathcal{L} = \{2\vec{l} : \vec{l} \in \mathcal{L}\}$. Then

$$v(2\mathcal{L}) = 4 \cdot v(\mathcal{L})$$

and it follows that

$$v(\mathcal{B}) > 4 \cdot v(\mathcal{L}) = v(2\mathcal{L}).$$

Hence according to the lemma, there exist two distinct vectors $\vec{x}$ and $\vec{y}$ in $\mathcal{B}$, such that $\vec{x} - \vec{y} \in 2\mathcal{L}$, or equivalently

$$\frac{1}{2}(\vec{x} - \vec{y}) \in \mathcal{L}.$$

But since $\mathcal{B}$ is symmetric, $\vec{y} \in \mathcal{B}$ implies that $-\vec{y} \in \mathcal{B}$, and therefore since $\mathcal{B}$ is convex we have

$$\frac{1}{2}\vec{x} + \frac{1}{2}(-\vec{y}) \in \mathcal{B}.$$

Hence

$$\frac{1}{2}(\vec{x} - \vec{y}) \in \mathcal{B} \cap \mathcal{L}$$

is a non-zero lattice point in $\mathcal{B}$. □

Now it is time to look at some applications.

7   Dirichlet's Theorem

One of the first theorems you will encounter in the area of Diophantine Approximation is a theorem of Dirichlet about the approximation of real numbers by rational numbers.

Theorem 5 (Dirichlet, 1842) *Let ξ be a real number and Q a positive integer, then there exist integers p and q with $1 \leq q \leq Q$ such that*

$$\left| \xi - \frac{p}{q} \right| < \frac{1}{qQ}.$$

Although it is not too hard to give a proof using the pigeonhole principle directly [5], the proof of this theorem using the tools from the geometry of numbers is very pretty. Note that if ξ is irrational Dirichlet's theorem implies that there are infinitely many rational numbers p/q with

$$\left| \xi - \frac{p}{q} \right| < \frac{1}{q^2}.$$

A stronger result can be established with the theory of continued fractions [7].

Proof We will take $\mathbf{Z}^2$ as our lattice $\mathcal{L}$ and let $\mathcal{B}$ be the following set

$$\left\{ (x,y) : |x - \xi y| < \frac{1}{Q}, |y| < Q + 1 \right\}.$$

Note that

$$v(\mathcal{B}) = \frac{2}{Q}\, 2(Q+1) > 4 = 4\, v(\mathcal{L}).$$

Hence $\mathcal{B}$ contains a non-zero lattice point $< p, q >$ for which

$$|p - \xi q| < \frac{1}{Q}$$

and $|q| < Q + 1$. Note that q cannot be zero, for else $|p| < 1/Q$ would imply that p is zero too. Hence the theorem follows. □

8   Sums of Squares

A famous theorem in number theory, which is attributed to Fermat, but was first proved by Euler, describes which primes can be written as the sum of two squares.

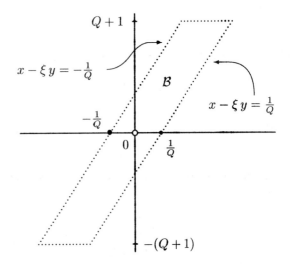

Figure 8: The set $\mathcal{B} = \{ (x,y) : |x - \xi y| < \frac{1}{Q}, |y| < Q + 1 \}$ in the real plane.

Theorem 6 (Fermat; Euler, 1754) *An odd prime p can be written as the sum of two squares*

$$p = x^2 + y^2$$

if and only if p is of the form $4k + 1$.

A classical proof of this result can be found in books on elementary number theory [7]. Most texts discussing the geometry of numbers will include the standard proof of this theorem using Minkowski's Convex Body Theorem [11]. Here, though, we will prove in a similar fashion a related result, again due to Fermat and proved by Euler.

Theorem 7 (Fermat; Euler, 1772) *Let p be a prime of the form $6k + 1$. Then there are integers x and y such that*

$$p = 3x^2 + y^2.$$

The general problem of which primes can be expressed in the form $x^2 + ny^2$ has been studied extensively in algebraic number theory [2]. Special cases can be handled using elementary techniques, but to settle the general case one needs higher reciprocity laws and class field theory.

Proof Let $p = 6k + 1$. From Gauss' theory of quadratic residues, a standard topic in any course on elementary number theory, we know that -3 is a quadratic residue of p ([7], Chapter 4), i.e. there exists an integer u, with $1 \leq u \leq p - 1$, such that

$$u^2 \equiv -3 \pmod{p}$$

or

$$3 + u^2 \equiv 0 \pmod{p}.$$

Now let $\mathcal{L}$ be the lattice generated by the vectors

$$\begin{pmatrix} 1 \\ u \end{pmatrix} \quad \text{and} \quad \begin{pmatrix} 0 \\ p \end{pmatrix}.$$

Then

$$v(\mathcal{L}) = \det \begin{pmatrix} 1 & 0 \\ u & p \end{pmatrix} = p.$$

A generic lattice point $<a, b>$ can thus be written as

$$\begin{aligned} \begin{pmatrix} a \\ b \end{pmatrix} &= a \begin{pmatrix} 1 \\ u \end{pmatrix} + x \begin{pmatrix} 0 \\ p \end{pmatrix} \\ &= \begin{pmatrix} 1 & 0 \\ u & p \end{pmatrix} \begin{pmatrix} a \\ x \end{pmatrix} \end{aligned}$$

for some integer x. In particular note that the last row implies

$$b \equiv au \pmod{p}.$$

Hence for lattice points $<a, b>$ we have

$$\begin{aligned} 3a^2 + b^2 &\equiv 3a^2 + (au)^2 \\ &\equiv a^2 (3 + u^2) \\ &\equiv 0 \pmod{p}, \end{aligned}$$

which means that p divides $3a^2 + b^2$.

Let $\mathcal{B}$ be the elliptic region

$$\mathcal{B} = \{(x, y) : 3x^2 + y^2 < R^2\}.$$

Then

$$v(\mathcal{B}) = \frac{\pi R^2}{\sqrt{3}}.$$

Hence if we take $R^2 = 2.5\, p$,

$$v(\mathcal{B}) = \frac{\pi(2.5\, p)}{\sqrt{3}} > 4p = 4\, v(\mathcal{L}).$$

Then, by Minkowski's Convex Body Theorem, $\mathcal{B}$ contains a non-zero lattice point $<a, b>$ with

$$0 < 3a^2 + b^2 < R^2 = 2.5p,$$

and since p divides $3a^2 + b^2$ we find

$$3a^2 + b^2 = p \quad \text{or} \quad 2p.$$

But $3a^2 + b^2 \equiv 0, 1,$ or $3 \pmod{4}$ and $2p \equiv 2 \pmod{4}$, so that

$$3a^2 + b^2 = p. \qquad \square$$

To illustrate the theorem, let $p = 7$. We can take $u = 5$, since then $u^2 + 3 \equiv 0 \pmod{7}$. If

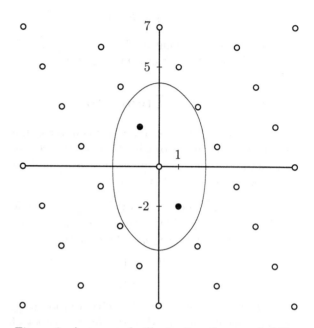

Figure 9: An example illustrating the proof of Fermat's Theorem for $p = 7$.

$\mathcal{L}$ is spanned by $\binom{1}{5}$ and $\binom{0}{7}$, $R = 4$, and $\mathcal{B} = \{(x, y) : 3x^2 + y^2 < 16\}$, it is easy to verify that $v(\mathcal{B}) > 4\,v(\mathcal{L})$. We see that $< 1, -2 >$ is a non-zero lattice point in $\mathcal{B}$ for which

$$7 = 3 \cdot 1^2 + (-2)^2.$$

Of course for any one particular prime of the form $p = 6k+1$ one could directly find x and y such that $p = 3x^2 + y^2$ by checking all x, y with $0 < x, y \leq \sqrt{p}$, but the theorem establishes the existence of such x, y for *all* these p.

9   Lagrange's Theorem

A famous and important theorem in additive number theory, which was first conjectured by Bachet in 1621, states that every positive integer can be written as the sum of four squares. Diophantus might have known the theorem, and Fermat claimed to have found a proof. Even Euler worked on it for a long time (and contributed the famous identity mentioned below), but it was Lagrange who finally proved the theorem in 1770. Here we will give a standard proof based on the geometry of numbers.

Theorem 8 (Lagrange, 1770)   *Every positive integer N can be written as:*

$$N = a^2 + b^2 + c^2 + d^2$$

for some integers a, b, c and d.

Proof Obviously the statement is true for $N = 1$ and 2, e.g.,

$$2 = 1^2 + 1^2 + 0^2 + 0^2.$$

We will establish the result for all odd primes first, and then for general numbers, using Euler's identity. Suppose $N = p$, an odd prime. We will construct a lattice with integers s and t for which

$$s^2 + t^2 + 1 \equiv 0 \pmod{p}.$$

To see that such integers exist, a simple counting argument suffices. Note that the two sets

$$\{\, s^2 : s = 0, \dots, (p-1)/2 \,\}$$

and

$$\{\, -(t^2 + 1) : t = 0, \dots, (p-1)/2 \,\}$$

each have $(p+1)/2$ distinct elements modulo p. Hence, since there are only p distinct values modulo p, by the pigeonhole principle, there are s and t with

$$s^2 \equiv -(t^2 + 1) \pmod{p}.$$

Let $\mathcal{L}$ be the lattice generated by

$$\begin{pmatrix} 1 \\ 0 \\ s \\ t \end{pmatrix}, \begin{pmatrix} 0 \\ 1 \\ -t \\ s \end{pmatrix}, \begin{pmatrix} 0 \\ 0 \\ p \\ 0 \end{pmatrix}, \begin{pmatrix} 0 \\ 0 \\ 0 \\ p \end{pmatrix}.$$

Then

$$v(\mathcal{L}) = \det \begin{pmatrix} 1 & 0 & 0 & 0 \\ 0 & 1 & 0 & 0 \\ s & -t & p & 0 \\ t & s & 0 & p \end{pmatrix} = p^2.$$

Now if $< a, b, c, d >$ is a lattice point, it can be written as

$$a \begin{pmatrix} 1 \\ 0 \\ s \\ t \end{pmatrix} + b \begin{pmatrix} 0 \\ 1 \\ -t \\ s \end{pmatrix} + x \begin{pmatrix} 0 \\ 0 \\ p \\ 0 \end{pmatrix} + y \begin{pmatrix} 0 \\ 0 \\ 0 \\ p \end{pmatrix}$$

for certain integers x and y; or more elegantly as:

$$\begin{pmatrix} a \\ b \\ c \\ d \end{pmatrix} = \begin{pmatrix} 1 & 0 & 0 & 0 \\ 0 & 1 & 0 & 0 \\ s & -t & p & 0 \\ t & s & 0 & p \end{pmatrix} \begin{pmatrix} a \\ b \\ x \\ y \end{pmatrix}.$$

The last two rows imply that

$$\begin{aligned} c &\equiv as - bt \pmod{p} \\ d &\equiv at + bs \pmod{p}. \end{aligned}$$

Hence for lattice points $< a, b, c, d >$ we have

$$\begin{aligned} & a^2 + b^2 + c^2 + d^2 \\ &\equiv a^2 + b^2 + (as - bt)^2 + (at + bs)^2 \\ &\equiv (a^2 + b^2)(1 + s^2 + t^2) \\ &\equiv 0 \pmod{p}, \end{aligned}$$

which means that p divides $a^2 + b^2 + c^2 + d^2$. Let $\mathcal{B}$ be the four-dimensional sphere of radius R, then

$$v(\mathcal{B}) = \frac{\pi^2}{2} R^4.$$

To see this, define

$$\mathcal{B}_R = \{(x, y, z, w) : x^2 + y^2 + z^2 + w^2 \le R\},$$

then $v(\mathcal{B}) = v(\mathcal{B}_R) = v(\mathcal{B}_1) R^4$, where $v(\mathcal{B}_1)$ can be calculated as follows:

$$\begin{aligned} & v(\mathcal{B}_1) \\ &= \int \cdots \int_{\mathcal{B}_1} 1 \, dx \, dy \, dz \, dw \\ &= \iint_{z^2 + w^2 \le 1} \left(\iint_{x^2 + y^2 \le 1 - (z^2 + w^2)} 1 \, dx \, dy \right) dz \, dw \\ &= \iint_{z^2 + w^2 \le 1} \pi \left(1 - (z^2 + w^2)\right) dz \, dw \\ &= \pi \int_0^1 \int_0^{2\pi} (1 - r^2) r \, d\varphi \, dr \\ &= \frac{\pi^2}{2}, \end{aligned}$$

which shows that $v(\mathcal{B}) = \frac{\pi^2}{2} R^4$.

If we now take $R^2 = 1.9 p$, then

$$v(\mathcal{B}) = \frac{\pi^2}{2}(1.9 p)^2 > 16 p^2 = 2^4 v(\mathcal{L}),$$

and by Minkowski's Convex Body Theorem $\mathcal{B}$ contains a non-zero lattice point $< a, b, c, d >$. For this point we have:

$$0 < a^2 + b^2 + c^2 + d^2 < R^2 = 1.9 p < 2p,$$

and since p divides $a^2 + b^2 + c^2 + d^2$ we find that

$$p = a^2 + b^2 + c^2 + d^2.$$

This establishes Lagrange's theorem for odd primes. To prove the theorem for general N it is sufficient to factor N into primes and use Euler's identity:

$$
\begin{aligned}
(a^2 &+ b^2 + c^2 + d^2)(A^2 + B^2 + C^2 + D^2) \\
= \quad &(aA + bB + cC + dD)^2 \\
&+ (aB - Ba + dC - cD)^2 \\
&+ (aC - cA + bD - dB)^2 \\
&+ (aD - dA + cB - bC)^2. \qquad \square
\end{aligned}
$$

10   Closing Remarks

We hope that the three applications of Minkowski's Convex Body Theorem we presented here have offered you an enlightening glimpse into the beautiful world of the geometry of numbers. Of course this chapter was just intended as an introduction. We've merely scratched the surface of this vast field of study. In fact we have only discussed applications of Minkowski's *first* theorem and haven't even mentioned two of his other famous theorems: the Successive Minima Theorem and the Linear Forms Theorem. Aside from the fact that the Geometry of Numbers has proven to be so useful in many areas of mathematics and in numerous applications, it is an absolutely fascinating area of mathematics to study in its own right. As David Hilbert, a close friend of Minkowski, stated in his memorial eulogy: "What fullness of content most alluring are the deep-lying arithmetical truths that are connected through a geometric bond in this great work of Minkowski." For those whose interest in the subject has been sparked by this discussion, we leave you with a few references that will surely delight you. Enjoy.

References

[1] J. W. S. Cassels. *An Introduction to the Geometry of Numbers*. Springer Verlag, 1971.

[2] D. A. Cox. *Primes of the Form $x^2 + ny^2$: Fermat, Class Field Theory and Complex Multiplication*. Wiley, 1989.

[3] P. M. Gruber and C. G. Lekkerkerker. *Geometry of Numbers*. North Holland, 1987.

[4] H. Hancock. *Development of the Minkowski Geometry of Numbers*. Macmillan, 1939.

[5] E. Hlawka, J. Schoißengeier, and R. Taschner. *Geometric and Analytic Number Theory*. Springer Verlag, 1991.

[6] H. Minkowski. *Geometrie der Zahlen*. Taubner, 1896 and 1910.

[7] H. E. Rose. *A Course in Number Theory*. Oxford Univ. Press, 1994.

[8] W. Scharlau and H. Opolka. *From Fermat to Minkowski: Lectures on the Theory of Numbers and Its Historical Development*. Springer Verlag, 1985.

[9] W. M. Schmidt. *Diophantine Approximation and Diophantine Equations*. LNM 1467, Springer Verlag, 1980.

[10] C. L. Siegel. *Lectures on the Geometry of Numbers*. Springer Verlag, 1989.

[11] I. N. Stewart and D. O. Tall. *Algebraic Number Theory*. Chapman and Hall, 1987.

[12] S. K. Stein and S. Szabó. *Algebra and Tiling: Homomorphisms in Service of Geometry*. MAA, 1994.

Statistical Symmetry

Charles Radin [*]
Mathematics Department
University of Texas
Austin, TX 78712
radin@math.utexas.edu

1 Introduction

After a short discussion of traditional ideas about the symmetries of patterns in the plane and space, we introduce a slight extension of the traditional notions. This new "statistical symmetry" is being used in various ways, from modeling quasicrystals to constructing new forms of graph paper.

2 Periodic Tilings

It is easy to determine the symmetries of a regular n-gon, for instance a regular octagon as in Figure 1 Such a figure coincides with itself after a rotation about its center by the angle $2\pi/8$ (radians), or any integer multiple of that angle such as $4\pi/8$ or $6\pi/8$. The figure also has the symmetry of reflection about any of the four lines which bisect a pair of adjacent sides, or the four which join opposite vertices. (The set of all symmetries of the octagon is called the dihedral group D_8.)

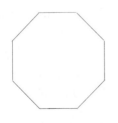

Figure 1: Octagon.

More interesting are the symmetries of patterns which fill the whole plane, for example the "bathroom tiling" shown in part in Figure 2, made of regular hexagons. It is an interesting exercise to

work out the symmetries of this pattern, which now includes some translations.

There has been a great deal of study of patterns in the plane such as that of Figure 2, that is, patterns in which there is some "unit cell" such that the whole pattern is obtained by translating the cell by integer multiples of vectors in two different directions. (We will call such patterns "periodic". For the hexagonal tiling we can use a hexagon for the unit cell and can take as vectors those perpendicular to two opposite sides of one of the hexagons, and $\sqrt{3}$ times their length.) It has been shown that there are precisely 17 different sorts of periodic patterns in the plane; that is, there are precisely 17 different sets of symmetries that can occur for periodic patterns. These 17 sets are called "the wallpaper groups". (As a general reference see [1].)

3 Aperiodic Tilings

There is a simple method for producing periodic tilings, namely first make a cell and then "grow" a larger and larger pattern by repeatedly adding on certain translations of the cell. We now con-

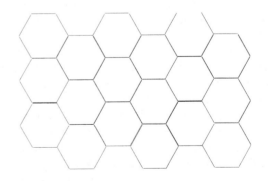

Figure 2: Tiling with hexagons.

[*]Research supported in part by NSF Grant No. DMS-9304269 and Texas ARP Grant 003658-113.

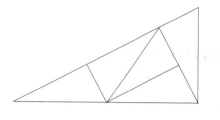

Figure 3: Hierarchy.

sider patterns made by a different sort of algorithm, called "hierarchical" (or fractal). An interesting example is produced by starting with a right triangle with legs of length one and two (and hypotenuse $\sqrt{5}$). Divide it into five smaller triangles as in Figure 3, noting that the five smaller triangles are congruent to each other and similar to the original large triangle, and then expand the collection of triangles by stretching in all directions by a factor $\sqrt{5}$, so again the triangles each have legs of size one and two.

Repeat this process in each of the five triangles, obtaining now twenty-five triangles each congruent to the one we started with. If you do this process five times you get a pattern like Figure 4. If you keep repeating this algorithm forever you get a pattern filling the plane, of which a part appears in Figure 5. If you look carefully at the twenty-five triangles after the second expansion, you will find one in the interior with edges parallel to the edges of the original triangle. Therefore if the expansions of our process are always taken about an appropriate point in this interior triangle, each two applications of the process amounts to simply adding more triangles around the original, and shows more explicitly in what way the process leads to a tiling of the plane. Such a tiling of the plane is called a "pinwheel" [3].

There are better known tilings of the plane, created by Roger Penrose, made out of two shapes with a similar hierarchical growth algorithm; Figure 6 contains part of such a Penrose tiling. For more on Penrose tilings, including the hierarchical algorithm, see [1].

These pinwheel and Penrose patterns and others like them, all of which we will call hierarchical, have symmetry properties different from the periodic patterns considered above; this new kind of symmetry is the main focus of this article. To be specific, we claim that in some appropriate sense a Penrose tiling is symmetric under rotation by the angle $2\pi/10$ about any point, and a pinwheel tiling is symmetric under rotation by any angle about any point. Clearly we have some explaining to do!

We begin our explanation by introducing a notion

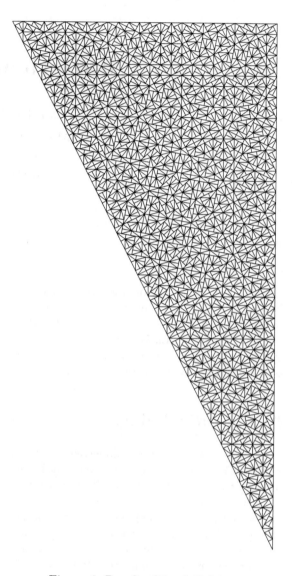

Figure 4: Result of 5 subdivisions.

of "frequencies" for patterns. Consider the Penrose tiling and fix some finite portion f of it, say the three tiles with heavy outline in Figure 7. Fix also some arc a of a unit circle, say the arc containing all angles from 0 to $\pi/20$ inclusive. Now take any large circle C in the tiling, of radius R and centered at a point P, and *count* the number of times the finite pattern f appears (completely) in the circle, where we only count occurrences of f for which the orientation is rotated from that of the original by some angle in a.

That means we count occurrences that have the original orientation as in Figure 7, or such an orientation rotated counterclockwise by any angle up to $\pi/20$. We claim that if one makes such a count, getting the number N, then one will find that $N/\pi R^2$ is approximately $\sqrt{5} - 2$, and this approximation

Figure 5: Pinwheel tiling.

Figure 6: Penrose tiling.

Figure 7: Tiling frequencies.

will be better and better the larger the size of the circle used. (We are assuming for definiteness that the area of the kite-like tile is 1.) We say that associated with the finite pattern f and arc a there is a frequency $F(f, a)$ which in this case is $\sqrt{5} - 2$. Of course this is just one example, and we are really claiming that there are analogous frequencies for each finite pattern f and arc a.

4   Statistical Symmetry

We are now ready to clarify the new "statistical" notion of symmetry. We say the Penrose tilings have the statistical symmetry of rotation by $2\pi/10$ because if such a tiling is rotated about any point by $2\pi/10$ *none of its frequencies change*. Note that in a Penrose tiling all the edges of the tiles are parallel to one of five lines, and the tiles only appear in ten different orientations. The statistical symmetry we have claimed means that the tiles appear *equally often* in each of these orientations; more generally, not just individual kites and darts but any finite pattern in the tiling has this property [3].

A more surprising fact is that for a pinwheel tiling, roughly speaking each finite part f appears equally often in all directions! In particular picking some elementary triangle in a pinwheel tiling, about one eighth of any large region is filled with copies of that triangle with orientation within $\pm 2\pi/8$ of the original.

This notion of symmetry is weaker than the usual one, in that if a pattern is invariant in the ordinary sense under a rotation by angle α about some point

(as is the hexagonal tiling under rotation by $2\pi/3$ about any vertex), then it is also statistically symmetric under rotation by α, while from the above examples the converse is not true in general. At this point we should give some indication why this notion of statistical symmetry is of interest.

5   Diffraction Patterns

Imagine the surface of a large lake, out of which is growing a collection of isolated, thin trees. If a wind created a sequence of parallel waves on the lake surface, as these waves struck each tree circular rings of waves would appear to come out of the tree; see Figure 8.

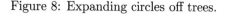

Figure 8: Expanding circles off trees.

The expanding rings from different trees eventually overlap and interfere with each other, producing complicated patterns of reinforcement, both positive (where the heights of two intersecting rings-waves are both high or both low) and negative (where one height is high and the other is low). This is similar to a mechanism which allows physicists to investigate the internal (atomic) structure of solids. To "visualize" their internal structure, parallel plane waves of very short wavelength (such as X-rays) are sent into the material, creating expanding spheres of waves issuing from each atom; the expanding spheres from different atoms interfere, and the patterns of interference are recorded on appropriate film. (The result is called a diffraction pattern for the material.)

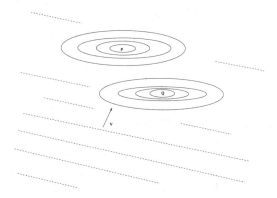

Figure 9: Expanding spheres off atoms.

We analyze the situation as follows. Imagine we have two atoms in the target material, at positions P and Q, and there are parallel waves hitting these atoms, the waves moving in the direction of the unit vector V; see Figures 9 and 10. (We use two-dimensional drawings for simplicity, with lines for the waves instead of planes, and circles instead of spheres.) The atoms emit expanding spheres of outgoing waves. At a point far away these spheres appear as plane waves, and we assume we will measure the size of the resulting waves at such a point in the direction of some unit vector W from the atoms. Comparing the distance traveled by the waves hitting the two atoms, we see there is an extra distance of size $V \cdot (P-Q) + (-W) \cdot (P-Q) = (V-W) \cdot (P-Q)$ for the wave going through P, as in Figure 10. Measuring distances by the natural scale of the wavelength of the waves, and recalling that waves reinforce optimally when corresponding maxima meet, we see that for maximum reinforcement this extra distance should be a multiple of a wavelength, which occurs only in special directions W.

So far we have considered how the spherical waves emanating from two atoms interfere. Now consider the total diffraction from the target, the sum of the contributions of all pairs of atoms. If the atoms in the target are arranged in a periodic array then there will be many copies of each vector of the sort $P - Q$ and one can then calculate precisely which directions W will show a large effect for given incoming beam direction V. (Each individual X-ray picture samples a small range of directions W; the bright spots are due to positive reinforcement from many sources (atoms). For a discussion of diffraction see [2, 5].)

The relevance of all this to statistical symmetry is the following. To get bright spots in a diffraction pattern of a material does not *require* that the material be a periodic crystal, all one needs is a lot of vectors $P - Q$ creating the same contribution. That is, the effect is a consequence of the appropriate statistical nature of the locations of the atoms in the material, namely the *frequencies of finite clusters of atoms*. And in particular, a diffraction pattern will show an ordinary rotational symmetry if the frequencies of finite sets of atomic locations have the corresponding statistical symmetry. (As noted above this is automatic if the configuration of atoms itself has the rotational symmetry in the ordinary sense, but this is not necessary to achieve statistical symmetry.)

About fifteen years ago certain metallic alloys, called quasicrystals, were discovered which exhibit unusual diffraction patterns, unusual in that they have rotational symmetries never seen before and which in fact were known to be impossible for any ordinary solid. The location of the atoms in an ordinary solid is periodic in the sense discussed above, and it has been known for many years what sort

Figure 10: Diffraction.

of symmetries could be produced by diffraction of any such periodic pattern. In an attempt to understand what sort of atomic structure could be producing these unusual diffraction patterns, physicists used three-dimensional versions of Penrose tilings, with what we would now call their *statistical symmetry of rotation by* $2\pi/10$, and showed that they could reproduce the unusual diffraction patterns with such models; see [5]. us understand previously unknown atomic structures of solids. And furthermore, the pinwheel with its complete rotational symmetry suggests that there are even wilder structures possible for the atoms in solids.

6   Conclusion

Although we were led to the pinwheel to understand the structure of materials, now that we have it before us we find that it has other uses. For instance, imagine a piece of graph paper, made up of many little squares all lined up. Such a pattern can be used as a model of a sort of (planar) discrete world, where you can only travel along the lines. (This is sometimes called a taxicab geometry, from analogy with the way a taxi travels along city streets.) There are all sorts of things one can imagine investigating about such a world. For a mathematician the isoperimetric problem comes to mind. That is, imagine we wanted to enclose a large region in such a world with the smallest possible amount of perimeter (fencing material), where the perimeter consists of lines in the pattern. What shape would be optimal for the region? The same problem in the ordinary plane, in which there are no constraining lines, goes back to the ancient Greeks and is called

Dido's problem; of course it has a circle for its solution. In taxicab geometry a little thought shows that for large regions the shape is not circular but square, oriented the same as the little squares. Of course what we are leading up to is: What is the optimal shape for pinwheel geometry? This is quite a bit harder, and has only recently been shown [4] to be—again a circle! (asymptotically, for large regions).

The pinwheel, with its statistical roundness, is a recent discovery and it must still be hiding lots of unknown but interesting properties. The above are just some of the first ones we have found. And then one can try to investigate statistical symmetry in three dimensions . . .

References

[1] B. Grünbaum and G.C. Shephard. *Tilings and Patterns*. Freeman, New York, 1986.

[2] A. Guinier. *X-ray diffraction in crystals, imperfect crystals, and amorphous bodies*. Freeman, San Francisco, 1963. Trans. P. Lorrain and D. Ste.-M. Lorrain.

[3] C. Radin. Symmetry and tilings. *Notices Amer. Math. Soc.*, 42:26–31, 1995.

[4] C. Radin and L. Sadun. The isoperimetric problem for pinwheel tilings. *Comm. Math. Phys.*, 177:255–263, 1996.

[5] C. Radin. *Miles of Tiles*. Student Math. Lib. vol. 1, Amer. Math. Soc., Providence, 1999.

Three-Dimensional Topology and Quantum Physics

Louis H. Kauffman

Department of Mathematics, Statistics and Computer Science

851 South Morgan Street

University of Illinois at Chicago

Chicago, Illinois 60607–7045

kauffman@math.uic.edu

1   Introduction

This paper is a quick introduction to key relationships between the theories of knots, links, three-manifold invariants, and the structure of quantum mechanics. In section 2 we review the basic ideas and principles of quantum mechanics. Section 3 shows how the idea of a quantum amplitude is applied to the construction of invariants of knots and links. Section 4 explains how the generalization of the Feynman integral to quantum fields led to invariants of knots, links, and three-manifolds. Section 5 is a summary.

This paper is a thumbnail sketch of recent developments in low-dimensional topology and physics. I recommend that the interested reader consult the references given here for further information, and I apologize to the many authors whose significant work was not mentioned here due to limitations of space and reference.

2   A Quick Review of Quantum Mechanics

To recall principles of quantum mechanics it is useful to have a quick historical recapitulation. Quantum mechanics really got started when DeBroglie introduced the fantastic notion that matter (such as an electron) is accompanied by a wave that guides its motion and produces interference phenomena just like the waves on the surface of the ocean or the diffraction effects of light going through a small aperture.

DeBroglie's idea was successful in explaining the properties of atomic spectra. In this domain, his wave hypothesis led to the correct orbits and spec-

tra of atoms, formally solving a puzzle that had been only described in ad hoc terms by the preceding theory of Niels Bohr. In Bohr's theory of the atom, the electrons are restricted to move only in certain elliptical orbits. These restrictions are placed in the theory to get agreement with the known atomic spectra and to avoid the paradox that arises if one thinks of the electron as a classical particle orbiting the nucleus of the atom. Such a particle is undergoing acceleration in order to move in its orbit. Accelerated charged particles emit radiation. Therefore the electron should radiate away its energy and spiral into the nucleus! Bohr commanded the electron to occupy only certain orbits and thereby avoided the spiral death of the atom—at the expense of logical consistency.

DeBroglie hypothesized a wave associated with the electron and he said that an integral multiple of the length of this wave must match the circumference of the electron orbit. Thus, not all orbits are possible, only those where the wave pattern can "bite its own tail". The mathematics works out, providing an alternative to Bohr's picture.

DeBroglie had waves, but he did not have an equation describing the spatial distribution and temporal evolution of these waves. Such an equation was discovered by Erwin Schrodinger. Schrodinger relied on inspired guesswork based on DeBroglie's hypothesis and produced a wave equation, known ever since as the Schrodinger equation. Schrodinger's equation was enormously successful, predicting fine structure of the spectrum of hydrogen and many other aspects of physics. Suddenly a new physics, *quantum mechanics*, was born from this musical hypothesis of DeBroglie.

Along with the successes of quantum mechanics came a host of extraordinary problems of interpretation. What is the status of this wave function of Schrodinger and DeBroglie? Does it connote a new element of physical reality? Is matter "nothing but" the patterning of waves in a continuum? How can the electron be a wave and still have the capacity to instantiate a very specific event at one place and one time (such as causing a bit of phosphor to glow there on your television screen)? It came to pass that Max Born developed a statistical interpretation of the wave-function wherein the wave determines a probability for the appearance of the localized particulate phenomenon that one wanted to call an "electron". In this story the wave function ψ takes values in the complex numbers and the associated probability is $\psi^*\psi$, where ψ^* denotes the complex conjugate of ψ. Mathematically, this is a satisfactory recipe for dealing with the theory, but it leads to further questions about the exact character of the statistics. If quantum theory is inherently statistical, then it can give no complete information about the motion of the electron. In fact, there may be no such complete information available even in principle. Electrons manifest as particles when they are observed in a certain manner and as waves when they are observed in another, complementary manner. This is a capsule summary of the view taken by Bohr, Heisenberg, and Born. Others, including DeBroglie, Einstein, and Schrodinger, hoped for a more direct and deterministic theory of nature.

As we shall see in the course of this essay, the statistical nature of quantum theory has a formal side that can be exploited to understand the topological properties of such mundane objects as knotted ropes in space and spaces constructed by identifying the sides of polyhedra. These topological applications of quantum mechanical ideas are exciting in their own right. They may shed light on the nature of quantum theory itself.

In this section we review a bit of the mathematics of quantum theory. Recall the equation for a wave:

$$f(x,t) = \sin(\frac{2\pi}{\lambda}(x - ct)).$$

With x interpreted as the position and t as the time, this function describes a sinusoidal wave travelling with velocity c. We define the wave number $k = 2\pi/\lambda$ and the frequency $\omega = 2\pi c/\lambda$ where λ is the wavelength. Thus we can write $f(x,t) = \sin(kx - \omega t)$. Note that the velocity, c, of the wave is given by the ratio of frequency to wave number, $c = \omega/k$.

DeBroglie hypothesized two fundamental relationships: between energy and frequency, and between momentum and wave number. These rela-

tionships are summarized in the equations

$$E = \hbar\omega$$
$$p = \hbar k$$

where E denotes the energy associated with a wave and p denotes the momentum associated with the wave. Here $\hbar = h/2\pi$ where h is Planck's constant.

For DeBroglie, the discrete energy levels of the orbits of electrons in an atom of hydrogen could be explained by restrictions on the vibrational modes of waves associated with the motion of the electron. His choices for the energy and the momentum in relation to a wave are not arbitrary. They are designed to be consistent with the notion that *the wave or wave packet moves along with the electron*. That is, the velocity of the wave-packet is designed to be the velocity of the "corresponding" material particle.

It is worth illustrating how DeBroglie's idea works. Consider two waves whose frequencies are very nearly the same. If we superimpose them (as a piano tuner superimposes his tuning fork over the vibration of the piano string), then there will be a new wave produced by the interference of the original waves. This new wave pattern will move at its own velocity, different (and generally smaller) than the velocity of the original waves. To be specific, let $f(x,t) = \sin(kx - \omega t)$ and $g(x,t) = \sin(k'x - \omega't)$. Let

$$\begin{aligned} h(x,t) &= [(\sin(kx - \omega t) + \sin(k'x - \omega't)] \\ &= f(x,t) + g(x,t). \end{aligned}$$

A little trigonometry shows that

$$\frac{1}{2}h(x,t) = \cos\left(\frac{k - k'}{2}x - \frac{\omega - \omega'}{2}t\right) \times$$
$$\sin\left(\frac{k + k'}{2}x - \frac{\omega + \omega'}{2}t\right). \quad (1)$$

If we assume that k and k' are very close and that ω and ω' are very close, then $(k + k')/2$ is approximately k, and $(\omega + \omega')/2$ is approximately ω. Thus $h(x,t)$ can be represented by

$$H(x,t) = \cos\left(\frac{\delta k}{2}x - \frac{\delta\omega}{2}t\right)f(x,t)$$

where $\delta k = (k - k')/2$ and $\delta\omega = (\omega - \omega')/2$. This means that the superposition, $H(x,t)$, behaves as the waveform $f(x,t)$ carrying a slower-moving "wave packet" $G(x,t) = \cos((\delta k/2)x - (\delta\omega/2)t)$. See Figure 1.

Since the wave packet (seen as the clumped oscillations in Figure 1) has the equation $G(x,t) =$

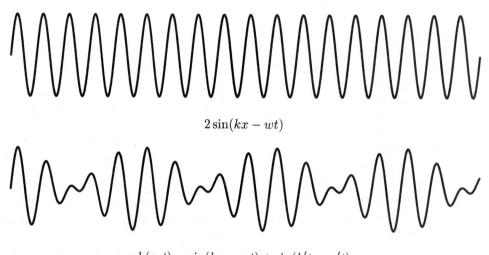

$$2\sin(kx - wt)$$

$$h(x,t) = \sin(kx - wt) + \sin(k't - w't)$$

Figure 1: Wave and wave packet.

$\cos((\delta k/2)x - (\delta\omega/2)t)$, we see that the velocity of this wave packet is $v_g = \delta\omega/\delta k$. Recall that wave velocity is the ratio of frequency to wave number. Now according to DeBroglie, $E = \hbar w$ and $p = \hbar k$, where E and p are the energy and momentum associated with this wave packet. Thus we get the formula $v_g = \delta E/\delta p$. In other words, *the velocity of the wave packet is the rate of change of its energy with respect to its momentum.* Now this is exactly in accord with the well-known classical laws for a material particle! For such a particle, $E = mv^2/2$ and $p = mv$. Thus $E = p^2/2m$ and $dE/dp = p/m = v$. It is this astonishing concordance between the simple wave model and the classical notions of energy and momentum that initiated the beginnings of quantum theory.

2.1 Schrodinger's Equation

Schrodinger answered the question: *Where is the wave equation for DeBroglie's waves?* Writing an elementary wave in complex form

$$\psi = \psi(x,t) = \exp(i(kx - \omega t)),$$

we see that we can extract DeBroglie's energy and momentum by differentiating:

$$i\hbar\frac{\partial\psi}{\partial t} = E\psi \text{ and } -i\hbar\frac{\partial\psi}{\delta x} = p\psi.$$

This led Schrodinger to postulate *the identification of dynamical variables with operators* so that the first equation,

$$i\hbar\frac{\partial\psi}{\partial t} = E\psi,$$

is promoted to the status of an equation of motion while the second equation becomes the definition of momentum as an operator:

$$p = -i\hbar\frac{\partial}{\partial x}.$$

Once p is identified as an operator, the numerical value of momentum is associated with an eigenvalue of this operator, just as in the example above. In our example $p\psi = \hbar k\psi$.

In this formulation, the position operator is just multiplication by x itself. Once we have fixed specific operators for position and momentum, the operators for other physical quantities can be expressed in terms of them. We obtain the energy operator by substitution of the momentum operator in the classical formula for the energy:

$$E = \frac{1}{2}mv^2 + V$$

$$E = \frac{p^2}{2m} + V$$

$$E = -\frac{\hbar^2}{2m}\frac{\partial^2}{\partial x^2} + V.$$

Here V is the potential energy, and its corresponding operator depends upon the details of the application.

With this operator identification for E, Schrodinger's equation

$$i\hbar\frac{\partial\psi}{\partial t} = -\frac{\hbar^2}{2m}\frac{\partial^2\psi}{\partial x^2} + V\psi$$

is an equation in the first derivatives of time and in second derivatives of space. In this form of the theory one considers general solutions to the differential equation and this in turn leads to excellent results in a myriad of applications.

In quantum theory, observation is modelled by the concept of eigenvalues for corresponding operators. *The quantum model of an observation is a projection of the wave function into an eigenstate.* An energy spectrum $\{E_k\}$ corresponds to wave functions ψ satisfying the Schrodinger equation, such that there are constants E_k with $E\psi = E_k\psi$. An *observable* (such as energy) E is a Hermitian operator on a Hilbert space of wave functions. A Hermitian operator acts on the infinite dimensional space of the wave functions and, like a Hermitian matrix, is equal to the conjugate of its transpose. Since Hermitian operators have real eigenvalues, this provides the link with measurement for the quantum theory.

It is important to notice that there is no mechanism postulated in this theory for how a wave function is "sent" into an eigenstate by an observable. Just as mathematical logic need not demand causality behind an implication between propositions, the logic of quantum mechanics does not demand a specified cause behind an observation. This absence of an assumption of causality in logic does not obviate the possibility of causality in the world. Similarly, the absence of causality in quantum observation does not obviate causality in the physical world. Nevertheless, the debate over the interpretation of quantum theory has often led its participants into asserting that causality has been demolished in physics.

Note that the operators for position and momentum satisfy the equation $xp - px = \hbar i$. This corresponds directly to the equation obtained by Heisenberg, on other grounds, that dynamical variables can no longer necessarily commute with one another. In this way, the points of view of DeBroglie, Schrodinger and Heisenberg came together, and quantum mechanics was born. In the course of this development, interpretations varied widely. Eventually, physicists came to regard the wave function not as a generalized wave packet, but as a carrier of information about possible observations. In this way of thinking, $\psi^*\psi$ represents the probability of finding the "particle" (an observable with local spatial characteristics) at a given point in spacetime.

2.2 Dirac Brackets

Recall Dirac's notation, $\langle a|b \rangle$, [6]. In this notation, $\langle a|$ and $|b \rangle$ are vectors and covectors respectively. Thus $\langle a|b \rangle$ is the evaluation of $\langle a|$ by $|b \rangle$, hence it is a scalar and in ordinary quantum mechanics it is a complex number. One can think of this as the amplitude for the state to begin in "a" and end in "b". That is, there is a process that can mediate a transition from state a to state b. Except for the fact that amplitudes are complex valued, they obey the usual laws of probability. This means that if the process can be factored into a set of all possible intermediate states $c_1, c_2, \ldots, c_n$, then the amplitude for $a \to b$ is the sum of the amplitudes for $a \to c_i \to b$. Meanwhile, the amplitude for $a \to c_i \to b$ is the product of the amplitudes of the two subconfigurations $a \to c_i$ and $c_i \to b$. Formally we have

$$\langle a|b \rangle = \sum_i \langle a|c_i \rangle \langle c_i|b \rangle$$

where the summation is over all the intermediate states $i = 1, \ldots, n$.

In general, the amplitude for mutually disjoint processes is the *sum* of the amplitudes of the individual processes. The amplitude for a configuration of disjoint processes is the *product* of their individual amplitudes.

Dirac's division of the amplitudes into *bras* $\langle a|$ and *kets* $|b \rangle$ is done mathematically by taking a vector space V (a Hilbert space, but it can be finite dimensional) for the bras; $\langle a|$ belongs to V. The dual space V^* is the home of the kets. Thus $|b \rangle$ belongs to V^* so that $|b \rangle$ is a linear mapping $|b \rangle : V \to C$ where C denotes the complex numbers. We restore symmetry to the definition by realizing that an element of a vector space V can be regarded as a mapping from the complex numbers to V. Given $\langle a| : C \to V$, the corresponding element of V is the image of 1 (in C) under this mapping. In other words, $\langle a|(1)$ is a member of V. Now we have $\langle a| : C \to V$ and $|b \rangle : V \to C$. The composition $\langle a||b \rangle = \langle a|b \rangle : C \to C$ is regarded as an element of C by taking the specific value $\langle a|b \rangle(1)$. The complex numbers are regarded as the "vacuum", and the entire amplitude $\langle a|b \rangle$ is a "vacuum to vacuum" amplitude for a process that includes the creation of the state a, its transition to b, and the annihilation of b to the vacuum once more.

Dirac notation has a life of its own. Let $P = |y \rangle \langle x|$ and let $\langle x| |y \rangle = \langle x|y \rangle$. Then

$$PP = |y \rangle \langle x||y \rangle \langle x| = |y \rangle \langle x|y \rangle \langle x| = \langle x|y \rangle P.$$

Up to a scalar multiple, P is a projection operator. That is, if we let $Q = P/\langle x|y \rangle$, then $QQ = PP/\langle x|y \rangle \langle x|y \rangle = \langle x|y \rangle P/\langle x|y \rangle \langle x|y \rangle = P/\langle x|y \rangle = Q$. Thus $QQ = Q$. In this language, the completeness of intermediate states becomes the statement that a certain sum of projections is equal to the identity: suppose that $\sum_i |c_i\rangle\langle c_i| = 1$ (summing over i) with $\langle c_i|c_i\rangle = 1$ for each i. Then

$$
\begin{aligned}
\langle a|b\rangle &= \langle a||b\rangle \\
&= \langle a|\sum_i |c_i\rangle\langle c_i||b\rangle \\
&= \sum_i \langle a||c_i\rangle\langle c_i||b\rangle \\
&= \sum_i \langle a|c_i\rangle\langle c_i|b\rangle.
\end{aligned}
$$

Iterating this principle of expansion over a complete set of states leads to the most primitive form of the Feynman integral [8]. Imagine that the initial and final states a and b are points on the vertical lines $x = 0$ and $x = n + 1$ respectively in the x-y plane, and that $(c(k)_{i(k)}, k)$ is a given point on the line $x = k$ for $0 < i(k) < m$. Suppose that the sum of projectors for each intermediate state is complete. That is, we assume that following sum is equal to one, for each k from 1 to n: $|c(k)_1\rangle\langle c(k)_1| + \cdots + |c(k)_m\rangle\langle c(k)_m| = 1$.

Applying the completeness iteratively, we obtain the following expression for the amplitude $\langle a|b\rangle$:

$$
\langle a|b\rangle = \\
\sum\sum \langle a|c(1)_{i(1)}\rangle\langle c(1)_{i(1)}|c(2)_{i(2)}\rangle \cdots \langle c(n)_{i(n)}|b\rangle
$$

where the sum is taken over all $i(k)$ ranging between 1 and m, and k ranging between 1 and n. Each term in this sum can be construed as a combinatorial path from a to b in the two-dimensional space of the x-y plane. Thus the amplitude for going from a to b is seen as a summation of contributions from all the "paths" connecting a to b. Feynman used this description to produce his famous path integral expression for amplitudes in quantum mechanics. His path integral takes the form

$$
\int dP \exp(iS)
$$

where i is the square root of minus one, the integral is taken over all paths from point a to point b, and S is the action for a particle to travel from a to b along a given path. For the quantum mechanics associated with a classical (Newtonian) particle the action S is given by the integral along the given path from a to b of the difference $T - V$ where T is the classical kinetic energy and V is the classical potential energy of the particle.

The beauty of Feynman's approach to quantum mechanics is that it shows the relationship between the classical and the quantum in a particularly transparent manner. Classical motion corresponds to those regions where all nearby paths contribute constructively to the summation. This classical path occurs when the variation of the action is null. To ask for those paths where the variation of the action is zero is a problem in the calculus of variations, and it leads directly to Newton's equations of motion. Thus with the appropriate choice of action, classical and quantum points of view are unified.

The drawback of this approach lies in the unavailability at the present time of an appropriate measure theory to support all cases of the Feynman integral.

To summarize, Dirac notation shows at once how the probabilistic interpretation for amplitudes is tied to the vector space structure of the space of states of the quantum mechanical system. Our strategy for bringing forth relations between quantum theory and topology is to pivot on the Dirac bracket. The Dirac bracket intermediates between notation and linear algebra. In a very real sense, *the connection of quantum mechanics with topology is an amplification of Dirac notation.*

The next two sections discuss how topological invariants in low-dimensional topology are related to amplitudes in quantum mechanics. In these cases the relationship with quantum mechanics is primarily mathematical. Ideas and techniques are borrowed. It is not yet clear what the effect of this interaction will be on the physics itself.

3   Knot Amplitudes

At the end of section 2 we said: *the connection of quantum mechanics with topology is an amplification of Dirac notation.*

Consider first a circle in a spacetime plane with time represented vertically and space horizontally, as shown in Figure 2. The circle represents a vacuum to vacuum process that includes the creation of two "particles" (Figure 3), and their subsequent annihilation (Figure 4).

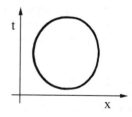

Figure 2: A circle in a spacetime plane.

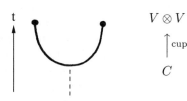

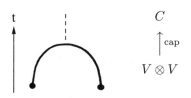

Figure 3: Creation of two particles.

Figure 4: Annihilation of two particles.

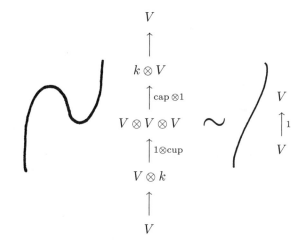

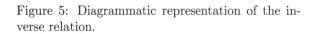

Figure 5: Diagrammatic representation of the inverse relation.

In accord with our previous description, we could divide the circle into these two parts (creation as shown in Figure 3 and annihilation as shown in Figure 4) and consider the amplitude $\langle a|b \rangle$. Since the diagram for the creation of the two particles ends in two separate points, it is natural to take a vector space of the form $V \otimes V$ as the target for the bra and as the domain of the ket.

We imagine at least one particle property being catalogued by each dimension of V. For example, a basis of V could enumerate the spins of the created particles. If $\{e_a\}$ is a basis for V then $\{e_a \otimes e_b\}$ forms a basis for $V \otimes V$. The elements of this new basis constitute all possible combinations of the particle properties. Since such combinations are multiplicative, the tensor product is the appropriate construction.

In this language the creation ket is a map cup,

$$\text{cup} = \langle a| : C \to V \otimes V,$$

and the annihilation bra is a mapping cap,

$$\text{cap} = |b\rangle : V \otimes V \to C.$$

The first hint of topology comes when we realize that it is possible to draw a much more complicated simple closed curve in the plane that is nevertheless decomposed with respect to the vertical direction into many cups and caps. In fact, any non-self-intersecting differentiable curve can be rigidly rotated until it is in general position with respect to the vertical. It will then be seen to be decomposed into these minima and maxima. Our prescriptions for amplitudes suggest that we regard

any such curve as an amplitude via its description as a mapping from C to C.

Each simple closed curve gives rise to an amplitude, but any simple closed curve in the plane is isotopic to a circle, by the Jordan Curve Theorem. If these are *topological amplitudes,* then they should all be equal to the original amplitude for the circle. Thus the question: What condition on creation and annihilation will insure topological amplitudes? The answer derives from the fact that all isotopies of the simple closed curves are generated by the cancellation of adjacent maxima and minima as illustrated in Figure 6.

In composing mappings it is necessary to use the identifications

$$(V \otimes V) \otimes V = V \otimes (V \otimes V)$$

and

$$V \otimes k = k \otimes V = V.$$

Thus in the illustration above, the composition on the left is given by

$$V = \quad V \otimes k \xrightarrow{1 \otimes \text{cup}} V \otimes (V \otimes V)$$
$$= (V \otimes V) \otimes V \xrightarrow{\text{cap} \otimes 1} k \otimes V = V.$$

This composition must equal the identity map on V (denoted 1 here) for the amplitudes to have a proper image of the topological cancellation. This condition is said very simply by taking a matrix representation for the corresponding operators.

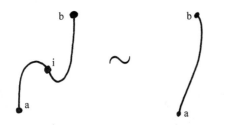

Figure 6: Cancellation of an adjacent maximum and minimum.

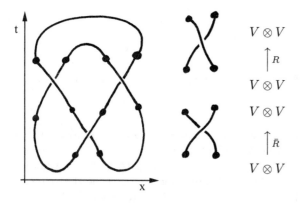

Figure 7: A knot with crossings.

Specifically, let $\{e_1, e_2, \ldots, e_n\}$ be a basis for V. Let $e_{ab} = e_a \otimes e_b$ denote the elements of the tensor basis for $V \otimes V$. Then there are matrices M_{ab} and M^{ab} such that $\text{cup}(1) = \sum M_{ab}e_{ab}$ with the summation taken over all values of a and b from 1 to n. Similarly, cap is described by $\text{cap}(e_{ab}) = M^{ab}$. Thus the amplitude for the circle is $\text{cap}[\text{cup}(1)] = \text{cap} \sum M_{ab}e_{ab} = \sum M_{ab}M^{ab}$. In general, the value of the amplitude on a simple closed curve is obtained by translating it into an "abstract tensor expression" in the M_{ab} and M^{ab} and then summing over these products for all cases of repeated indices.

Returning to the topological conditions we see that they are just that the matrices (M_{ab}) and (M^{ab}) are inverses in the sense that $\sum M_{ai}M^{ib} = I_a^b$ and $\sum M^{ai}M_{ib} = I_b^a$ are the identity matrices.

In Figure 6, we show the diagrammatic representative of the equation $\sum M_{ai}M^{ib} = I_a^b$.

In the simplest case, cup and cap are represented by 2×2 matrices. The topological condition implies that these matrices are inverses of each other. Thus the problem of the existence of topological amplitudes is very easily solved for simple closed curves in the plane.

Now we go to knots and links. Any knot or

link can be represented by a picture that is configured with respect to a vertical direction in the plane. The picture will decompose into minima (creations), maxima (annihilations), and crossings of the two types shown in Figure 8. (Here I consider knots and links that are unoriented. They do not have an intrinsic preferred direction of travel.) In Figure 7, next to each of the crossings we have indicated mappings of $V \otimes V$ to itself, called R and $\bar{R}$ respectively. These mappings represent the transitions corresponding to these elementary configurations.

That R and $\bar{R}$ really must be inverses follows from the isotopy shown in Figure 8. (This is the second Reidemeister move.)

We now have the vocabulary of cup, cap, R, and $\bar{R}$. Any knot or link can be written as a composition of these fragments, and consequently a choice of such mappings determines an amplitude for knots and links. In order for such an amplitude to be topological, we want it to be invariant under the list of local moves on the diagrams shown in Figure 10. These moves are an augmented list of the Reidemeister moves, adjusted to take care of the fact that the diagrams are arranged with respect to a given direction in the plane. The equivalence relation generated by these moves is called regular isotopy. It is one move short of the relation known as *ambient isotopy*. The missing move is the first Reidemeister move shown in Figure 9.

In the first Reidemeister move, a curl in the diagram is created or destroyed. Ambient isotopy (generated by all the Reidemeister moves) corresponds to knots and links embedded in three-dimensional space. Two link diagrams are ambient isotopic via the Reidemeister moves if and only if there is a continuous family of embeddings in three dimensions leading from one link to the other. The moves give us a combinatorial reformulation of the spatial topology of knots and links.

By ignoring the first Reidemeister move, we allow the possibility that these diagrams can model framed links, that is links with a normal vector field or, equivalently, embeddings of curves that are thickened into bands. It turns out to be fruitful to study invariants of regular isotopy. In fact, one can usually normalize an invariant of regular isotopy to obtain an invariant of ambient isotopy. We shall see an example of this phenomenon with the bracket polynomial in a few paragraphs.

As the reader can see, we have already discussed the algebraic meaning of moves 0 and 2. The other moves translate into very interesting algebra. Move 3, when translated into algebra, is the

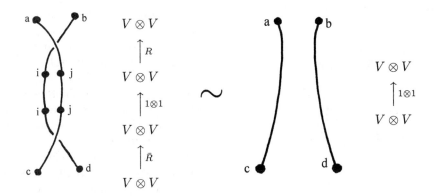

Figure 8: The second Reidemeister move. $R_{ij}^{ab} \bar{R}_{cd}^{ij} = I_c^a I_c^b$.

Figure 9: The first Reidemeister move.

famous Yang-Baxter equation. The Yang-Baxter equation occurred for the first time in problems related to exactly solved models in statistical mechanics (See [18]). All the moves taken together are directly related to the axioms for a quasi-triangular Hopf algebra (aka quantum group). We shall not go into this connection here.

There is an intimate connection between knot invariants and the structure of generalized amplitudes, as we have described them in terms of vector space mappings associated with link diagrams. This strategy for the construction of invariants is directly motivated by the concept of an amplitude in quantum mechanics. It turns out that the invariants that can actually be produced by this means (that is, by assigning finite-dimensional matrices to the caps, cups and crossings) are incredibly rich. They encompass, at present, all of the known invariants of polynomial type (Alexander polynomial, Jones polynomial and their generalizations).

It is now possible to indicate the construction of the Jones polynomial via the bracket polynomial as an amplitude, by specifying its matrices.

The cups and the caps are defined by $(M_{ab}) = (M^{ab}) = M$ where M is the 2×2 matrix (with $ii = -1$)

$$M = \begin{bmatrix} 0 & iA \\ -iA^{-1} & 0 \end{bmatrix}.$$

Note that $MM = I$ where I is the identity matrix.

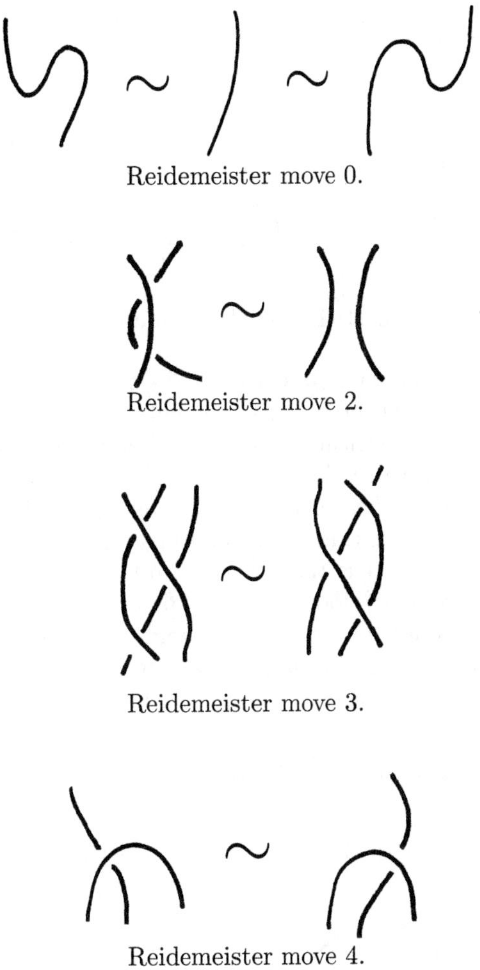

Reidemeister move 0.

Reidemeister move 2.

Reidemeister move 3.

Reidemeister move 4.

Figure 10: Local moves under which amplitudes should be invariant.

Note also that the amplitude for the circles is

$$
\begin{aligned}
\sum M_{ab} M^{ab} &= \sum M_{ab} M_{ab} \\
&= \sum (M_{ab})^2 \\
&= (iA)^2 + (-iA^{-1})^2 \\
&= -A^2 - A^{-2}.
\end{aligned}
$$

The matrix R is then defined by the equation

$$
R_{cd}^{ab} = AM^{ab} M_{cd} + AI_c^a I_d^b,
$$

or symbolically by

$$
R_{\square\square}^{\square\square} = AM^{\square\square} M_{\square\square} + A^{-1} I_\square^\square I_\square^\square.
$$

For example, we have the specific evaluation

$$
\begin{aligned}
R_{12}^{12} &= AM^{12} M_{12} + A^{-1} I_1^1 I_2^2 \\
&= A(iA)(iA) + A^{-1}(1)(1).
\end{aligned}
$$

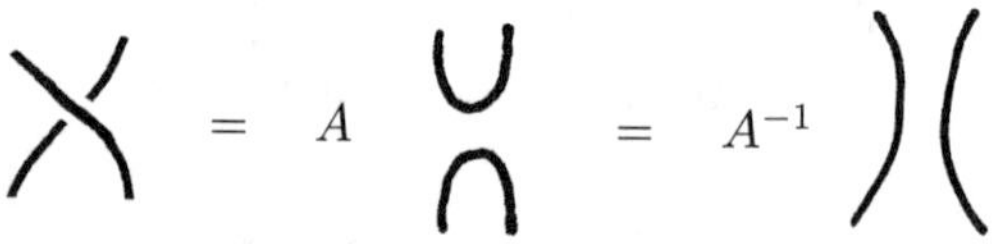

$$
\bigtimes \;=\; A \;\; \bigcup_{\bigcap} \;=\; A^{-1} \;\;)(
$$

Figure 11: Loop value of a crossing.

Since, diagrammatically, we identify R with a (right-handed) crossing, this equation can be written diagrammatically as shown in Figure 11. Taken together with the loop value of $-A^2 - A^{-2}$, we get Figure 12. These equations can be regarded as a recursive algorithm for computing the amplitude. This algorithm is the bracket state model for the (unnormalized) Jones polynomial [13]. This model can be studied on its own grounds. We end this section with some comments about this algorithm and its properties.

3.1   The Bracket Model

If we were to start with just the calculational formulas as indicated above but with arbitrary coefficients

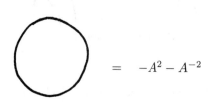

$$
= \;\; -A^2 - A^{-2}
$$

Figure 12: Loop value of $-A^2 - A^{-2}$.

A and B for the two smoothings, and an arbitrary loop value d, then it is easy to see that the resulting method of calculating a three-variable polynomial (in the commuting variables A, B, and d) from a link diagram is well defined, although not necessarily invariant under the Reidemeister moves. It is then an interesting exercise to see that asking for invariance under just the second Reidemeister move essentially forces $B = A^{-1}$ and $d = -A^2 - A^{-2}$. Thus the parameters arising from the algebra that we have sketched actually come directly from the topology. It is equally easy to see the resulting Laurent polynomial is a well-defined invariant of regular isotopy. Let's denote that invariant by $\langle K \rangle$, the (unnormalized) bracket polynomial of K. In this version of the bracket we have $\langle O \rangle = -A^2 - A^{-2}$ where O denotes a circle in the plane. If we define $f_K(A) = (-A^3)^{-w(K)} \langle K \rangle / \langle O \rangle$ where $w(K)$ denotes the sum of the signs of the crossings in an oriented link K (See [13] or [16].), then $f_K(A)$ is an invariant of ambient isotopy and the original Jones polynomial [11] $V_K(t)$ is given by the formula

$$
V_K(t) = f_K(t^{-1/4}).
$$

The bracket model for the Jones polynomial is quite useful both theoretically and in terms of practical computations. One of the neatest applications is to simply compute $f_K(A)$ for the trefoil knot T (see Figure 12) and determine that $f_K(A)$ is not equal to $f_K(A^{-1})$. This shows that the trefoil is not ambient isotopic to its mirror image, a fact that is quite tricky to prove by classical methods.

4   Topological Quantum Field Theory—First Steps

In order to further justify this idea of the amplification of Dirac notation, consider the following scenario. Let M be a 3-dimensional manifold. Suppose that F is a closed orientable surface inside M dividing M into two pieces, M_1 and M_2. These pieces are 3-manifolds with boundary. They meet along the surface F. Now consider an amplitude $\langle M_1 | M_2 \rangle = Z(M)$. The form of this amplitude generalizes our previous considerations, with the surface F constituting the distinction between the "preparation" M_1 and the "detection" M_2. This generalization of the Dirac amplitude $\langle a | b \rangle$ amplifies the notational distinction consisting in the vertical line of the bracket to a topological distinction in a space M. The amplitude $Z(M)$ will be said to be a *topological amplitude* for M if it is a topological invariant of the 3-manifold M. Note that a topological amplitude does not depend upon the choice of surface F that divides M.

From a physical point of view the independence of the topological amplitude of the particular surface that divides the 3-manifold is the most important property. An amplitude arises in the condition of one part of the distinction carved in the 3-manifold acting as "the observed" and the other part of the distinction acting as "the observer". If the amplitude is to reflect physical (read topological) information about the underlying manifold, then it should not depend upon this particular decomposition into observer and observed. The same remarks apply to 4-manifolds and interface with ideas in relativity. We mention 3-manifolds because it is possible to describe many examples of topological amplitudes in three dimensions. The matter of 4-dimensional amplitudes is a topic of current research. The notion that an amplitude be independent of the distinction producing it is prior to topology. Topological invariance of the amplitude is a convenient and fundamental way to produce such an independence.

This sudden jump to topological amplitudes has its counterpart in mathematical physics. In [21] Edward Witten proposed a formulation of a class of 3-manifold invariants as generalized Feynman integrals taking the form $Z(M)$ where

$$Z(M) = \int dA \exp\left(\frac{ik}{4\pi}S(M, A)\right).$$

Here M denotes a 3-manifold without boundary and A is a gauge field (also called a gauge potential or gauge connection) defined on M. The gauge field is a one-form on M with values in a representation of a Lie algebra. The group corresponding to this Lie algebra is said to be the *gauge group* for this particular field. In this integral the "action" $S(M, A)$ is taken to be the integral over M of the trace of the Chern-Simons three-form $CS = AdA + (2/3)AAA$. (The product is the wedge product of differential forms.)

Instead of integrating over paths, the integral $Z(M)$ integrates over all gauge fields modulo gauge equivalence. This generalization from paths to fields is characteristic of quantum field theory. Quantum field theory was designed in order to accomplish the quantization of electromagnetism. In quantum electrodynamics the classical entity is the electromagnetic field. The question posed in this domain is to find the value of an amplitude for starting with one field configuration and ending with another. The analogue of all paths from point a to point b is "all fields from field A to field B".

Witten's integral $Z(M)$ is, in its form, a typical integral in quantum field theory. In its content $Z(M)$ is highly unusual. The formalism of the integral and its internal logic support the existence of a large class of topological invariants of 3-manifolds and associated invariants of knots and links in these manifolds.

Invariants of three-manifolds were initiated by Witten as functional integrals in [21] and at the same time defined in a combinatorial way by Reshetikhin and Turaev in [20]. The Reshetikhin-Turaev definition proceeds in a way that is quite similar to the definition that we gave for the bracket model for the Jones polynomial in section 2. It is an amazing fact that Witten's definition seems to give the very same invariants. We are not in a position to go into the details of this correspondence here. However, one theme is worth mentioning: For k large, the Witten integral is approximated by those connections A for which $S(M, A)$ has zero variation with respect to change in A. These are the so-called *flat connections*. It is possible in many examples to calculate this contribution via both the functional integral and by the combinatorial definition of Reshetikhin and Turaev. In all cases, the two methods agree (see, e.g., [9]). This is one of the pieces of evidence in a puzzle that everyone expects will eventually justify the formalism of the functional integral. Note how this case corresponds exactly to the relation of classical and quantum physics as it was discussed in Section 2.

In order to obtain invariants of knots and links from Witten's integral, one adds an extra bit of machinery to the brew. The new machinery is the *Wilson loop*. The Wilson loop is an exponentiated version of integrating the gauge field along a loop K. We take this loop K in three space to be an embedding (a knot) or a curve with transversal self-intersections. It is usually indicated by the symbolism $tr(P\exp(\int_K A))$. Here the P denotes *path ordered integration*—that is, we are integrating and exponentiating matrix valued functions, and one must keep track of the order of the operations. The symbol tr denotes the trace of the resulting matrix.

With the help of the Wilson loop function on knots and links, Witten [21] writes down a functional integral for link invariants in a 3-manifold M:

$$Z(M, K) = \int dA \exp\left(\frac{ik}{4p}S(M, A)\right) tr\left(P\exp\left(\int_K A\right)\right).$$

Here S(M,A) is the Chern-Simons Lagrangian, as in the previous discussion.

If one takes the standard representation of the Lie algebra of $SU(2)$ as 2×2 complex matrices, then it is a fascinating exercise to see that the formalism of $Z(S^3, K)$ (where S^3 denotes the three-dimensional sphere) yields up the original Jones polynomial with the basic properties as discussed in section 2. See Witten's paper [21], or [16] or [17] for discussions of this part of the heuristics.

This approach to link invariants crosses boundaries between different methods. There are close relations between $Z(S^3, K)$ and the invariants defined by Vassiliev (See [3, 17].), to name one facet of this complex crystal.

This deep relationship between topological invariants in low-dimensional topology and quantum field theory in the sense of Witten's functional integral is really still in its infancy. There will be many surprises in the future as we discover that what has so far been uncovered is only the tip of an iceberg.

5 Caboose

We have, in this short paper, given an almost unbroken line of argument from the beginnings of quantum mechanics to the construction of topological quantum field theories and link invariants associated with quantum amplitudes. While the approach using the formalism of the functional integral gives a direct route into the heart of the subject, it involves a significant number of leaps of faith. These leaps are slowly being filled by rigorous mathematics. The algebraic approach via amplitudes is rigorous in its inception and has given rise to beautiful new theories of invariants of knots, links, and three-manifolds.

One of the most exciting prospects for these new invariants is the possibility of their application in quantum gravity. See [2] for an account of these developments. Many other applications are possible, and the subject is just beginning. For a survey of past and present applications of knots and links we refer the reader to [7].

References

[1] M. F. Atiyah. *The Geometry and Physics of Knots*. Cambridge University Press, 1990.

[2] J. Baez and J. P. Muniain. *Gauge Fields, Knots and Gravity*. World Sci. Press, 1994.

[3] D. Bar-Natan. On the Vassiliev knot invariants. *Topology*, 34(2):423–472, 1995.

[4] R. J. Baxter. *Exactly Solved Models in Statistical Mechanics*. Acad. Press, 1982.

[5] L. C. Biedenharn and J.D. Louck. *Angular Momentum in Quantum Physics—Theory and Application*. Cambridge University Press, 1979.

[6] P. A .M. Dirac. *Principles of Quantum Mechanics*. Oxford University Press, 1958.

[7] L. H. Kauffman (Editor). *Knots and Applications*. World Scientific Pub. Co., 1995.

[8] R. Feynman and A. R. Hibbs. *Quantum Mechanics and Path Integrals*. McGraw Hill, 1965.

[9] D. S. Freed and R. E. Gompf. Computer calculation of Witten's 3-manifold invariant. *Commun. Math. Phys.*, 141:79–117, 1991.

[10] V. F. R. Jones. A polynomial invariant for links via von Neumann algebras. *Bull. Amer. Math. Soc.*, 129:103–112, 1985.

[11] V. F. R. Jones. A new knot polynomial and von Neumann algebras. *Notices of AMS*, 33:219–225, 1986.

[12] L. H. Kauffman. *On Knots*. Number 115 in Annals of Mathematics Studies. Princeton University Press, 1987.

[13] L. H. Kauffman. State models and the Jones polynomial. *Topology*, 26:395–407, 1987.

[14] L. H. Kauffman. New invariants in the theory of knots. *Amer. Math. Monthly*, 95(3):195–242, March 1988.

[15] L. H. Kauffman. Statistical mechanics and the Jones polynomial. *AMS Contemp. Math. Series*, 78:263–297, 1989.

[16] L. H. Kauffman. *Knots and Physics*. World Scientific Pub., 1991 and 1993.

[17] L. H. Kauffman. Functional integration and the theory of knots. *J. Math. Phys.*, 36(5):2402–2429, 1995.

[18] L. H. Kauffman. Knots and statistical mechanics. In L. H. Kauffman, editor, *Proceedings of Symposia in Applied Mathematics—The Interface of Knots and Physics*, volume 51, pages 1–87, 1996.

[19] L. H. Kauffman and D. E. Radford. Invariants of 3-manifolds derived from finite-dimensional Hopf algebras. *Journal of Knot Theory and its Ramifications*, 4(1):131–162, 1995.

[20] N. Yu. Reshetikhin and V. G. Turaev. Invariants of 3-manifolds via link polynomials and quantum groups. *Invent. Math*, 103:547–597, 1991.

[21] E. Witten. Quantum field theory and the Jones polynomial. *Commun. Math. Phys.*, 121:351–399, 1989.

Bridges between Geometry and Graph Theory

Tomaž Pisanski*
Department of Mathematics
University of Ljubljana,
Ljubljana, Slovenia
Tomaz.Pisanski@fmf.uni-lj.si

Milan Randić
Drake University
Department of Mathematics
and Computer Science
Des Moines, Iowa

Abstract

Graph theory owes many powerful ideas and constructions to geometry. Several well-known families of graphs arise as intersection graphs of certain geometric objects. Skeleta of polyhedra are natural sources of graphs. Operations on polyhedra and maps give rise to various interesting graphs. Another source of graphs are geometric configurations where the relation of incidence determines the adjacency in the graph. Interesting graphs possess some inner structure which allows them to be described by labeling smaller graphs. The notion of covering graphs is explored.

1 Introduction

We assume that the reader is familiar with basic mathematics. For instance, we will not give any geometric introduction. We expect that the reader is familiar with concepts such as group, graph, matrix, and permutation, but we do not require any advanced knowledge of any of these topics. We do not give any rigorous definition of surface or map on a surface.

Books listed among our references, [5, 6, 7, 8, 11, 13, 14, 15, 18, 28, 29, 31, 32, 35, 37, 59, 66, 67, 68] provide a spectrum of background material ranging from motivating and introductory chapters for general readership up to advanced and rigorous monographs that can be used as a follow-up for those readers who would like to specialize their knowledge in a particular theme that we touch here.

*Work supported in part by grants J1-6161 and J2-6193 of the Ministry of Science and Technology of Slovenia.

2 Intersection Graphs

Given a set of n points $V = \{v_1, v_2, \ldots, v_n\}$ in some metric space and a positive number $r > 0$, we may draw n closed balls $B_i := B(v_i, r), i = 1, 2, \ldots, n$, each ball B_i centered at v_i and having radius r. Define a graph $G(V, r)$ as follows: The vertices are the n selected points. Two vertices v_i and v_j are adjacent iff the corresponding balls intersect, i.e., if $B_i \cap B_j \neq \emptyset$. The radius r will be called a *unit* and the graph the *unit sphere graph*.

We will use the notation 2D to represent the standard Euclidean plane and 3D to represent Euclidean 3-dimensional space; 1D can be used to denote the real line.

If we take the metric space 2D or 3D, we can easily represent a unit sphere graph G in the space itself. We represent the vertices by points and draw a line segment from v_i to v_j for any pair of adjacent vertices of the graph G.

Here are some specific examples:

Example 1 *Let us select the following points in the Euclidean plane: (x, y), $x \in \{1, 2, \ldots, a\}$, $y \in \{1, 2, \ldots, b\}$. Hence $n = ab$. Let $r = 0.5$. The unit sphere graph is the well-known $a \times b$ grid graph $Gr(a, b)$. The grid graph is also known in graph theory as the Cartesian product of the paths P_a and P_b. Figure 1 shows the case for $a = 5$ and $b = 3$.*

In order to understand better the nature of unit sphere graphs, one can do at least two experiments. One can keep the set of vertices V fixed and change the unit r. For $0 < r < 0.5$ we get the graph consisting of ab isolated vertices and no edges. For $0.5 \leq r < \sqrt{2}/2$ we get the grid graph (Figure 1). For $\sqrt{2}/2 \leq r < 2$ we get the graph that is obtained from $Gr(a, b)$ by adding the pair of diagonals in each unit square (Figure 2).

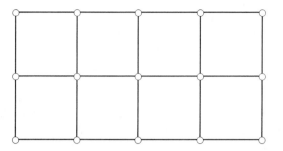

Figure 1: The grid graph $Gr(5,3)$. This is the same as the Cartesian product of paths P_5 and P_3.

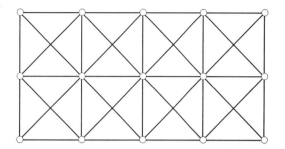

Figure 2: The strong product of paths P_5 and P_3.

This graph is also known as the *strong product* of paths P_a and P_b. Obviously a unit sphere graph with a smaller value of r is a subgraph of the unit sphere graph with a larger unit sphere value. However, to determine the number of distinct unit sphere graphs even for regular configurations of points such as the grid configuration seems to be nontrivial. For a given pair of integers $0 \le a \le b$ one can write a computer program that will determine the number $t(a,b)$ of distinct unit sphere graphs. For a given set V one can do the same thing and determine the number of distinct unit sphere graphs $t(V)$. One way of getting the result is to calculate the $n \times n$ symmetric distance matrix $D(V)$ with 0 on the main diagonal in which the entry in the i-th row and j-th column is given as the distance between v_i and v_j. Then $t(V)$ is simply the number of distinct values in $D(V)$. Since there are at most $n(n-1)/2$ edges in a graph with n vertices, there are at most $1+n(n-1)/2$ distinct unit sphere graphs. If we select n points in the real line 1D with coordinates $1, 2, 4, 8, \ldots, 2^{n-1}$, we can see that all $n(n-1)/2$ differences are distinct. The points give rise to the maximum possible number of unit sphere graphs. If we select n equidistant points on the line, we get only n distinct unit sphere graphs. The reader may prove that we cannot reduce this number. In 2D we can do much better if we select the points as the vertices of the regular n-gon. There are only $1+\lfloor n/2 \rfloor$ unit sphere graphs. For a given n, find the minimum in 2D and also in 3D.

Unit sphere graphs admit various generalizations and modifications. The most obvious one would be that each point v_i carries its own radius r_i. The balls $B_i := B(v_i, r_i)$ now have different radii; however, the condition remains the same. One can consider the case of n radio stations with different ranges broadcasting on various frequencies. There is interference if a listener can listen to two dif-

ferent radio stations broadcasting on the same frequency. We may model the situation by a sphere graph. Each vertex of the graph is assigned a frequency (color). The broadcasting has no interference if and only if the corresponding graph is properly colored (no two adjacent vertices are assigned the same color). Minimizing the number of frequencies is equivalent to finding the chromatic number of the graph (i.e., the least number of colors used in a proper coloring of the graph).

When setting up the model one has to select the radii. There is a natural choice for the radius of a vertex; let r_i of the vertex v_i be the distance to the nearest neighbor of v_i. This is how we get the *nearest neighbor* graphs.

Here is another possible generalization: To each point v_i we associate two radii $r_i \le R_i$ and two balls $b_i := B(v_i, r_i)$, $B_i := B(v_i, R_i)$. Let b_i^0 denote the interior of the ball b_i. The adjacency condition now reads:

$$v_i \equiv v_j \text{ if and only if } b_i^0 \cap b_j^0 = \emptyset \text{ and } B_i \cap B_j \neq \emptyset$$

Here are two examples:

Example 2 Chemistry. *We may model some molecules by assigning radii to various atoms. By placing atoms in a plane (or space) we then automatically get the graph of the molecule. Several computer programs for molecular dynamics such as RasMol [61] use such models; see also [55]. It can be shown that all benzenoid graphs can be described as unit sphere graphs in 2D. A graph is called a benzenoid graph[1] if it can be obtained by selecting a connected subset of hexagons in an infinite planar hexagonal lattice (representing graphite).*

---

[1] In chemistry a benzenoid graph is sometimes defined in a slightly different way. Namely, it is required to have at least one Kekule structure, i.e., a perfect matching. Also, a long string of hexagons that winds in a spiral may be considered a benzenoid even though distinct hexagons may project to the same hexagon in the graphite lattice.

Example 3 Touching coins and touching pennies, [31]. *Given a set of coins* $c_1, c_2, \ldots, c_n$ *in the plane, we may define the graph that has the n vertices in the centers of* $c_1, c_2, \ldots, c_n$ *and the i-th and j-th vertices are adjacent if and only if the coins* c_i *and* c_j *touch. In 1935 it was shown by Koebe that any planar graph can be realized as a coin graph. If we impose an additional condition that all coins have equal radius we obtain the so-called penny graph.*

Unit sphere graphs are a special case of intersection graphs. Given a family of sets $\{S_1, S_2, \ldots, S_n\}$ we define a graph on n vertices as follows: the vertex set is $\{S_1, S_2, \ldots, S_n\}$. Two vertices S_i and S_j are adjacent iff $S_i \cap S_j \neq \emptyset$. By selecting various geometric objects we get interesting families of graphs. For instance, the so-called *interval graphs* are intersection graphs of finite families of line segments in the 1D line.

If we select a direction on each edge of graph G the so-called *directed graph* or *digraph D* is obtained. Each digraph D corresponds to a binary relation R on the set of vertices of D. Two vertices a and b are related by R if and only if there is a directed arc in D from a to b. Graph G has a *transitive orientation* if the binary relation corresponding to the digraph is transitive.

Interval graphs are characterized by the following structural theorem, where a cycle in a graph is called *chordless* if it is an induced subgraph: no diagonal is an edge of the graph.

Theorem 1 *G is an interval graph if and only if it has no chordless cycle* C_n, *for* $n > 3$ *and its complement admits a transitive orientation.*

Unit sphere graphs in 1D are called unit interval graphs. They were characterized by F. S. Roberts, [60].

Theorem 2 *G is a unit interval graph if and only if it is an interval graph and has no induced* $K_{1,3}$ *subgraph, where* $K_{n,m}$ *is the complete bipartite graph with n vertices in one part and m vertices in the other part.*

It would be interesting and useful to characterize the unit sphere graphs in 2D and 3D. For instance, all platonic graphs arise as unit sphere graphs in 3D. One has to take the vertices of the corresponding platonic solid and radius r equal to half of the edge length. It appears that one cannot get the cube graph Q_3 as a unit sphere graph in 2D.

Finally, we defined a unit sphere graph for any metric space. Any connected graph admits the structure of a metric space. The points are vertices and the distance is the length of the shortest path from one vertex to another. It would be interesting to see what unit sphere graphs arise from a given connected graph. Let G be a graph and let G^i denote the graph obtained from G by joining two vertices u and v by an edge if and only if the distance between u and v in G is at most i. It can be shown that the unit sphere graphs arising from G are precisely all induced subgraphs of $G^i, i = 1, 2, \ldots$. Even among specialists in graph theory it is not widely known that a graph can allow additional metrics. Besides the common shortest path metric we also have the *resistance distance metric* and other metric alternatives [40, 41].

3 Polyhedra and Graphs

In the previous section we found a way from a polyhedron to a graph via unit sphere graphs. There is a much easier way of getting a graph from a polyhedron, obtained by taking the one-skeleton, i.e., the graph composed of vertices and edges of the polyhedron. We keep the vertices and the edges and forget all the other information (e.g., the facial structure or metric structure.) This route is quite interesting and gives among other things a very important family of graphs. A graph G is 3-connected if deletion of any pair of vertices results in a connected subgraph. It is planar if it can be drawn in the plane without crossings. Let us call a graph *polyhedral* if it is a one-skeleton of a convex polyhedron.

Theorem 3 *(Steinitz [63].) The one-skeleton of an arbitrary convex polyhedron is a planar 3-connected graph and each planar 3-connected graph is polyhedral.*

Example 4 *The one-skeleton of the octahedron is the complete tripartite graph* $K_{2,2,2}$ *on 6 ver-*

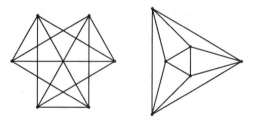

Figure 3: Two alternative drawings of $K_{2,2,2}$ (the graph of the octahedron).

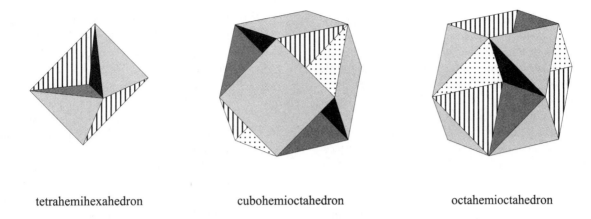

tetrahemihexahedron   cubohemioctahedron   octahemioctahedron

Figure 4: The heptahedron (left) is not convex, is self-intersecting, and even is non-orientable as a map in the projective plane. Its one-skeleton is the graph in Figure 3. The octahemioctahedron and cubohemioctahedron share with the cuboctahedron the same one-skeleton which is shown in Figure 10.

tices. One way of drawing this graph is to take the so-called Schlegel diagram of the octahedron. A Schlegel diagram of a convex polyhedron is obtained by first selecting a face and then using a stereographic projection from the center of that face onto a plane. Thus the selected face becomes the infinite face of the plane graph, which is the Schlegel diagram. For aesthetic reasons the resulting drawing is then homeomorphically changed in such a way that the faces are not accumulated too much in the center. However, we should always bear in mind that the graph G does not carry explicit information about the position of its vertices. We can represent the same graph by different drawings; see Figure 3.

The route back from a graph to a polyhedron is not so obvious. Recovering hidden or missing information is never as easy and obvious as throwing away information.

If we allow non-convex polyhedra we may get different polyhedra giving the same one-skeleton. The *uniform polyhedron* on the left in Figure 4 is a model of a projective plane. In [66] it is called a *heptahedron*. However, it is better known as a *tetrahemihexahedron*; see [49], where you can find more information about and illustrations of uniform polyhedra. It has 6 vertices, 12 edges, 4 triangles, and 3 squares. Its one-skeleton is again $K_{2,2,2}$; see Figure 3.

Figure 5 shows its embedding in the projective plane. The antipodal points of the disk are identified.

If we start with a cuboctahedron, we can obtain two different polyhedra depending on what types of faces are replaced by hexagons.

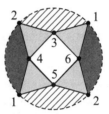

Figure 5: $K_{2,2,2}$ embedded in the projective plane with 4 triangles and 3 squares. The antipodal points on the dotted border are identified.

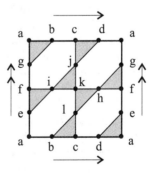

Figure 6: The map of the octahemioctahedron in the torus with 8 triangles and 4 hexagons. The arrows and double arrows show how to identify the sides of the rectangle in order to form the torus.

By keeping all triangles and replacing the quadrilaterals by main hexagons, we get an *octahemioctahedron* whose map on the torus is shown in Figure 6. On the other hand one may keep quadrilaterals and remove triangles. The polyhedron is called a *cubohemioctahedron* [49]. It consists of 6 quadrilaterals and 4 hexagons.

By keeping information about faces, we get the so-called *2-skeleton* of a polyhedron. It has vertices, edges, and faces and all the information about how to glue the pieces together.

Informally, a *map* is a collection of fused polygons. More rigorously, a *map* is a collection of polygons with directed sides, such that the total number of sides is even, together with an involution on the set of sides without fixed points, which determines how the sides are pairwise glued (respecting the orientation). If the resulting complex is connected, we get a surface. (For an exact definition and more examples, see the book by Ringel [59]; see also [67].)

4 Operations on Maps

A map is a combinatorial representation of a closed surface. There are several equivalent combinatorial descriptions available for maps. In the computer package Vega [57], we implemented a series of operations on maps. This enables us to produce new maps from old ones. In turn, we can get new polyhedra or new graphs.

Here we present some operations on maps; all of them are explained via examples in the figures. We consider only connected maps.

- *Du*: *Dual.* This operation is well known for planar graphs. However, it can be generalized for maps in other surfaces. It is also known as the *Poincaré dual*. The dual map $Du(M)$ is built from the original map M as follows: we put a vertex of $Du(M)$ in the center of each face of M. For each edge e of M we produce its dual $Du(e)$ so that $Du(e)$ connects the vertices corresponding to the faces of M that have e on the common boundary. We place the vertex $Du(M)$ in the same surface as M in such a way that the faces of $Du(M)$ correspond to the vertices of M. This means that the dual edges are traversed along faces in the same cyclic order as the original edges are traversed cyclically around a vertex. It can be shown that $Du(Du(M)) = M$.

    Let us consider three examples that will serve us also for other operations.

1. The cube Q_3 in the sphere: $v = 8$, $e = 12$, $f = 6$. The dual is the octahedral graph $K_{2,2,2}$ in the sphere: $v = 6$, $e = 12$, $f = 8$; see Figure 8.

    A projection of a sphere-like polyhedron on a plane is sometimes called a *Schlegel diagram*. Such a projection is similar to the so-called stereographic projection in which exactly one point on a sphere, called the *center*, is mapped to infinity. In a polyhedron the center is usually taken either at a vertex, at the center of an edge or at a center of a face, see Figure 7.

2. The *bouquet* B_n of n circles is a graph with n loops attached to a single vertex. The bouquet of one circle B_1 in the projective plane: $v = 1$, $e = 1$, $f = 1$. It is self-dual: $v = 1$, $e = 1$, $f = 1$.

3. Another example of a self-dual map is the tetrahedron K_4 in the sphere.

4. K_4 in the torus: $v = 4$, $e = 6$, $f = 2$. One face is a quadrilateral and the other one is a hexagon. The dual graph has two vertices; one vertex has a double loop and there are four parallel edges between the two vertices: $v = 2$, $e = 6$, $f = 4$. All faces are triangles.

If v, e, f are the parameters of the original and v', e', f' are the parameters of the transformed map we have the relations:

$$v' = f$$
$$e' = e$$
$$f' = v.$$

- *Su1*: *1-dimensional subdivision.* This is the simplest of all. We subdivide each edge by inserting a vertex at the midpoint of each edge thus splitting the edge in two; see Figure 9.

1. The cube Q_3 in the sphere: $v = 8$, $e = 12$, $f = 6$. The 1-dimensional subdivision has the parameters: $v = 20$, $e = 24$, $f = 6$.

2. The bouquet of one circle B_1 in the projective plane: $v = 1$, $e = 1$, $f = 1$. Its subdivision $Su1$ is the cycle C_2 in the same surface: $v = 2$, $e = 2$, $f = 1$.

3. K_4 in the torus: $v = 4$, $e = 6$, $f = 2$. The subdivision graph has parameters $v = 10$, $e = 12$, $f = 2$. All faces have twice as many edges as before.

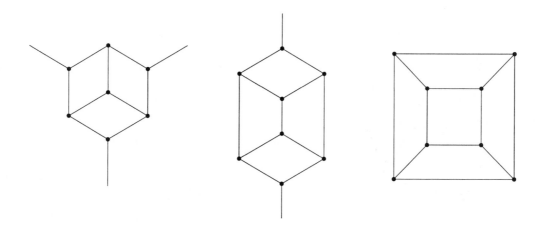

Figure 7: Schlegel diagrams of cube centered at (a) a vertex, (b) an edge, (c) at a face.

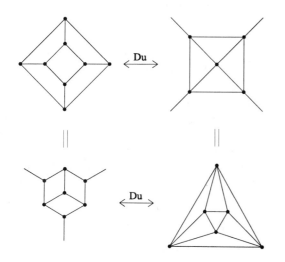

Figure 8: The cube and its dual, the octahedron.

If v, e, f are the parameters of the original and v', e', f' are the parameters of the transformed map we have the relations:

$$v' = v + e$$
$$e' = 2e$$
$$f' = f.$$

- *Pa*: *Parallelization*. Here we replace each edge by a pair of parallel edges forming a digon on the surface. Instead of giving this description we could define Parallelization formally as: $Pa(M) := Du(Su1(Du(M)))$. The parame-

ters are transformed as follows:

$$v' = v$$
$$e' = 2e$$
$$f' = e + f.$$

- *Si*: *Simplification*. This operation is in a sense the inverse to the operation of $Su1$. It removes all vertices of valence 2 but leaves the graph (and the map) topologically the same. The only exception is the bouquet B_1 which does not allow further reductions.

- *PSi*: *Parallel Simplification*. This operation is in a sense the inverse to the operation of Pa. It removes all digons by changing them into a single edge. $PSi(M) := Du(Si(Du(M)))$

- *Me*: *Medial*. This is probably the most important transformation of a map and is not so simple. The new vertices are the midpoints of the original edges. New vertices are adjacent if and only if the two original edges span an angle; i.e., the two edges must be incident and consecutive when traversing a rotation path about their common vertex in the map. The medial graph is thus a subgraph of the *line-graph*. For the definition of line-graph, see for instance, [68]. In the line-graph each original vertex gives rise to a complete graph; in the medial graph it only gives rise to a cycle.

If v, e, f are the parameters of the original and v', e', f' are the parameters of the transformed

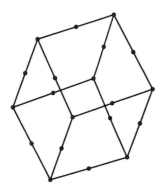

Figure 9: The subdivided cube.

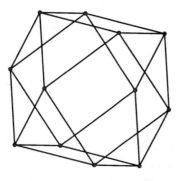

Figure 10: The cuboctahedron, the medial of the cube.

map we have the relations:

$$v' = e$$
$$e' = 2e$$
$$f' = v + f.$$

1. The cube Q_3 in the sphere: $v = 8$, $e = 12$, $f = 6$. Its medial is the cuboctahedron and has the parameters: $v = 12$, $e = 24$, $f = 14$; see Figure 10.

2. The bouquet of one circle B_1 in the projective plane: $v = 1$, $e = 1$, $f = 1$. Its medial is B_2 in the same surface: $v = 1$, $e = 2$, $f = 2$.

3. K_4 in the torus: $v = 4$, $e = 6$, $f = 2$. The medial graph Me is obtained from $K_{2,2,2}$ by doubling 8 edges with one endpoint in the same color class with parameters $v =$

6, $e = 12$, $f = 6$. There are 4 triangles, one quadrilateral and one octagon.

Medials have interesting properties:

- Each one is isomorphic to the medial of the dual:

$$Me(G) = Me(Du(G))$$

- All are 4-valent and their duals are bipartite.

- The structure of the map and its dual are visible in the medial.

- Face lengths and vertex valencies are readily visible in the medial.

- The map and its dual occur symmetrically in the medial.

- *Tr: Truncation*

  Truncation can be first described intuitively. We cut off the neighborhood of each vertex by a plane "close" to the vertex that meets each edge incident to the vertex. Using transformations from above we may define truncation Tr to be $Tr(G) = PSi(Me(Su1(G)))$. The medial of a 1-subdivided map would introduce parallel edges and digons, so we insist that the operation is followed by parallel simplification. Compare [20].

- *Su2: 2-dimensional subdivision*

  The 2-dimensional subdivision of a graph is obtained by adding a vertex in the center of each face and joining it by edges to the vertices of the original face; see Figure 11.

  There is also an interesting connection between truncation and the 2-dimensional subdivision, namely, $Tr(Du(G)) = Du(Su2(G))$; see for instance [40]. Therefore, $Su2$ can be defined in terms of truncation:

$$Su2(G) = Du(Tr(Du(G)))$$
$$= Du(PSi(Me(Su1(Du(G))))).$$

- BS: *Barycentric subdivision*

  The barycentric subdivision is a composite operation: It is a 2-dimensional subdivision of the 1-dimensional subdivison.

$$BS(G) = Su2(Su1(G))$$
$$= Du(PSi(Me(Su1(Du(Su1(G)))))).$$

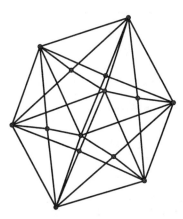

Figure 11: The two-dimensional subdivision of the cube.

The barycentric subdivision has an important role in topology and its dual is also very interesting, as it is a cubic graph and thus much easier to visualize.

- Pe: *Petrie Dual*

The Petrie dual is the first operation that keeps the 1-skeleton fixed and only changes the faces of the map. New faces are composed of all left-right walks on the surface: every two consecutive edges of the new faces are also consecutive edges of the old faces. However, no three consecutive edges of new faces form consecutive edges of the old faces. It can be shown that it is indeed a "dual" in the sense that $Pe(Pe(G)) = G$. The Petrie dual is a powerful operation as it can turn an orientable surface into a nonorientable one and vice versa.

For instance, the Petrie dual of the ordinary cube Q_3 gives its embedding in the torus with all hexagonal faces. However, there is another, different hexagonal embedding of Q_3 in the torus; see Figure 12. This may run counter to our intuition, since Q_3 in particular has only one drawing in the sphere.

- R2: *Rotation Square*

The rotation square changes faces. It is well-defined only for graphs that have all vertices of odd valence. The map is completely defined if we specify its *rotation scheme*, i.e., at each vertex we give the cyclic order of incident edges. Rotation square takes the square of the overall permutation. If $(e_1, e_2, \ldots, e_s)$ is the local rotation at a vertex, then we have in the rotation square the following sequence:

$(e_1, e_3, e_5, \ldots, e_2, e_4, \ldots)$. That is why we need s to be odd.

- S2: *Embedded Square*

The idea behind the embedded square is similar to the idea of the rotation square; however, the construction is quite different and the resulting graph is quite different from the original. The graph is obtained by keeping the vertices of the original and adding an edge at each angle. If we have a face in the original that has $(v_1, v_2, \ldots, v_s)$ as a sequence of vertices then the new edges are $v_1 - v_3, v_2 - v_4, v_3 - v_5, \ldots$.

- Sn1 and Sn2: *Snub*

There are two snub operations. First we take two consecutive medials $Me(Me(G))$. The resulting map is 4-valent and also equipped with a collection of quadrilaterals, arising from vertices of the first medial operation $Me(G)$, such that each vertex of $Me(Me(G))$ belongs to two quadrilaterals. By inscribing a diagonal in one of the quadrilaterals one induces diagonals in all remaining quadrilaterals. If the other initial diagonal is selected, the resulting map is $Sn2(G)$ instead of $Sn1(G)$. When $Sn1$ and $Sn2$ are isomorphic we simply have a snub Sn. For instance, for a tetrahedron T, $Sn(T) = Sn1(T) = Sn2(T)$ is topologically equivalent to the icosahedron I but $Sn1(I) \neq Sn2(I)$. There are five Platonic polyhedra, with all vertices, all edges, and all faces mutually equivalent. There are 13 Archimedean polyhedra in which all vertices are equivalent but neither the edges nor the faces are. All Platonic and Archimedean polyhedra can be obtained from the tetrahedron T by some sequence of operations that we have introduced; see Figures 13 and 14.

- Le: *Leapfrog*

Leapfrog [20] is a term coined by chemists. It represents a composite operation. It is the truncation of the dual:

$$
\begin{aligned}
Le(G) &= Tr(Du(G)) \\
&= PSi(Me(Su1(Du(G)))).
\end{aligned}
$$

It can be described in an intuitive way. Recall how we envision truncation. The process of truncation involves a collection of planes, one for each vertex, that cut off parts of the polyhedron close to each vertex. If we "move" these planes towards the center of polyhedron, each

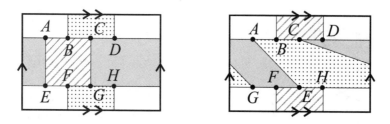

Figure 12: Two all-hexagonal drawings of the cube Q_3 on the torus.

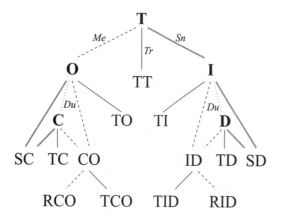

Figure 13: Evolution of the Platonic and Archimedean polyhedra from the tetrahedron.

original edge is diminished and is at a certain moment reduced to a single point, the midpoint. The polyhedron obtained at that moment is the medial. If we continue the process beyond that point, we obtain the leapfrog. Similar intuitive processes can be applied to an arbitrary map yielding the same result.

Patchwork

Consider the polyhedron with the seams of the traditional soccer ball as edges. This is the truncated icosahedron, one of the trivalent Archimedean polyhedra, having 60 vertices and faces that are pentagons and hexagons. Furthermore, the pentagons are colored black and the hexagons white as in the soccer ball. The black faces are vertex disjoint but each of the 60 vertices lies on the boundary of a black face. Such a collection of faces is called a *patchwork*. There are many trivalent polyhedra with pentagonal and hexagonal faces. They are called *fullerenes*. The name was suggested by Harry Kroto, one of the 1996 Nobel prize winners and discoverers of C_{60}, which is a novel allotropic form of carbon different from the well-known forms

graphite and diamond. They are named after the architect Buckminster Fuller, in appreciation for his pioneering work on design and construction of huge domes based on geometrical considerations. A fullerene with a patchwork is called a *generalized soccer ball*. Note that in generalized soccer balls in addition to black pentagons there will be some black hexagons. The following is a characterization [20] of generalized soccer balls:

Theorem 4 *A fullerene is a generalized soccer ball if and only if it is a leapfrog of another fullerene.*

It is known that fullerenes with abutted pentagons are not chemically stable and are therefore of little interest while generalized soccer balls have no abutted pentagons.

Triangulations

If a polyhedron has some non-triangular faces, we may introduce additional diagonals until all faces are triangles. The corresponding graph is known as a *maximally planar graph* or *triangulation*. Note that the introduction of any additional edge to such a graph necessarily produces a non-planar graph, i.e., a graph that cannot be drawn in the plane without crossing of lines.

Diagonal flips [52]

It can be shown that any two triangulations on the same number of vertices are equivalent under diagonal flips: this means that one can start with one triangulation and obtain any other by a series of diagonal flips.

The dual of a triangulation is a cubic polyhedron. The dual operation of the diagonal flip is sometimes referred to in chemistry as the Stone-Wales transformation.

Before we leave the subject let us think a bit about the analogy of a self-intersecting projective planar polyhedron and a drawing of a non-planar

Symbol	Polyhedron	Operation	Formula
T	Tetrahedron	Primitive	—
O	Octahedron	Medial	$Me(T)$
C	Cube(Hexahedron)	Dual	$Du(O) = Du(Me(T))$
I	Icosahedron	Snub	$Sn(T)$
D	Dodecahedron	Dual	$Du(Sn(T))$
TT	Truncated tetrahedron	Truncation	$Tr(T)$
TO	Truncated octahedron	Truncation	$Tr(O) = Tr(Me(T))$
TC	Truncated cube	Truncation	$Tr(C) = Tr(Du(Me(T)))$
TI	Truncated icosahedron	Truncation	$Tr(I) = Tr(Sn(T))$
TD	Truncated dodecahedron	Truncation	$Tr(D) = Tr(Du(Sn(T)))$
CO	Cuboctahedron	Medial	$Me(C) = Me(Me(T))$
ID	Icosidodecahedron	Medial	$Me(I) = Me(Sn(T))$
TCO	Truncated cuboctahedron	Truncation	$Tr(CO) = Tr(Me(Me(T)))$
TID	Truncated icosidodecahedron	Truncation	$Tr(ID) = Tr(Me(Sn(T)))$
RCO	Rhombicuboctahedron	Medial	$Me(CO) = Me(Me(Me(T)))$
RID	Rhombicosidodecahedron	Medial	$Me(ID) = Me(Me(Sn(T)))$
SC	Snub cube	Snub	$Sn(C) = Sn(Du(Me(T)))$
SD	Snub dodecahedron	Snub	$Sn(D) = Sn(Du(Sn(T)))$

Figure 14: Derivation of the Platonic and Archimedean polyhedra from the tetrahedron.

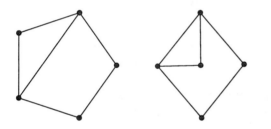

Figure 15: Two embeddings of a graph in the plane.

graph in the plane. In both cases topologists would call such a phenomenon an *immersion*. If there are no singularities (no self-intersections) then we speak about *embeddings*. Clearly only planar graphs allow embeddings in the plane. Actually each planar graph without loops and parallel edges admits a straight-line embedding in the plane—this is known as Farey's Theorem—although the embedding is not necessarily unique as shown in Figure 15.[2]

On the other hand, every graph can be embedded in 3D using only straight-line segments as edges. We can pose the following optimization problem. Let us take the three-dimensional grid of the shape of the cube with $n \times n \times n$ points. The problem is to select k out of the n^3 points in such a way that the vertices of the complete graph K_k can be placed

on the selected points and that would result in the embedding of K_k in 3D with line-segments representing edges meeting only at points representing vertices. For a given n what is the largest possible k? For $n = 1$ it is obviously $k = 1$ and for $n = 2$ we get $k = 5$. It seems that for $n = 3$ we get $k = 6$. What is the value of k for $n = 4$?

5   Symmetry and Orbits

Platonic and Archimedean polyhedra are highly symmetric. Each of them has indistinguishable vertices. However, the edges of Archimedean polyhedra are not all alike. This phenomenon can be understood by the concept of *orbits*. We will explain this concept first in graphs.

Let $G = (V, E)$ be a graph with the vertex set V and edge set E. A permutation $\pi : V \to V$ is an *automorphism* of G if it preserves adjacencies: $(u, v) \in E$ if and only if $(\pi(u), \pi(v)) \in E$. The automorphisms of G form a group that we denote by $Aut(G)$. Two vertices u and v of G are *indistinguishable* if one can be mapped to the other by some graph automorphism. The set of indistinguishable vertices form an *orbit* of $Aut(G)$. If all vertices of G are indistinguishable we say that G is *vertex-transitive*. We may also view automorphisms of G acting on the edges of G.

The graph of a regular prism is obviously vertex-transitive but in general not edge-transitive. It is impossible to distinguish among the vertices, but one can tell lateral edges apart from the base edges

---

[2]By Whitney's theorem, any planar 3-connected graph admits a unique embedding on the sphere. A graph G other than the triangle K_3 is 3-connected if the graph obtained by deleting any two vertices from G remains connected.

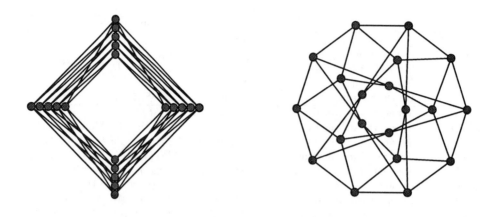

Figure 16: The standard drawing of the Folkman graph and another drawing of the same graph.

unless the prism is a cube.

There are graphs that have all edges indistinguishable but do not have all the vertices alike. One of the graphs is the so-called Folkman graph. The Folkman graph is therefore edge-transitive but not vertex-transitive. See Figure 16.

There is another notion that we will use later; it is called arc-transitivity. A graph G is called *arc-transitive* if for any pair of edges $(u, v) \in E$, $(u', v') \in E$ one can find two automorphisms π and π' such that $\pi(u) = u', \pi(v) = v'$ and $\pi'(u) = v'$ and $\pi'(v) = u'$. Hence the notion of arc-transitivity is stronger than edge-transitivity. It means that we can map any edge to any other edge in an arbitrary direction. Clearly arc-transitivity implies edge-transitivity.

6   Configurations and Levi Graphs

A *configuration* of type (n_r, b_k) is an ordered pair $(P, \mathcal{B})$, consisting of a set of n *points* $P = \{p_1, \ldots, p_n\}$ and a collection of b *lines (blocks)* $\mathcal{B} = \{B_1, \ldots, B_b\}$ with $B_i \subseteq P$, for each $i = 1, \ldots, n$, and $\pi(p_j) := \{B \in \mathcal{B} | p_j \in B\}$, for each $j = 1, \ldots, n$, such that

1. Each line contains k points: $|B_i| = k$, for each $i = 1, \ldots, b$;

2. Each point lies on r lines: $|\pi(p_j)| = r$, for each $j = 1, \ldots, n$;

3. Two different points are connected by at most one line: $|B_i \cap B_j| \leq 1$, for each $i \neq j, i = 1 \ldots, b, j = 1, \ldots, b$.

The ordered pair (p, B) with $p \in P, B \in \mathcal{B}$ and $p \in B$ is called a *flag*. If $k = r$ and hence $n = b$ (see, for instance [18]) the configuration is called *symmetric* and its type is denoted by n_k.

Let $\Pi := \{\pi(p) | p \in P\}$, where π is defined above. Clearly the ordered pair $(\mathcal{B}, \Pi)$ forms a configuration which is called the *dual* of $(P, \mathcal{B})$. In the dual configuration we only reverse the role of points and lines. Since there is a natural bijection $\pi : P \to \Pi$ we can write $(\mathcal{B}, P)$ instead. The term duality is appropriate here since applying duality twice in a row gives a configuration that is isomorphic to the original one. We can state that more formally as follows.

Let $C = (P, \mathcal{B})$ be a configuration. A map $\alpha : (P, \mathcal{B}) \to (P, \mathcal{B})$ with $\alpha(P) = \mathcal{B}$ and $\alpha(\mathcal{B}) = P$ which respects the incidence structure, namely for each $p \in P$ and for each $B \in \mathcal{B}$ there is $p \in B$ if and only if $\alpha(B) \in \alpha(p)$, is called an *anti-automorphism*. A configuration which admits an *anti-automorphism* is called *self-dual*. An anti-automorphism of order 2 is called a *polarity*. A configuration which admits a polarity is called *self-polar*.

Let C be a configuration; then $A(C)$ denotes the group of all its automorphisms and anti-automorphisms, while $Aut(C)$ denotes the group of its automorphisms. Then $Aut(C)$ is a subgroup of $A(C)$. If C is self-dual, it is a proper subgroup; otherwise, $Aut(C) = A(C)$.

If $Aut(C)$ is transitive on points P, lines $\mathcal{B}$, or flags, C is called *point-*, *line-*, or *flag-transitive* respectively.

A *triangle* of a configuration consists of three points, say a, b, and c, such that each of the three pairs $\{a, b\}$, $\{b, c\}$, and $\{a, c\}$ is contained in a line. A configuration that has no triangles is called a

Configuration $(P, \mathcal{B})$	Levi graph $L(P, \mathcal{B})$
incidence structure	bipartite graph
1-configuration [27]	girth ≥ 6
n_r, b_k	semi-regular, girth ≥ 6
n_3	bipartite, cubic, girth ≥ 6
points	black vertices
lines (blocks)	white vertices
no triangles	girth ≥ 8
self-dual	$\gamma \in Aut(L)$ swaps black and white vertices
self-polar	$\gamma \in Aut(L)$ of order 2 swaps black and white vertices
indecomposable	connected
incidence matrix	bi-adjacency matrix
point-transitive	all black vertices in the same orbit under $Aut(L)$
line-transitive	all white vertices in the same orbit under $Aut(L)$
flag-transitive	edge-transitive
flag-transitive, self-dual	arc-transitive
point-transitive, self-dual	vertex-transitive
cyclic	Haar graph of girth ≥ 6

Figure 17: Properties of configurations and their Levi graphs.

triangle-free configuration.

If there is a cyclic subgroup of $Aut(C)$ which acts transitively on P, the configuration C is called *cyclic*. Cyclic configurations can be described in a concise way. We only have to specify the number of points n and provide the 0-th line. The i-th line is then obtained by adding $i \pmod n$ to the index of each element of the 0-th line.

The *Levi graph* $L = L(P, \mathcal{B})$ of a configuration was introduced by Coxeter in 1950, see [12]. It is a bipartite graph with "black" vertices P and "white" vertices $\mathcal{B}$ and with an edge between $p \in P$ and $B \in \mathcal{B}$ if and only if $p \in B$. Note that dual configurations have the same Levi graph but the roles of black and white vertices are interchanged.

Complete information about a configuration can be recovered from its Levi graph with a given black and white coloring of vertices. Hence counting configurations can be done via careful counting of Levi graphs. Two configurations $(P, \mathcal{B})$ and $(P', \mathcal{B}')$ are isomorphic if and only if there is an isomorphism between their Levi graphs $L(P, \mathcal{B})$ and $L(P', \mathcal{B}')$ that maps black vertices into black vertices.

It is well known [12] that a graph G is a Levi graph of some n_3 configuration if and only if:

1. G is cubic;

2. G is bipartite;

3. The length of the shortest cycle in G, the girth, is at least 6.

Figure 17 shows the correspondence between properties of configurations and their Levi graphs.

Recently the number of n_3 configurations up to $n \leq 18$ was computed [4].

Example 5 *Each regular graph of valence k without loops and multiple edges can be viewed as an (n_k, b_2) configuration of points and lines. Vertices correspond to points and edges correspond to lines (line segments). The corresponding Levi graph is the subdivision graph of the original graph.*

Example 6 *Take the equilateral triangle with all three altitudes and the inscribed circle. The configuration consists of the 7 points, 6 line segments and the circle; see Figure 18. The configuration graph is the well-known Heawood graph; see Figure 19. This graph on 14 vertices has shortest cycle of length 6. The configuration is interesting as it represents the smallest model of a finite projective plane, the well-known Fano plane. This is the unique and smallest configuration of points and lines.*

Any abstract configuration can be described as a collection of sets. The elements represent points and the sets represent lines. The Fano configuration can thus be described as:

$$\{\{1, 2, 6\}, \{2, 3, 4\}, \{1, 3, 5\}, \{1, 4, 7\},$$
$$\{2, 5, 7\}, \{3, 6, 7\}, \{4, 5, 6\}\}$$

Clearly it is isomorphic to the configuration:

$$\{\{1, 2, 3\}, \{1, 4, 5\}, \{1, 6, 7\}, \{2, 4, 6\},$$
$$\{2, 5, 7\}, \{3, 4, 7\}, \{3, 5, 6\}\}$$

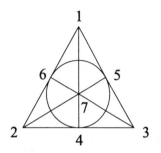

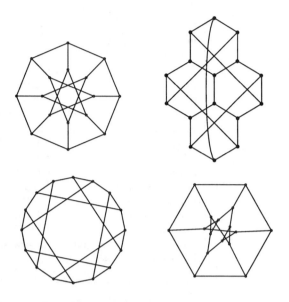

Figure 18: The Fano configuration with 7 points and 7 lines. The seventh line is drawn as a circle.

Figure 20: Four views of the Möbius-Kantor graph $GP(8,3) = H(133)$. A modification of the upper right-hand drawing can be found in [36].

Figure 19: The Levi graph of the Fano configuration, also known as the Heawood graph or the 6-cage.

7   Generalized Petersen Graphs and Haar Graphs

Since the Fano configuration is cyclic, it can also be generated by $n = 7$ and 0-th row $\{0, 2, 6\}$.

$$\{\{0,2,6\}, \{1,3,0\}, \{2,4,1\}, \{3,5,2\},$$
$$\{4,6,3\}, \{5,0,4\}, \{6,1,5\}\}.$$

It can be verified that the Fano configuration is self-dual; actually it is self-polar.

Example 7 *Take three lines a, b, and c forming a triangle ABC. The corresponding configuration graph is the complete bipartite graph $K_{3,3}$. In general, each n-gon defines a self-dual configuration of type n_2.*

Here is another example.

The Möbius-Kantor configuration is the only 8_3 configuration. It is also transitive and self-dual. It can be shown that the Möbius-Kantor configuration cannot be realized in the real projective plane. It can be realized in the complex projective plane. Figure 20 shows four drawings of its Levi graph. The top left drawing reminds us of the well-known Petersen graph; see Figure 26. That is why we introduce the family of generalized Petersen graphs.

For a positive integer $n \geq 3$ and $1 \leq r < n/2$, *the generalized Petersen graph $GP(n,r)$ has vertex set* $\{u_0, u_1, \ldots, u_{n-1}, v_0, v_1, \ldots, v_{n-1}\}$ *and edges of the form $u_i v_i, u_i u_{i+1}, v_i v_{i+r}$, where $i \in \{0, 1, \ldots, n-1\}$ with arithmetic modulo n.*

In [21] the automorphism group of $GP(n,r)$ was determined for each n and r. With the exception of the dodecahedron $GP(10,2)$, the generalized Petersen graph $GP(n,r)$ is vertex-transitive, if and only if $r^2 \equiv \pm 1 (\text{mod } n)$; see also [43]. It was also shown [21] that $GP(n,r)$ is arc-transitive if and only if

$$(n,r) \in \{(4,1), (5,2), (8,3), (10,2),$$
$$(10,3), (12,5), (24,5)\}.$$

Note that $GP(4,1)$ is the cube Q_3. On the other hand, $GP(5,2)$ is the Petersen graph and $GP(8,3)$ is known as the Möbius-Kantor graph [8], since it is the Levi graph of the unique 8_3 configuration. Similarly, $GP(10,3)$ is the Levi graph of the Desargues configuration 10_3 and $GP(12,5)$ is the Levi graph of one of the 229 configurations of type 12_3 [27].

Recall that any cyclic configuration of type n_k can be described by n and the first line $T \subseteq \{0, 1, \ldots, n-1\}, |T| = k, 0 \in T$. This, in turn, can be put in one-to-one correspondence with a positive

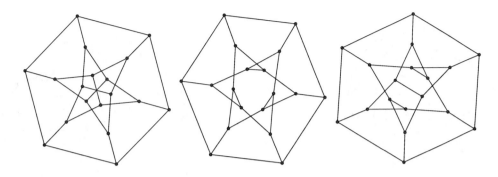

Figure 21: Levi graphs for the three 9_3 configurations. Left, the Pappus configuration; center a Haar graph $H(261)$.

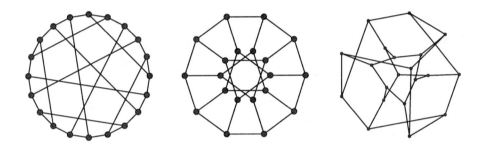

Figure 22: Levi graph of the Desargues configuration. This is the same as $GP(10,3)$. See also Figure 28.

integer N via its binary notation,

$$N = b_0 2^{n-1} + \cdots + b_{n-2} 2 + b_{n-1},$$

by letting $t \in T$ if and only if $b_t = 1$. Hence the complete information about a cyclic configuration and its Levi graph can be encoded by a positive integer N. In this way we get a graph $H(N)$ for each integer N which is called the *Haar graph* of N [33, 34]. If its girth is at least 6 then $H(N)$ is a Levi graph of a cyclic configuration.

In order to make the definition clear, let us construct the Heawood graph as the Haar graph $H(69)$. Since 69 is written in binary as 1000101 we conclude that $H(69)$ contains 7 "black" vertices, say, $v_0, v_1, \ldots, v_6$ and seven "white" vertices $u_0, u_1, \ldots, u_6$. Furthermore, vertex v_0 is adjacent to vertices u_0, u_4, and u_6 and vertex v_i is adjacent to vertices u_{0+i}, u_{4+i}, and u_{6+i}, where addition is taken mod 7. Note that we constructed in passing the Fano configuration:

$$\{\{0, 4, 6\}, \{1, 5, 0\}, \{2, 6, 1\}, \{3, 0, 2\},$$
$$\{4, 1, 3\}, \{5, 2, 4\}, \{6, 3, 0\}\}.$$

Of course $GP(2m + 1, 1)$ does not have a Haar graph representation, whereas $GP(2m, 1) =$

$H(2^{2m-1} + 3)$ and $GP(8, 3)$, the only other generalized Petersen graph that is a Haar graph, is isomorphic to $H(133)$. Its automorphism group is of interest in topological graph theory [65].

We specialize the result from [45] to Levi graphs and combine it with common knowledge.

Proposition 1 *The graph $GP(8,3)$ is the only generalized Petersen graph that is a Levi graph of a cyclic configuration [45]. The other two generalized Petersen graphs that are Levi graphs of point-transitive, self-dual configurations are $GP(10,3)$ and $GP(12,5)$.*

There are three 9_3 configurations of points and lines of type 9_3; see Figure 21. The most famous, called the Pappus configuration, is transitive and self-dual.

There are 10 configurations of points and lines of type 10_3. Only one of the 10 configurations is non-realizable in 3D. (It is realizable over the quaternions!) It is given as follows:

$$\{\{1, 2, 3\}, \{1, 6, 10\}, \{1, 7, 9\}, \{2, 4, 10\},$$
$$\{2, 5, 6\}, \{3, 4, 9\}, \{3, 5, 7\},$$
$$\{4, 5, 8\}, \{5, 7, 8\}, \{8, 9, 10\}\}.$$

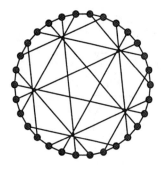

Figure 23: The Levi graph of the Cremona-Richmond configuration.

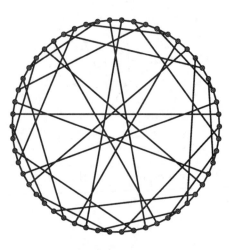

Figure 24: The Gray graph, which is the Levi graph of a flag-transitive non-self-dual 27_3 configuration.

The most important of the ten configurations is the Desargues configuration which is also transitive and self-dual; see Figure 22.

There are 31 configurations of type 11_3 and 229 of type 12_3. These facts were established more than hundred years ago by Daublebsky von Sterneck, [16], [17], although he missed one 12_3 configuration. Only recently it has been shown [64] that all these configurations admit realizations in 3D using integer coordinates.

The Cremona-Richmond configuration with 15 points and lines is the smallest n_3 configuration with no triangles [66]; see Figure 23.

For a more thorough introduction to the interesting area of configurations the reader is referred to the work of Harald Gropp; see, for instance, [24], [23], [26], [25]. Algorithmic aspects are covered in [6] and [64].

Configurations used to play important role in geometry. For instance, the following theorem is one of the main results of the PhD thesis of renowned German geometer E. Steinitz [62]:

Theorem 5 *Each n_3 configuration can be drawn in a plane with at most one curved line.*

For instance, the Fano configuration cannot be drawn with all lines straight, but not more than one curved line is needed; see Figure 18. A modern version of Theorem 5 can be found in algorithmic form in [6] and has also been implemented as a computer program [57].

Alspach and Zhang [1] proved that every cubic Cayley graph of a dihedral group is Hamiltonian. Compare also [2]. This result covers also cubic Haar graphs. Hence it applies to cyclic configurations.

Proposition 2 (Alspach and Zhang) *A Levi graph of a cyclic n_3 configuration is Hamiltonian.*

One can show that all Levi graphs for cyclic n_r configurations where $r > 2$ have girth 6. In other words, all non-trivial cyclic configurations contain triangles.

It is perhaps of interest to note that there exist flag-transitive, non-self-dual configurations. Figure 24 depicts the Levi graph of one such configuration, known as the Gray configuration of type 27_3. Since it has all flags indistinguishable it must have all points alike and all lines alike. However, if we interchange the role of points and lines we do not get the same configuration.

As we mentioned earlier, the generalized Petersen graph $G(10,3)$ is the Levi graph of the Desargues configuration. It is interesting that it appears in a different context.

Let us start with the 5-element set $\{A, B, C, D, E\}$ partitioned into a 2-element set and a 3-element set, say AB-CDE. Two partitions are adjacent, if and only if they have disjoint 2-element sets. For instance AB-CDE is adjacent to CD-ABE, CE-ABD, and DE-ABC. The resulting graph shown in Figure 25 is the Petersen graph.

We may repeat this process but now we label the vertices of a trigonal bipyramid. If we use labels AB for the axial position and CDE for the equatorial position we construct the Petersen graph again. This mechanism also allows for the occurrence of the enantiomers (mirror image). If we want to differentiate the two we obtain $GP(10,3)$, the Levi graph of the Desargues configuration; see Figure 28.

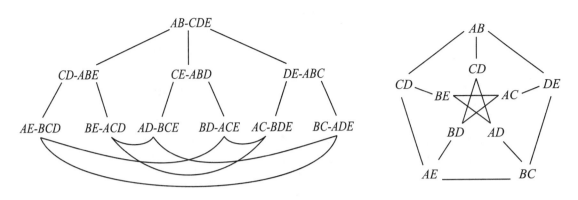

Figure 25: The Petersen graph $GP(5,2)$.

8   Cages and Covering Graphs

The length of the shortest cycle in a graph G is called the girth of G. For a simple graph the girth is at least 3. The smallest trivalent graph of girth g is called a g-cage. Obviously K_4 is the unique 3-cage and $K_{3,3}$ is the only 4-cage. The Petersen graph is the only 5-cage.

It turns out that the 6-cage is the Heawood graph; see Figure 19. The 7-cage has 24 vertices and is depicted in Figure 30. It was Tutte who proved that the 8-cage is the graph of the Cremona-Richmond configuration. That is why this graph is more often called the Tutte 8-cage.

It is interesting that the 9-cage was not found until quite recently. The search for the 9-cage involved a lot of computer checking and the result came as a surprise. There are 18 non-isomorphic 9-cages. All smaller cages have regular structure and all are unique. However, the 9-cages do not show any apparent structure; they are computed in [10].

Balaban found one of the three 10-cages which is shown in Figure 27. It is perhaps of interest to note that the 10-cages were known before all the 9-cages were computed. The reason is simply in the fact that the gap between easily proven lower bound and the actual size of the cage is larger for the 9-cage than for the 10-cage. One can prove that there is no trivalent graph of girth 9 on 46 or fewer vertices and that there is no such graph of girth 10 on 62 or fewer vertices. Since the 9-cage has 58 vertices and the 10-cage has 70 vertices the respective gaps are 12 for the 9-cage and only 8 for the 10-cage. For a survey of cages, see [69].

There is a way of describing certain large graphs using labels on smaller ones. We will introduce this method using the cages as examples.

Let us start with the *5-cage*, the Petersen graph. We will project the Petersen graph onto a handcuff

graph. The *handcuff graph* is not simple. It has two loops attached to the endpoints of a single edge. The outer circle is projected to one loop, the inner pentagram is projected to the second loop and the rims are projected to the edge between the loops. We may direct the edges of the handcuff graph and assign permutations on the edges as shown in Figure 26. The first loop gets permutation $(1,2,3,4,5)$, the second loop gets $(1,3,5,2,4)$ and the edge gets the identity $(1)(2)(3)(4)(5)$. The assignment of the permutations on the edges of the handcuff graph is called the *permutation voltage assignment*. Now we will show how to construct the permutation-derived graph. Above each vertex of the handcuff graph we place five vertices, one on each layer. Above each edge of the handcuff graph we place five edges. However, the edge that was directed from a to b and having voltage p runs from the vertex above a on layer i to the vertex above b on layer $p(i)$. The resulting graph is clearly the Petersen graph. In this case we have a shortcut. Instead of permutations we can assign group elements from Z_5. The first loop gets 1, the second loop 3 and the edge 0. This is called the *ordinary voltage assignment*.

Now we label the layers by group elements and perform the operation again, getting the same result. The *voltage graph* is also called the *base graph* and the derived graph is called the *covering graph*. The terminology follows the one of covering spaces in algebraic topology. The reader is referred to the book by Gross and Tucker [28] for further information about graph coverings. The Petersen graph obtained in this way can be labeled as $GP(5,3)$. This means that we are working in the group Z_5 and that the voltage on the second loop is 3. Note that this process can be generalized to Petersen graphs $GP(n,k)$ with group Z_n and voltage k.

If the covering graph is trivalent then the base graph must be trivalent as well. It may have loops

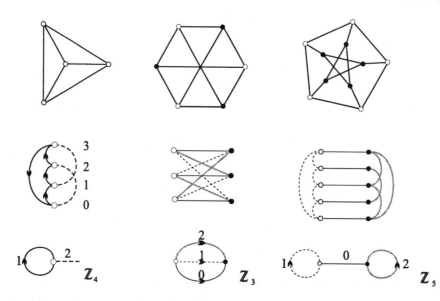

Figure 26: The 3-, 4- and 5-cages as covering graphs, with the corresponding voltage graphs below.

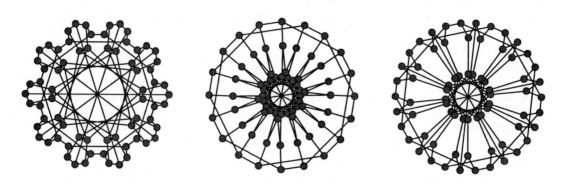

Figure 27: Three views of Balaban's 10-cage.

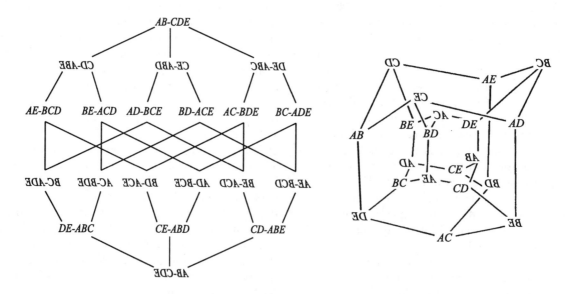

Figure 28: $GP(10,3)$ is the Kronecker double cover of $GP(5,2)$.

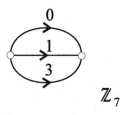

Figure 29: The theta graph, equipped with voltages from Z_7 for the 6-cage.

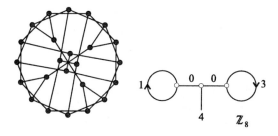

Figure 30: The 7-cage with its voltage graph. This is also known as the McGee graph.

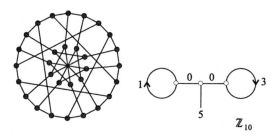

Figure 31: The 8-cage with its voltage graph. This is also known as the Cremona-Richmond graph.

and parallel edges. Even though we assigned voltages to directed edges, we get the same result if we change the direction at an edge. We only have to assign the inverse voltage to the reversely directed edge. We may also have to allow *half-edges* in the base graph. The only condition is the half-edge must have an involutory voltage.

The *3-cage* is the covering graph over the graph with a single vertex with a loop and a half-edge. The voltages are taken in Z_4. The loop gets voltage 1 and the half-edge gets voltage 2. The *4-cage* is the 3-fold covering graph over the theta graph (with voltages 0,1,2 in Z_3.

Taking the theta graph and the voltages 0, 1, and 3 in Z_4 we get the graph Q_3, but the same voltages in Z_7 (see Figure 29) define the *6-cage*,

i.e., the Heawood graph. The *7-cage* is an 8-fold covering graph over the voltage graph on 3 vertices; see Figure 30. The *8-cage* can be represented as a 5-fold covering graph over a graph on 6 vertices; see Figure 31.

We should mention two special constructions. One is called the *Kronecker double cover*. For an arbitrary graph G we may label all the edges with 1, the non-trivial element of the two-element group Z_2. This defines the double cover graph, $G(2)$ also known as the tensor product by K_2. Here are some examples:

Graph	Kronecker double cover
P_n	$2P_n$
C_{2n}	$2C_{2n}$
C_{2n+1}	C_{4n+2}
K_4	Q_3
G - bipartite	2G
$GP(5,2)$	$GP(10,3)$

Recall that $GP(4,1)$ is the cube Q_3. It turns out that $GP(8,3)$, $GP(12,5)$, and $GP(24,5)$ are its covers, [8]. On the other hand, $GP(5,2)$ is the Petersen graph whose canonical (Kronecker) double cover is $GP(10,3)$, as clearly seen in Figures 25 and 28, while $GP(10,2)$, the skeleton of the dodecahedron, arises as a double cover of its pentagonal embedding in the projective plane. An interesting project would be to determine the Kronecker double covers of all generalized Petersen graphs $GP(n,k)$.

The simplest voltage graph G has the voltage group Z_n and has only two types of voltage labels: the trivial 0 and the non-trivial 1. In this case the covering graph is called a *rotagraph*. The subgraph M of G composed of those edges of G with trivial voltages is called a *monograph*; see [56] for more information on rotagraphs.

Many well-known graphs can be described as rotagraphs, for example the prisms and antiprisms. However, the Petersen graph is not a rotagraph although it has a rotational symmetry. In this case we need jumps of size one as well as jumps of size 2. Hence, we will call the covering graph obtained by the cyclic group Z_n a generalized rotagraph. The main idea about generalized rotagraphs is that we may easily keep the voltages and vary n. So we always get not only one graph but an infinite family of graphs of the same type. Viewing the Petersen graph as a generalized rotagraph, we immediately get the whole family $GP(n,2)$.

The notion of cages can be generalized to other regular graphs. A (d,g)-cage is the smallest regular d-valent graph with girth g.

Figure 32: The (4,5)-cage.

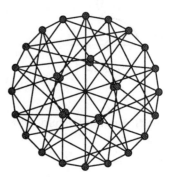

Figure 34: The fourth (5,5)-cage.

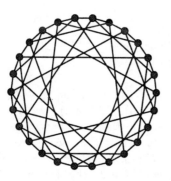

Figure 33: The (4,6)-cage.

In Figures 27, 32, 33, and 34 are given some known examples, perhaps drawn in a new way.

Recently Markus Meringer showed by computer that there are exactly 4 (5,5)-cages. His computer program `genreg`, which can generate all connected k-regular graphs with a given minimum girth, found the fourth (5,5)-cage and proved that there are no other ones [50].

9 Final Comment

Most of the figures in this paper were created by programs that are integrated in the system Vega [57]. For graph drawings, variations of the so-called spring embedder were used.

References

[1] B. Alspach and C. Q. Zhang. Hamilton cycles in cubic Cayley graphs on dihedral groups. *Ars. Combin.*, 28:101–108, 1989.

[2] B. Alspach, C. C. Chen, and K. McAvaney. On a class of Hamilton laceable 3-regular graphs. *Discr. Math.*, 151:19–38, 1996.

[3] Anton Betten and Dieter Betten. Regular Linear Spaces. *Beiträge zur Algebra und Geometrie*, 38:1, 111–124, 1997.

[4] A. Betten, G. Brinkmann, and T. Pisanski. Counting symmetric configurations v_3, submitted.

[5] N. Biggs. *Algebraic Graph Theory*, Second Edition. Cambridge Univ. Press, Cambridge, 1993.

[6] J. Bokowski and B. Strumfels. *Computational Synthetic Geometry*. LNM 1355, Springer-Verlag, 1989.

[7] J. A. Bondy and U. S. R. Murty. *Graph Theory with Applications*. American Elsevier Publishing Co., 1976.

[8] I. Z. Bouwer et al, editors. *The Foster Census*. The Charles Babbage Research Centre, Winnipeg, Canada, 1988.

[9] G. Brinkmann. Fast Generation of Cubic Graphs. *J. Graph Theory*, 23:139–149, 1996.

[10] G. Brinkmann, B. D. McKay, and C. Saager. The smallest cubic graph of girth 9. *Combinatorics, Probability and Computing*, 5:1–13, 1995.

[11] H. S. M. Coxeter. *Regular Polytopes*, Third Edition. Dover Publications Inc., New York, 1973.

[12] H. S. M. Coxeter. Self-dual configurations and regular graphs. *Bull. Amer. Math. Soc.*, 56:413–455, 1950.

[13] H. S. M. Coxeter, R. Frucht, and D. L. Powers. *Zero-Symmetric Graphs.* Academic Press, New York, 1981.

[14] H. S. M. Coxeter and W. O. J. Moser. *Generators and Relators for Discrete Groups*, Fourth Edition. Springer-Verlag, 1980.

[15] H. M. Cundy and A. R. Rollett. *Mathematical Models*, Third Edition. Tarquin Publications, Norfolk, 1981.

[16] R. Daublebsky von Sterneck. Die Configurationen 11_3. *Monatsh. Math. Physik*, 5:325–330, 1894.

[17] R. Daublebsky von Sterneck. Die Configurationen 12_3. *Monatsh. Math. Physik*, 6:223–255, 1895.

[18] P. Dembowski. *Finite Geometries.* Ergebnisse der Mathematik und ihre Greenzgebiete, Bd. 44, Springer-Verlag, New York, 1968.

[19] S. F. Du, D. Marušič, and A. O. Waller. On 2-arc-transitive covers of complete graphs, submitted.

[20] P. Fowler and T. Pisanski. Leapfrog Transformation and Polyhedra of Clar Type. *J. Chem. Soc. Faraday Trans.*, 90:2865–2871, 1994.

[21] R. Frucht, J. E. Graver, and M. E. Watkins. The groups of the generalized Petersen graphs. *Proc. Cambridge Philos. Soc.*, 70:211–218, 1971.

[22] M. C. Golumbic. *Algorithmic Graph Theory and Perfect Graphs.* Academic Press, New York 1980.

[23] H. Gropp. On the history of configurations. In A. Deza, J. Echeverria and A. Ibarra, editors, *Internat. Symposium on Structures in Math. Theories*, 263–268, Bilbao, 1990.

[24] H. Gropp. Configurations and Graphs. *Discrete Math.*, 111:269–276, 1993.

[25] H. Gropp. The Construction of All Configurations $(12_4, 16_3)$. In J. Nešetřil and M. Fiedler, editors, *Fourth Czechoslovak Symposium on Combinatorics, Graphs and Complexity*, 85–91, Elsevier, 1992.

[26] H. Gropp. Configurations and their realizations. *Discrete Math.*, 174:137–151, 1997.

[27] H. Gropp. Configurations. In C. J. Colburn and J. H. Dinitz, editors, *The CRC Handbook of Combinatorial Designs*, 253–255, CRC Press, 1996.

[28] J. L. Gross and T. W. Tucker. *Topological Graph Theory.* Wiley Interscience, 1987.

[29] L. Grossman and W. Magnus. *Groups and their graphs.* Random House, New York, 1966.

[30] T. F. Havel. The Combinatorial Distance Geometry Approach to the Calculation of Molecular Conformation, Ph. D. Thesis (Biophysics), Berkeley, 1982.

[31] N. Hartsfield and G. Ringel. *Pearls in Graph Theory*, Revised Edition. Academic Press, New York, 1994.

[32] D. Hilbert and S. Cohn-Vossen. *Geometry and the Imagination.* Chelsea Publishing Co., New York, 1952.

[33] M. Hladnik and T. Pisanski. Schur Norms of Haar Graphs, work in progress.

[34] M. Hladnik, D. Marušič, and T. Pisanski. Haar graphs, work in progress.

[35] D. A. Holton and J. Sheehan. *The Petersen Graph.* Cambridge Univ. Press, 1993.

[36] H. Hosoya, Y. Okuma, Y. Tsukano, and K. Nakada. Multilayered Cyclic Fence Graphs: Novel Cubic Graphs Related to the Graphite Network. *J. Chem. Inf. Comp. Sci.*, 35:351–356, 1995.

[37] A. Jurišić. Antipodal Covers, Ph.D. Thesis, University of Waterloo, Canada, 1995.

[38] M. Kaufman, T. Pisanski, D. Lukman, B. Borštnik, and A. Graovac. Graph-drawing algorithms geometries versus molecular mechanics in fullerenes. *Chem. Phys. Letters*, 259:420–424, 1996.

[39] S. Klavžar and M. Petkovšek. Intersection graphs of halflines and halfplanes. *Discrete Math.*, 66:133–137, 1987.

[40] D. J. Klein. Graph geometry, graph metrics, and Wiener. *MATCH*, 35:7–27, 1997.

[41] D. J. Klein, M. Randić. Resistance Distance. *J. Math. Chemistry*, 12:81–95, 1993.

[42] D. J. Klein, H. Zhu. All-Conjugated Carbon Species. In A. T. Balaban, editor, *From Chemical Topology to Three-Dimensional Geometry*, 297–341, Plenum Press, New York, 1997.

[43] M. Lovrečič-Sarazin. A note on the generalized Petersen graphs that are also Cayley graphs. *J. Combin. Theory (B)*, 69: 226–229, 1997.

[44] M. Lovrečič-Sarazin. Cayleyevi in Petersenovi grafi. *Obzornik Mat. fiz.*, 42:193–198, 1995.

[45] D. Marušič and T. Pisanski. The remarkable generalized Petersen graph $G(8,3)$, submitted.

[46] H. Maehara. Space graphs and sphericity. *Discrete Appl. Math.*, 7:55–64, 1984.

[47] H. Maehara. On the sphericity for the join of many graphs. *Discrete Math.*, 49:311–313, 1984.

[48] H. Maehara. On the sphericity of semiregular polyhedra. *Discrete Math.*, 58:311–315, 1986.

[49] R. E. Maeder. Uniform Polyhedra. *The Mathematica Journal*, 3(4):48–57, 1993.

[50] M. Meringer. Computer Program `genreg`, `ftp://132.180.16.20`

[51] M. Meringer: Erzeugung regulärer Graphen, Master's thesis, Universität Bayreuth, January 1996.

[52] S. Negami. Diagonal Flips in Triangulations of Surfaces. *Discr. Math.*, 135:225–232, 1994.

[53] T. Pisanski, B. Plestenjak, and A. Graovac. NiceGraph Program and its application in Chemistry. *Croatica Chemica Acta*, 68:283–292, 1995.

[54] T. Pisanski, B. Plestenjak, and A. Graovac. Generating Fullerenes at Random. *J. Chem. Inf. Comp. Sci.*, 36:825–828, 1996.

[55] T. Pisanski, M. Razinger, and A. Graovac. Geometry versus Topology: Testing Self-consistency of the NiceGraph Program. *Croatica Chemica Acta*, 69 (in press), 1996.

[56] T. Pisanski, A. Žitnik, A. Graovac, and A. Baumgartner. Rotagraphs and their generalizations. *J. Chem. Inf. Comp. Sci.*, 34:1090–1093, 1994.

[57] T. Pisanski, editor. *Vega Version 0.2; Quick Reference Manual and Vega Graph Gallery*. IMFM, Ljubljana, 1995. `http://vega.ijp.si/`

[58] M. Randić. Symmetry properties of graphs of interest in chemistry. II. Desargues-Levi Graph. *Int. J. Quant. Chem.*, 15:663–682, 1979.

[59] G. Ringel. *Map Color Theorem*. Springer-Verlag, Berlin, 1974.

[60] F. S. Roberts. Indifference graphs. In F. Harary, editor, *Proof Techniques in Graph Theory*, 139–146, Academic Press, 1969.

[61] R. Sayle. RasMol V2.5, `http://www.bio.-cam.ac.uk/doc/rasmol.html`

[62] E. Steinitz. *Über die Construction der Configurationen n_3*. Dissertation, Breslau, 1894.

[63] E. Steinitz (with H. Rademacher). *Vorlesungen über die Theorie der Polyeder*. Springer-Verlag, Berlin, 1934.

[64] B. Sturmfels and N. White. All 11_3 and 12_3 configurations are rational. *Aequationes Mathematicae*, 39:254–260, 1990.

[65] T. W. Tucker. There is only one group of genus two. *J. Combin. Theory B*, 36:269–275, 1984.

[66] D. Wells. *The Penguin Dictionary of Curious and Interesting Geometry*. Penguin Books, London, 1991.

[67] A. White. *Graphs, Groups, and Surfaces*. North-Holland Pub. Co., Amsterdam, 1973.

[68] R. Wilson. *Introduction to Graph Theory*. Academic Press, New York, 1979.

[69] P. K. Wong. Cages—a survey. *J. Graph Theory*, 6:1–22, 1982.

Polytopes in Combinatorial Optimization

Thomas Burger
FB IV, Mathematik, Universität Trier
D-54286 Trier, Germany
burger@dm7.uni-trier.de

Peter Gritzmann[*]
Center for Mathematical Sciences
Arcisstr. 21
D-80333 Munich
gritzman@mathematik.tu-muenchen.de

1   Introduction

For millennia, polytopes played an important role in mathematics and mythology. The ancient Greeks associated four of the five three-dimensional regular polytopes with the four basic "elements":

fire	⟷	tetrahedron
earth	⟷	cube
air	⟷	octahedron
water	⟷	icosahedron.

Having one regular polytope left, they imagined the dodecahedron as the shape that envelopes the universe. The name *platonic solid* indicates the fact that the tetrahedron, the cube, the octahedron, the dodecahedron and the icosahedron were already known to Plato.

The mystical and theological aspects were still present in the seventeenth century. Even someone like Kepler was so much captured by the fascination of the five three-dimensional regular polytopes that he derived a divine reasoning for the existence and completeness of the then known five planets [14]:

> *God ... has based the design of the orbits of the planets upon those five regular bodies .... The Earth is the measure for all other orbits. Its orbit circumscribes a dodecahedron; its circumsphere is Mars. Mars' orbit circumscribes a tetrahedron; its circumsphere is Jupiter. Jupiter's orbit circumscribes a cube; its circumsphere is Saturn. Now inscribe an icosahedron in Earth's orbit; its insphere is Venus. In Venus' orbit inscribe an octahedron; its insphere is Mercury. There you have the reason for the number of the planets.*[1]

The mystical approach to life has mostly disappeared in our intellectual world, and certainly the mythological aspects of polytopes are not central anymore. Nonetheless, polytopes have proved useful in making certain structures visible and thus better understandable. As a matter of fact, polytopes have become powerful tools for solving very difficult but very important practical problems, and it will be this role of polytopes that we will demonstrate here.

The three examples we are going to give are the *matching polytope*, which is a polytope that is associated with the problem of finding maximum matchings in graphs, the *traveling salesman polytope* which comes up in the context of finding shortest tours in graphs, and the *order polytope* which is related to partially ordered sets.

The first two examples belong to the realm of *polyhedral combinatorics*, an important tool for solving hard combinatorial optimization problems. The feasible solutions of a given problem—the matching problem and the traveling salesman problem in our case—are encoded as the vertices of an associated polytope, and the aim is to understand the facial structure of this polytope well enough to apply linear programming based techniques; the underlying branch-and-cut algorithm will be sketched in Section 3. Section 2 uses the matching problem to introduce and develop the concept of polyhedral

[*]Research of the second author was supported in part by a Max Planck Research Award.

[1]Full original quote: Creator Optimus maximus, in cre- atione Mundi huius mobilis, et dispositione Coelorum, ad illa quinque regularia corpora, inde a PYTHAGORA et PLATONE, ad nos vsque, celebratissima respexerit, atque ad illorum naturam coelorum numerum, proportiones, et motuum rationem accomodauerit ([14, p. 9]). ... Terra est Circulus mensor omnium: Illi circumscribe Dodecaedron: Circulus hoc comprehendens erit Mars. Marti circumscribe Tetraedron: Circulus hoc comprehendens erit Jupiter. Ioui circumscribe Cubum: Circulus hunc comprehendens erit Saturnus. Iam terrae inscribe Icosaedron: Illi inscriptus Circulus erit Venus: Veneri inscribe Octaedron: Illi inscriptus Circulus erit Mercurius. Habes rationem numeri planetarum ([14, p. 13]).

combinatorics; it is a rare case where a complete set of linear inequalities is known that fully describes the associated polytope. The traveling salesman problem is much harder, since only partial information about the facial structure of the associated polytope is available; it will be studied in Section 4.

The final example that we give is of a different flavor. It is the order polytope which encodes the linear extensions of a given partial order as the simplices of one of its triangulations. This "geometrization" of the underlying combinatorial problem is useful for various algorithmic tasks, some of which are outlined in Section 5. We include it here to indicate that polyhedral methods can be utilized in more than one way, and that polytopes arise naturally in a variety of different applications.

2   The Matching Polytope

Imagine yourself as the manager of a dating service. Saturday night is coming up, your database contains applications by lonely hearts containing information about age, profession, hobbies, etc., and you wonder whom to pair up with whom. From your experience you know that for most people it is most pleasant to have a partner whose interests are closely related. By comparing each possible pair you can thus derive a "similarity level" for each, say on a scale from 0 to 10. You have compiled your results in the following picture.

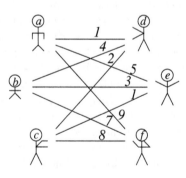

Figure 1: The "dating graph" for this weekend. (Business is slow.)

Note that according to your database your clients belong to the two different classes $\{a, b, c\}$ and $\{d, e, f\}$ and pairs within a class are not allowed. By exhaustive search, you find that the combination (a, e), (b, d), (c, f) yields a maximum combined similarity of 17. In the absence of better criteria you decide that this is the best matching combination.

Were this the only application of the matching problem one might wonder—given the fact that "success of a relationship" is not a well-understood concept—how relevant the application of mathematical tools really is. However, the same problem occurs in the context of personnel assignment, chromosome identification, air traffic control, job scheduling on parallel machines and many more; see, e.g., [1]. Moreover, algorithms for solving the general matching problem are important tools in many hard combinatorial optimization problems.

In many of the above mentioned applications, the matching problem involves thousands of objects, and then methods for solving these problems are needed that are much more efficient than "checking all possibilities." We will come back to this point later; let us first formally introduce the problem in its natural environment, *graph theory*.

A *graph* $G = (V, E)$ is a pair consisting of a set V and a set E of 2-element subsets of V. The elements of V are called *nodes*, the elements of E *edges*.[2] If $e = \{v, w\} \in E$ is an edge of G we also write $e = vw$ and say that v and w are *incident* to e. In the special case that E is the full set of all possible 2-element subsets of V, the graph is called *complete*. Note that the concept of a graph is purely combinatorial. So, we may, for instance, speak of *the* complete graph on $|V|$ nodes, or we may choose node sets of the form $V = \{1, 2, \ldots, n\}$ to simplify the notation. A *weighted graph* is a triple $G = (V, E, c)$ where (V, E) is a graph and c is a real-valued function defined on the edge set. So our dating graph of Figure 1 is a weighted graph on six nodes; the nine edges are labeled by their weights.

A set of edges of a graph with the property that no two have a node in common is called a *matching*. For a subset $M \subset E$, we define the *weight* of M as $c(M) = \sum_{e \in M} c(e)$. In particular, this defines a weight of a matching. Then our matching problem is the following task:

MATCHING.

Input: *A weighted graph* $G = (V, E, c)$.

Task: *Find a matching of* G *with maximum weight.*

---

[2]From the definition it is certainly allowed that V or E is empty. For our purposes the case that no edge is present is not particularly interesting, so we will tacitly assume throughout the paper that E (whence also V) is nonempty. For a picture of the empty graph and a discussion of its relevance see F. HARARY and R. C. READ, *Is the null-graph a pointless concept in graphs and combinatorics*, Lecture Notes in Math. **406**, Springer, 1974, p. 37.

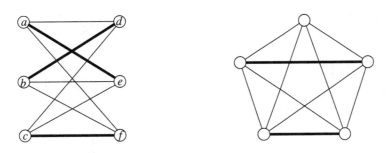

Figure 2: *Left:* The maximum matching in the "dating graph." *Right:* The complete graph on five vertices. The weights (which are not depicted) are all 1. Hence all matchings of cardinality two are maximal. One of them is highlighted.

Figure 2 shows the maximum matching in the "dating graph" (that we found by exhaustive search), and also a maximum matching in a (weighted) complete graph on five nodes.

Recall that our "dating graph" has a special structure: its nodes can be partitioned into two sets such that all edges have exactly one endpoint in each set. Such graphs are called *bipartite*. It turns out that the special case of bipartite graphs can be handled more easily than the general case. But, how can we solve this problem of BIPARTITE MATCHING efficiently, and what does "efficiently" mean?

For the purpose of this article, *efficiently* simply means that we can reduce the problem to linear programming (or at least that the problem is not much harder than linear programming). This informal "measure" of difficulty of a problem is used to save the reader from a technical dive into the theory of computational complexity. It does not seem appropriate here to burden the reader with models of computations and beasts from the zoo of complexity classes. For those readers who are familiar with complexity theory: the adjective *efficient* will always indicate polynomial-time, while *hard* or *intractable* refers to **NP**- or even #**P**-hardness. Readers who want to see formal definitions for these classes (and much more) may consult [8].

Recall that a typical *linear program* looks as follows:

> Given $m, d \in \mathbf{N}$, $c, a_1, \ldots, a_m \in \mathbf{R}^d$,
> $b_1, \ldots, b_m \in \mathbf{R}$,
>
> maximize $c^T x$ for $x \in \mathbf{R}^d$
>
> with respect to $a_i^T x \le b_i$, $i = 1, \ldots, m$.
>
> (2.1)

(Note that we identify $\mathbf{R}^d$ with the set of all d-dimensional column vectors. Hence $x \in \mathbf{R}^d$ is a

column vector, c^T is a row vector, and $c^T x$ is therefore a real number.) Geometrically, the *set of feasible solutions*,

$$P = \left\{ x \in \mathbf{R}^d : a_i^T x \le b_i \text{ for all } i = 1, \ldots, m \right\},$$

of the linear program (2.1) is just the intersection of a finite number of closed halfspaces, a *polyhedron*. In all our applications the set of feasible solutions will also be bounded; then P is called a *polytope*. Equivalently, a polytope is the convex hull of a finite point set. (The proof of this fact is not trivial, see, e.g., [33].) The dimension of the smallest affine subspace that contains a given polytope is called the *dimension* of P, and denoted by $\dim(P)$. A *face* of a polytope P is a proper subset of P that is the intersection of P with some hyperplane such that P lies completely on one side of the hyperplane. In particular, a face is again a polytope.[3] Formulated in terms of linear programming, a subset Z of a d-dimensional polytope P in $\mathbf{R}^d$ is a face of P if and only if there exists a vector $c \in \mathbf{R}^d \setminus \{0\}$ such that Z coincides with the set of optimal solutions of the linear program $\max_{x \in P} c^T x$. The faces of dimension 0 are called *vertices*, the faces of dimension 1 *edges*, and the faces of dimension $\dim(P) - 1$ *facets* of P; see Figure 3 (left). Note that if the feasible region P of (2.1) is nonempty and bounded, then the optimum of the objective function is finite and attained at a vertex of P.

Throughout this article we will assume (without going into details) that an efficient procedure for finding an optimal vertex of a linear programming problem (with nonempty and bounded feasible region P) is available. This assumption is well justified [25, 32]: huge linear programs can be solved

---

[3]Often, $\emptyset$ and P itself (even if it is fulldimensional) are also regarded as faces of P; they are sometimes called *improper* in contrast to the proper faces that we are interested in.

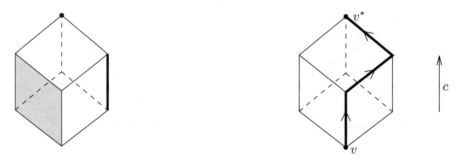

Figure 3: *Left:* A polytope P with a facet, an edge and a vertex highlighted. *Right:* An objective function $x \mapsto c^T x$, two vertices v and v^*, the latter being maximal with respect to the objective function, and a path as constructed by the simplex algorithm.

quickly, and linear programming has become a major part of everyone's optimization toolbox. In particular, the essential idea behind the *simplex algorithm*, introduced by Dantzig (see [25, 29, 32]) is to construct a path in the union of P's edges ending in an optimal vertex, on which the objective function is strictly increasing; see Figure 3 (right). For an account of linear programming from a geometric point of view see [9].

So, how can we formulate our matching problem as a linear program? Let's try! Let $G = (V, E)$ be a given graph, and let us introduce a variable x_e for each edge in G. The underlying rationale is that the variable x_e should take only the values 0 and 1, and in a feasible solution all edges e with $x_e = 1$ form a matching in G. How can we characterize such sequences $(x_e : e \in E)$ of 0-1-entries? Clearly, the definition of a matching requires that no node of G be contained in more than one edge. Hence denoting, for a node $v \in V$, the set of all edges of G that contain v by $\delta(v)$ and writing $c_e = c(e)$ for all $e \in E$, we obtain the following formulation of MATCHING:

$$\text{maximize} \quad \sum_{e \in E} c_e x_e$$

with respect to

$$\left\{ \begin{array}{ll} \sum_{e \in \delta(v)} x_e \leq 1 & \text{for all} \quad v \in V, \\ x_e \in \{0, 1\} & \text{for all} \quad e \in E. \end{array} \right. \tag{2.2}$$

We can see that (2.2) is indeed equivalent to MATCHING as follows: A feasible solution has only components 0 or 1, and for each node there is at most one edge e such that $x_e = 1$. Hence every feasible solution corresponds to a matching. On the other hand, given a matching M of G and setting the variables x_e to 0 or 1, depending on whether

the edge e is contained in M, we obtain a solution that is feasible for (2.2). Hence the set of feasible points of (2.2) corresponds precisely to the set of matchings of G, and the value of the objective function $\sum_{e \in E} c_e x_e$ is just the weight of the associated matching. This shows that MATCHING is equivalent to (2.2).

It is desirable to collect all values of the variables x_e in a vector $x = (x_e : e \in E)^T$. To be able to do this in an unambiguous way we need to assume that the edges of E are indexed in an arbitrary but fixed way—say $E = \{e_1, e_2, \ldots, e_{|E|}\}$. Then we can simply associate the ith coordinate of a vector $(x_1, \ldots, x_{|E|})^T$ with the value of x_{e_i}. In particular, for a matching M, the 0-1-vector with $x_e = 1$ if and only if $e \in M$ is called the *incidence vector* of the matching M. (As a way of getting familiar with the optimization formulation of (2.2), the reader may write down the corresponding conditions for the "dating graph" from Figure 1.)

Unfortunately, (2.2) is not a linear program, since it contains the integrality conditions $x_e \in \{0, 1\}$. But is this integrality condition needed? If we simply replace the conditions $x_e \in \{0, 1\}$ by the linear constraints $0 \leq x_e \leq 1$, or—in view of the first set of constraints—equivalently by $0 \leq x_e$ for all $e \in E$ we obtain the ordinary linear program:

$$\text{maximize} \quad \sum_{e \in E} c_e x_e$$

with respect to

$$\left\{ \begin{array}{ll} \sum_{e \in \delta(v)} x_e \leq 1 & \text{for all} \quad v \in V, \\ -x_e \leq 0 & \text{for all} \quad e \in E. \end{array} \right. \tag{2.3}$$

What have we lost? The set of feasible points of the linear program (2.3) is now a polyhedron P, not a finite set of points anymore as is the case in (2.2).

Certainly, P is bounded (it is contained in the standard cube $[0, 1]^{|E|}$) and nonempty (since $0 \in P$), hence the maximum of the objective function is obtained at a vertex of P. So the remaining question is: Are all vertices of P already 0-1-vectors? Recall that the optimum of the linear program (2.3) is attained at a vertex of P and, in particular, all incidence vectors of matchings are vertices of P. But are there others? If not, P is merely the convex hull of all feasible points of (2.2), whence (2.3) is then essentially a linear reformulation of (2.2).

It turns out that this is indeed so in the bipartite case (see [11, 32]). This means that problems having the special structure of our (bipartite) dating problem can be solved just by linear programming techniques. The reason is that the coefficient matrix A of the linear program (2.3) (when written in the form $\max c^T x$ subject to $Ax \le \mathbf{1}$, $x \ge 0$ with $\mathbf{1}$ being the vector of $\mathbf{R}^{|V|}$ with all components 1) is the *incidence matrix* of a bipartite graph, and such a matrix is known to be *totally unimodular*, meaning that each square submatrix has determinant -1, 0, or 1. It is actually not hard to see that this is true, and it is equally simple to deduce then that all vertices of P are integral; we leave it as an exercise for the highly motivated reader; others may consult, e.g., [32].

But what happens if the graph is not necessarily bipartite? This really happens in many relevant applications! In terms of our initial example this means that the group of people to be matched does not fall into two separate groups with no intergroupal match, respectively. Consider the complete graph K_3 on 3 nodes (i.e., a "triangle") with edges labeled 1, 2, 3. Then the constraints in (2.2) are equivalent to

$$
\begin{array}{rrrcl}
x_1 & + & x_2 & & \le & 1, \\
x_1 & & + & x_3 & \le & 1, \\
& x_2 & + & x_3 & \le & 1, \\
-x_1, & -x_2, & -x_3 & & \le & 0, \\
x_1, & x_2, & x_3 & & \in & \mathbf{Z}.
\end{array} \qquad (2.4)
$$

These conditions are satisfied by the following set of incidence vectors:

$$
\left\{ \begin{pmatrix} 0 \\ 0 \\ 0 \end{pmatrix}, \begin{pmatrix} 1 \\ 0 \\ 0 \end{pmatrix}, \begin{pmatrix} 0 \\ 1 \\ 0 \end{pmatrix}, \begin{pmatrix} 0 \\ 0 \\ 1 \end{pmatrix} \right\}.
$$

On the other hand, after dropping the integrality

condition, (2.4) describes the polytope

$$
P = \operatorname{conv} \left\{ \begin{pmatrix} 0 \\ 0 \\ 0 \end{pmatrix}, \begin{pmatrix} 1 \\ 0 \\ 0 \end{pmatrix}, \begin{pmatrix} 0 \\ 1 \\ 0 \end{pmatrix}, \right.
$$
$$
\left. \begin{pmatrix} 0 \\ 0 \\ 1 \end{pmatrix}, \begin{pmatrix} 1/2 \\ 1/2 \\ 1/2 \end{pmatrix} \right\} \qquad (2.5)
$$

depicted in Figure 4.

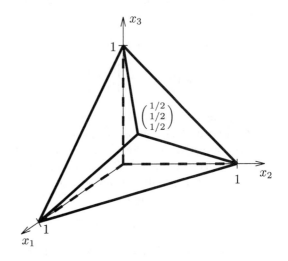

Figure 4: The polytope of the linear constraints (2.4) of K_3.

Clearly, the vertex $(1/2, 1/2, 1/2)^T$ does not correspond to a matching. Therefore, the integrality condition cannot be discarded. In particular, if the weights on the edges all equal 1, then all optimum matchings have weight 1 while (2.4) without the integrality conditions has optimum value $3/2$.

So, if we want to formulate the general matching problem as a linear program we need to add more linear constraints that do render the integrality condition redundant. In the above example we could add the condition

$$
x_1 + x_2 + x_3 \le 1. \qquad (2.6)
$$

The integrality condition would then automatically be satisfied. But how do we find such additional inequalities for graphs more interesting than K_3? Edmonds' Matching Theorem (see Theorem 2.1 below) gives a satisfactory answer. But before we state the result let us take a step back and look at the general concept underlying our approach.

We may regard MATCHING and many other important problems in combinatorial optimization (including the traveling salesman problem that will be studied in Section 4) as being given in terms of a triple $(S, \mathcal{F}, c)$ consisting of a finite set S, a family $\mathcal{F}$ of subsets of S, and a real-valued weight function c defined on S. The aim is to find a member F of $\mathcal{F}$ of maximum (or, in other cases: minimum) weight, where the weight of F is defined as $\sum_{s \in F} c(s)$. In case of MATCHING, S is the set of edges of the given graph G, $\mathcal{F}$ is the set of matchings in G, and the function c specifies the weights on the edges.

For each $F \in \mathcal{F}$, let x^F denote the *incidence vector* of F (based again on an "indexing" of the elements of S):

$$x^F = (x_s^F : s \in S)^T \in \{0, 1\}^{|S|}$$

$$\text{with} \quad x_s^F = \begin{cases} 1, & \text{if } s \in F, \\ 0, & \text{if } s \notin F. \end{cases} \quad (2.7)$$

Writing again $c_s = c(s)$, for $s \in S$, the original combinatorial problem becomes that of optimizing the linear objective function $\sum_{s \in F} c_s x_s^F$ over the set of incidence vectors x^F, $F \in \mathcal{F}$. Aiming at a formulation of this problem as a linear program we consider the *incidence polytope*

$$P(\mathcal{F}) = \text{conv}\{x^F : F \in \mathcal{F}\}$$

and optimize over $P(\mathcal{F})$ instead. In order to apply linear programming techniques we need to find a description of $P(\mathcal{F})$ in terms of linear inequalities.

For the matching problem, the incidence polytope is called the *matching polytope* and denoted by $\mathbf{M}_G$. What we have seen so far in this section reads now as follows:

- The inequalities $\sum_{e \in \delta(v)} x_e \leq 1$ (for all $v \in V$) and $0 \leq x_e$ (for all $e \in E$) provide a full description (in terms of inequalities) of the matching polytope $\mathbf{M}_G$ of a bipartite graph G.

- In general, the matching polytope is a proper subset of the polytope specified by the above inequalities.

To see that the linear description of the matching polytope is really useful consider the *complete bipartite graph* $K_{n,n}$ whose node set falls into two disjoint sets V_1 and V_2 of cardinality n each, and whose edge set is $\{vw : v \in V_1, w \in V_2\}$; note that our dating graph was just $K_{3,3}$. It is a simple counting exercise to show that

$$\text{"Number of matchings in } K_{n,n}\text{"} = \sum_{k=0}^{n} \binom{n}{k}^2 k!$$

which is exponential in n, while the above linear description of the corresponding polytope involves only n^2 variables and $2n + n^2$ inequalities. In all typical problems the set $\mathcal{F}$ is very large, so it is totally impractical to try to find the extremum by complete evaluation of the objective function at all feasible solutions. It should be pointed out, though, that the small number of inequalities that describe the bipartite matching polytope is not typical either. Usually the number of facets of incidence polytopes $P(\mathcal{F})$ associated with a combinatorial optimization problem is extremely large too, so a complete linear description of $P(\mathcal{F})$ is not possible. However, even partial information can usually be utilized. We will come back to this point in Section 3.

Let us now return to the problem of finding a complete linear description of the matching polytope for a general graph $G = (V, E)$. In order to state and prove Edmonds' Theorem we need two more definitions from graph theory.

For a subset $U \subset V$ let $E(U)$ denote the set of edges connecting two nodes in U, and let $\delta(U)$ be the set of edges connecting a node in U with a node in $V \setminus U$. More formally,

$$\begin{aligned} E(U) &= \{vw \in E : v, w \in U\}, \\ \delta(U) &= \{vw \in E : v \in U, w \in V \setminus U\}. \end{aligned}$$

Note that for singletons $U = \{v\}$ the new notation $\delta(U) = \delta(\{v\})$ coincides, up to the curly brackets, with our older notation $\delta(v)$, which we will continue to use.

Let us now formulate the fundamental result of Edmonds [5], [6] that really was the starting point of polyhedral combinatorics.

Theorem 2.1 (Edmonds' Matching Theorem)
The matching polytope $\mathbf{M}_G$ *of a graph* $G = (V, E)$ *is an* $|E|$-*dimensional polytope in* $\mathbf{R}^{|E|}$ *determined by the following inequalities:*

$$\sum_{e \in \delta(v)} x_e \leq 1 \qquad \text{for all } v \in V, \qquad (2.8)$$

$$\sum_{e \in E(U)} x_e \leq \left\lfloor \frac{|U|}{2} \right\rfloor \qquad \text{for all } U \subset V \qquad (2.9)$$

$$\text{with } |U| \text{ odd and } |U| \geq 3,$$

$$-x_e \leq 0 \qquad \text{for all } e \in E. \qquad (2.10)$$

Recall that, for $x \in \mathbf{R}$, $\lfloor x \rfloor$ denotes the floor function, i.e., the largest integer not exceeding x.

We knew already that the *degree constraints* (2.8) and the *nonnegativity constraints* (2.10) are reasonable; the *odd set constraints* (2.9) are new. (Not completely new: for K_3 there is only one additional inequality of the form (2.9), and this is just the

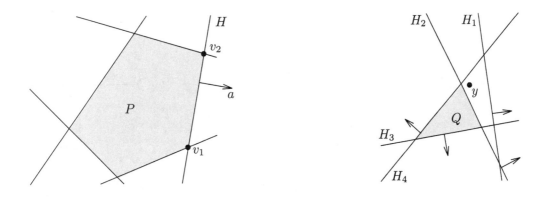

Figure 5: *Left:* A polytope $P \subset \mathbf{R}^2$ with the hyperplane $H = \{x \in \mathbf{R}^2 : a^T x = b\}$; the inequality $a^T x \le b$ is facet inducing (as are the others) because H contains the two affinely independent vertices v_1, v_2 of P. *Right:* A polytope $Q \subset \mathbf{R}^2$ given as the intersection of halfspaces $H_i^- = \{a_i^T x \le b_i\}$ with bounding hyperplanes H_i, $i = 1, \ldots, 4$; the hyperplane H_2 is facet inducing because $y \in H_1^- \cap H_3^- \cap H_4^-$, but $y \notin H_2^-$; the hyperplane H_1 is redundant.

constraint (2.6).) It is not hard to see that all inequalities stated in Theorem 2.1 are satisfied by all elements of $\mathbf{M}_G$; the nontrivial part of the proof is to show that all inequalities necessary to describe $\mathbf{M}_G$ are among them. The proof falls into a more combinatorial part that detects certain properties of maximal matchings and a more geometric part that utilizes properties of facets of polytopes. We will give both parts in detail. However, before we do so let us accept that Theorem 2.1 is correct and discuss the given constraints in more detail: Which of them describe facets? How can we find out?

Let us introduce some relevant geometric notation first. Given a polytope P in $\mathbf{R}^d$, an inequality $a^T x \le b$ is called *valid* for P if P is contained in the closed halfspace $H^- = \{x \in \mathbf{R}^d : a^T x \le b\}$. Obviously there is a whole continuum of valid inequalities so we are actually aiming at a small such system. In practice it is not always easy (or even desirable) to really shoot for a minimal system since redundancy might not always be easy to detect. The inequality $a^T x \le b$ is called *facet inducing* for P if it is valid for P and if the hyperplane $H = \{x \in \mathbf{R}^d : a^T x = b\}$ intersects P in a facet. If P is fulldimensional, i.e., $\dim(P) = d$, it is clear that there is precisely one facet inducing inequality per facet, and all such facet inducing inequalities have to be contained in every finite set of linear inequalities that determines the polytope. It is, further, not hard to see (for the "expert": via polarity or the appropriate separation theorem) that all inequalities which are not facet inducing are redundant: each d-dimensional polytope (or *d-polytope*) in $\mathbf{R}^d$ is the set of feasible solutions of the system of all facet inducing inequalities.

Suppose we are given a d-polytope P in $\mathbf{R}^d$ and an inequality $a^T x \le b$. How can we check if it is facet inducing? By definition an inequality is facet inducing for P if and only if it is valid for P and there are d affinely independent vertices of P that satisfy the inequality with equality. Finding such a set of vertices is the first way of proving that the given inequality is facet inducing. A second method is to use the property of any finite linear description of P that it contains each facet inducing inequality: the inequality $a^T x \le b$ is facet inducing if and only if it is irredundant; i.e., it cannot be omitted without changing the polytope. This means we need to find a point y that satisfies all other inequalities of our given system of valid inequalities for P but violates the "test inequality," i.e., $a^T y > b$. See Figure 5.

Let us try these tests for the inequalities specified in Theorem 2.1 for some graph $G = (V, E)$, beginning with the nonnegativity constraints (2.10). For the first method note that for each $e \in E$ the set $\{e\}$ is a matching in G and so is $\emptyset$. The corresponding incidence vectors are the standard unit vectors of $\mathbf{R}^{|E|}$ and 0. Whence $\mathbf{M}_G$ is of dimension $|E|$, and each coordinate hyperplane contains $|E|$ affinely independent incidence vectors of matchings. So the nonnegativity constraints (2.10) are all facet inducing. Now, let's see how the second method works for the same set of constraints. In fact, if we consider for a fixed edge e of G the vector whose coordinates are all 0 except for the value -1 of the coordinate associated with e, then this vector satisfies all the constraints in (2.8)–(2.10) except for the inequality $x_e \ge 0$ which is violated. Hence it follows again that the nonnegativity constraints are all irredundant.

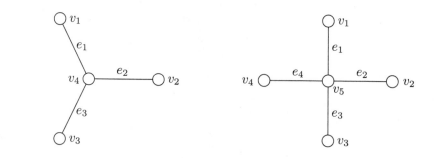

Figure 6: K_2 and two "stars."

So we have two proofs showing that all inequal-ities (2.10) are facet inducing for *any* graph G. Is the same true also for the other inequalities? Let us look at the constraints (2.8). For the complete graph K_2 depicted on the left of Figure 6 (two nodes v_1, v_2 and one edge e) we have two constraints of this form, namely

$$\sum_{e \in \delta(v_1)} x_e \leq 1 \qquad \text{and} \qquad \sum_{e \in \delta(v_2)} x_e \leq 1,$$

but they both read $x_e \leq 1$. So they are identi-cal and one is certainly redundant. If this example looks "too degenerate" look at the "star" in the center of Figure 6. Because of the nonnegativity constraints, of the four conditions of type (2.8),

$$x_{e_1} \leq 1, \quad x_{e_2} \leq 1, \quad x_{e_3} \leq 1,$$

$$x_{e_1} + x_{e_2} + x_{e_3} \leq 1,$$

only the last is facet inducing. A similar situation occurs with the constraints (2.9): some of them may be redundant, see Figure 6 (right).

Is it maybe even true that some of the constraints are generally redundant, i.e., redundant for any graph, and we can hence discard them once and for all? Unfortunately this is not the case: A constraint that is redundant for some graph may be facet in-ducing for another. In particular, all constraints mentioned in Theorem 2.1 are facet inducing for the complete graph K_n, at least whenever $n \geq 4$. A refinement of this statement is the content of the following remark.

Remark 2.2 *(i) The nonnegativity constraints (2.10) are all facet inducing for $\mathbf{M}_G$ for every graph G.*

(ii) The degree constraints (2.8) are all facet in-ducing for $\mathbf{M}_{K_n}$ when $n \geq 4$.

(iii) The odd set constraints (2.9) are all facet in-ducing for $\mathbf{M}_{K_n}$ when $n \geq 3$.

Proof. The first statement was shown before, so let us turn directly to (ii). Let $n \geq 4$, $K_n = (V, E)$, $v \in V$, and set

$$x_e = \begin{cases} \frac{1}{n-2}, & \text{if } e \in \delta(v), \\ 0, & \text{if } e \notin \delta(v). \end{cases}$$

Then

$$\sum_{e \in \delta(v)} x_e = \frac{n-1}{n-2} > 1,$$

hence the corresponding constraint (2.8) is violated. However, all other constraints are still satisfied. This is trivial for the constraints (2.10). For the other inequalities (2.8) it follows from the fact that $\delta(v) \cap \delta(w) = \{vw\}$ for each $w \in V \setminus \{v\}$ whence

$$\sum_{e \in \delta(w)} x_e = \frac{1}{n-2} < 1.$$

Since $\lfloor |U|/2 \rfloor \geq 1$ for all $U \subset V$ with $|U| \geq 3$, an inequality of type (2.9) could only be violated by x_e if the whole set $\delta(v)$ was involved, i.e., $\delta(v) \subset E(U)$. But this means that $U = V$, and then, because $n \geq 4$,

$$\sum_{e \in E(U)} x_e = \frac{n-1}{n-2} \leq \frac{n-1}{2} \leq \left\lfloor \frac{|U|}{2} \right\rfloor.$$

Finally, let's deal with (iii). Let $U \subset V$ with $k = |U|$ odd and $k \geq 3$. Obviously, $\left\lfloor \frac{|U|}{2} \right\rfloor = \frac{k-1}{2}$, and $E(U)$ is the (edge set of the) complete graph on k vertices, and is therefore of cardinality $\frac{k(k-1)}{2}$. Now set

$$x_e = \begin{cases} \frac{1}{k-1}, & \text{if } e \in E(U), \\ 0, & \text{if } e \notin E(U). \end{cases}$$

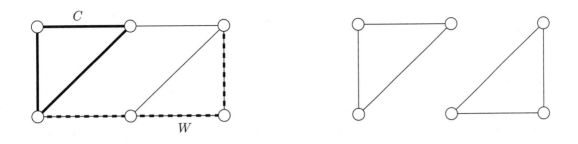

Figure 7: *Left:* A connected graph, with a 3-path W (the broken edges) and a 3-cycle C highlighted. *Right:* A graph that is not connected.

Then we have

$$\sum_{e \in E(U)} x_e = \frac{k(k-1)}{2} \frac{1}{k-1}$$

$$= \frac{k}{k-1} \left\lfloor \frac{|U|}{2} \right\rfloor > \left\lfloor \frac{|U|}{2} \right\rfloor, \quad (2.11)$$

whence the corresponding constraint (2.9) is violated. Clearly, the nonnegativity constraints are all satisfied, and so are the inequalities (2.8) since

$$\sum_{e \in \delta(v)} x_e \le \frac{1}{k-1} (|U| - 1) = 1,$$

$$\text{for all } v \in V. \quad (2.12)$$

Finally, let us check that all other inequalities of type (2.9) are satisfied. So, let $W \subset V$, $W \neq U$, with $l = |W| \ge 3$ odd, and let $m = |U \cap W|$. We have $m \le k, l$ and, because $W \neq U$, $m < k$ or $m < l$. Thus, $m(m-1) \le (k-1)(l-1)$, and we have

$$\sum_{e \in E(W)} x_e = \frac{m(m-1)}{2} \frac{1}{k-1}$$

$$= \frac{m(m-1)}{(l-1)(k-1)} \left\lfloor \frac{|W|}{2} \right\rfloor \le \left\lfloor \frac{|W|}{2} \right\rfloor. \quad (2.13)$$

Therefore all other constraints are satisfied, and the proof is completed. □

Now we turn to the proof of Edmonds' Theorem. (Readers in a great hurry may skip the rest of this section and continue with Section 3; but be warned: you don't know what you miss then.) As indicated before, we begin with a combinatorial "structural lemma." For a graph G let $\mathcal{M} = \mathcal{M}_G$ denote the set of all matchings in G, and for every weight function $c \colon E \mapsto \mathbf{R}$ let, in addition,

$$c^* = \max\{c(M) \colon M \in \mathcal{M}\},$$
$$\mathcal{M}^*(c) = \{M \in \mathcal{M} \colon c(M) = c^*\}.$$

Note that, in geometric terms, $\mathcal{M}^*(c)$ corresponds to (the vertices of) a face of $\mathbf{M}_G$, and each face of $\mathbf{M}_G$ corresponds to a set $\mathcal{M}^*(c)$ for some weight function c. A k-*path* or a path of cardinality k in a graph G is a subset W of E with the following property: there exists a sequence $(v_0, v_1, \ldots, v_k)$ of pairwise different nodes such that $W = \{v_{i-1}v_i \colon i = 1, \ldots, k\}$. Usually we do not distinguish between the path W and the graph on $\{v_0, v_1, \ldots, v_k\}$ whose edge set is W; so such a graph will be called a path too. The nodes v_0 and v_k are called the *endpoints* of the path, and the path is said to *connect* v_0 and v_k. The *distance* of two different nodes is the smallest k such that there exists a k-path connecting them, if such a path exists. A graph G is called *connected* if any two different nodes are connected by a path in G. Note that a graph G is connected if and only if $\delta(U)$ is nonempty for every proper subset U of V. A subset C of E of cardinality k is called a k-*cycle* if every $(k-1)$-element subset of C is a $(k-1)$-path. See Figure 7 for examples of these notions.

Lemma 2.3 *Let the weighted graph $G = (V, E, c)$ be connected, let $c(e) > 0$ for all $e \in E$, and suppose there is a matching $M^* \in \mathcal{M}^*(c)$ such that $|M^*| < \lfloor |V|/2 \rfloor$. Then there exists a node $u \in V$ such that $\delta(u) \cap M \neq \emptyset$ for all $M \in \mathcal{M}^*(c)$; see Figure 8.*

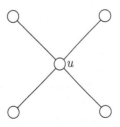

Figure 8: A weighted graph with all weights equal to 1 for which Lemma 2.3 is valid: the node u is contained in all maximum weight matchings.

Proof. Since $|M^*| < \lfloor |V|/2 \rfloor$ there are at least two different nodes v^* and w^* that are not contained in any edge of M^*. Since G is connected, v^* and w^* are the endpoints of some path in G. There is only a finite number of choices $(M; v, w)$ of matchings $M \in \mathcal{M}^*(c)$ with $|M| < \lfloor |V|/2 \rfloor$, and nodes v and w that are not contained in any edge of M, and we may assume that among all these choices, $(M^*; v^*, w^*)$ is such that the distance k of v^* and w^* is minimal. Let W be a k-path in G connecting v^* and w^*. Clearly, $k > 1$, since otherwise $v^*w^* \in E$, so $\hat{M} = M^* \cup \{v^*w^*\} \in \mathcal{M}$, but $c(\hat{M}) = c(M^*) + c(v^*w^*) > c(M^*)$, contradicting the maximality of M^*. Hence there is another node incident to W, say u. By the minimum distance choice of v^* and w^*, there is an edge of M^* that contains u. If u is incident to all $M \in \mathcal{M}^*(c)$ we are done. It remains to show that this is always the case.

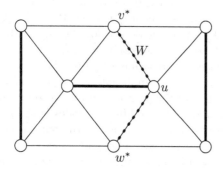

Figure 9: A graph G with a matching M^* (heavy lines), two vertices v^*, w^* not contained in M^*, and a vertex u lying on some shortest path W connecting v^* and w^* (dotted).

Suppose there is a matching $M' \in \mathcal{M}^*(c)$ such that $\delta(u) \cap M' = \emptyset$. Let us consider the graph $H = (V, M^* \cup M')$; see Figure 10 (left). No node of H lies in more than two edges, hence H can be partitioned into some s node-disjoint graphs $H_1, \ldots, H_s$, the edge set of each being either empty, a path, or a cycle. Suppose that u is a node of $H_1 = (V_1, E_1)$; then what kind of graph is H_1? Since u is incident to M^* but not to M', H_1 must be a path, and u is one of its endpoints. Since both M^* and M' are matchings, the edges of H_1 alternate between M^* and M', hence we can construct two new matchings by replacing the part of M^* in E_1 by $M' \cap E_1$, and vice versa, see Figure 10 (right). More formally, define

$$\hat{M}^* = (M^* \setminus E_1) \cup (M' \cap E_1)$$

and

$$\hat{M}' = (M' \setminus E_1) \cup (M^* \cap E_1).$$

Clearly, $\hat{M}^*$ and $\hat{M}'$ are matchings and

$$c(\hat{M}^*) + c(\hat{M}') = c(M^*) + c(M') = 2c^*;$$

whence $c(\hat{M}^*) = c^*$ and $c(\hat{M}') = c^*$, i.e.,

$$\hat{M}^*, \hat{M}' \in \mathcal{M}^*(c).$$

The vertex u is not incident to an edge of $\hat{M}^*$. If $|\hat{M}^*| = \lfloor |V|/2 \rfloor$, then v^* and w^* are incident to an edge of $\hat{M}^*$. If, on the other hand, $|\hat{M}^*| < \lfloor |V|/2 \rfloor$ then we can draw the same conclusion from the minimum distance choice of v^* and w^* since the distances of v^* and u and also of u and w^* are both smaller than the distance of v^* and w^*. So in any case, both nodes, v^* and w^*, are incident to an edge of $\hat{M}^*$. But is this really possible? By the definition of $\hat{M}^*$ this can only happen if $v^*, w^* \in V_1$. But since v^* and w^* are not contained in any edge of M^* none of them can be incident to two edges of H, let alone H_1. Since, except for its endpoints, each node in a path is contained in two different edges both nodes v^* and w^* must therefore be endpoints of H_1. But we know already that u is an endpoint of H_1. Since paths cannot have three endpoints, this is a contradiction. Hence our assumption that there is a matching $M' \in \mathcal{M}^*(c)$ such that $\delta(u) \cap M' = \emptyset$ must be wrong. So $\delta(u) \cap M \neq \emptyset$ for all $M \in \mathcal{M}^*(c)$ as claimed. $\square$

Let us remark that the main idea of the proof of Lemma 2.3, the "flipping of edges," along paths is a special case of the "augmenting path technique" which serves so prominently in devising efficient algorithms for a variety of combinatorial optimization problems; see [1, 24, 29].

With the aid of the combinatorial Lemma 2.3 we are now able to verify the description of the matching polytope given in Edmonds' Theorem.

Proof of Theorem 2.1 Let $\mathbf{M}_G$ be the matching polytope of the graph $G = (V, E)$. As we have seen already, $\mathbf{M}_G$ is of dimension $|E|$. Now let $d = |E|$ and $E = \{e_1, \ldots, e_d\}$, and suppose that

$$\sum_{i=1}^{d} a_i x_i \leq a_0 \qquad (2.14)$$

is a facet inducing inequality for $\mathbf{M}_G$; let Z denote this facet, and let $\mathcal{M}^*$ be the set of all matchings whose incidence vectors lie in Z. Clearly, $Z = \text{conv}\{x^M : M \in \mathcal{M}^*\} = \mathbf{M}_G \cap H$, where H is the hyperplane $\{x \in \mathbf{R}^d : \sum_{i=1}^{d} a_i x_i = a_0\}$. Further, note that, conversely, the system of linear equalities

$$\sum_{i=1}^{d} a_i x_i^M = a_0, \qquad M \in \mathcal{M}^*, \qquad (2.15)$$

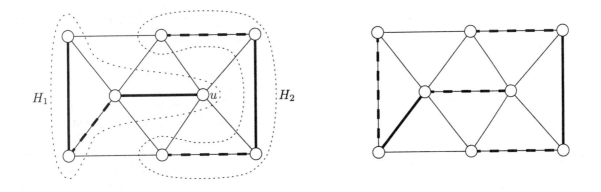

Figure 10: *Left:* A graph G with the subgraph $H = (V, M^* \cup M')$ highlighted (M^* is the set of thick solid edges, M' the set of thick dashed edges). *Right:* The corresponding flipping $\hat{M}^*$, $\hat{M}'$ in G ($\hat{M}^*$, heavy lines; $\hat{M}'$, dashed).

can be regarded as a (homogeneous) system of linear equalities in the variables $a_0, a_1, \ldots, a_d$. The rows of the coefficient matrix are the vectors $\binom{-1}{x^M}^T$, $M \in \mathcal{M}^*$. Since Z is a facet, d of the vectors x^M, $M \in \mathcal{M}^*$, are affinely independent, whence the system (2.15) is of rank d. So its solution space is 1-dimensional: the coefficient vector $(a_0, a_1, \ldots, a_d)$ of (2.14) is determined by (2.15) up to a scalar. We are going to show that (2.14) coincides with one of the inequalities (2.8)–(2.10) (up to multiplication by a scalar).

Let us first dispose of some simple cases. Is it possible that a_0 is negative? No, clearly not, since $\emptyset \in \mathcal{M}$ and $x^\emptyset = 0 \in \mathbf{R}^d$, 0 then would violate the inequality. So, $a_0 \geq 0$. Next, what happens if, for some j, the coefficient a_j is negative, after normalization, say, -1? Suppose there is a matching $M \in \mathcal{M}^*$ that contains the edge e_j, and let $\hat{M} = M \setminus \{e_j\}$. Since $\sum_{i=1}^d a_i x_i^M = a_0$, we have $\sum_{i=1}^d a_i x_i^{\hat{M}} = a_0 + 1 > a_0$, a contradiction. This means that no matching of $\mathcal{M}^*$ contains the edge e_j. Hence Z is contained in the coordinate hyperplane given by the equation $x_j = 0$, and since $\dim(Z) = d - 1$, its affine hull H actually is this coordinate hyperplane. Therefore the inequality (2.14) is just the nonnegativity constraint $-x_j \leq 0$, i.e., of the form (2.10). Hence, we may assume in the following that $a_i \geq 0$ for all $i = 1, \ldots, d$. Since these coefficients are not all 0, this implies further that $a_0 > 0$.

Now, suppose that there is $u \in V$ such that $|\delta(u) \cap M| = 1$ for all $M \in \mathcal{M}^*$. Then a solution to (2.15) is $a_0 = 1$, $a_i = 1$ for $e_i \in \delta(u)$ and $a_i = 0$ otherwise, $i = 1, \ldots, d$. Hence (2.14) is of the form (2.8). So, suppose that no such u exists, and consider the graph $G' = (V', E')$ that is obtained from

G by first deleting all edges e_i such that $a_i = 0$ and then deleting all isolated nodes, i.e., all nodes that are not incident to any edge. Note that our inequality cannot discriminate anyway whether, in a matching M, the deleted edges e_i are present; H is parallel to all corresponding coordinate axes, and the deletion of edges is geometrically just an orthogonal projection on the remaining coordinate space X. So X is an embedding of $\mathbf{R}^{|E'|}$ into $\mathbf{R}^{|E|}$; its dimension is $|E'|$. The formal reason for considering G' is that we want to apply Lemma 2.3. Turning from G to G' is a matter of "forcing" the positivity condition for the coefficients of the inequality. What about connectivity? Suppose G' is not connected. Then there is a nonempty proper subset V_1' of V' such that $\delta(V_1') = \emptyset$, i.e., no edges cross over from V_1' to $V_2' = V' \setminus V_1'$. Let $G_1' = (V_1', E(V_1'))$ and $G_2' = (V_2', E(V_2'))$, and note that both edge sets are nonempty. Let M' be a matching of G' such that $x^{M'}$ satisfies (2.14) with equality. Then

$$a_0 = \sum_{e_i \in E'} a_i x_i^{M'}$$
$$= \sum_{e_i \in E(V_1')} a_i x_i^{M'} + \sum_{e_i \in E(V_2')} a_i x_i^{M'} = b_1 + b_2,$$

(2.16)

where

$$b_1 = \max_{M''} \sum_{e_i \in E(V_1')} a_i x_i^{M''}$$

and

$$b_2 = \max_{M''} \sum_{e_i \in E(V_2')} a_i x_i^{M''},$$

and the maximum is taken over all matchings M'' of G', respectively. This is a consequence of the fact that the matching property in G_1' is independent from the matching property in G_2'. Hence $x^{M'}$ lies in the intersection of the two hyperplanes H_1' and H_2' of X, where

$$H_j' = \{x \in X: \sum_{e_i \in E(V_j')} a_i x_i = b_j\}, \qquad j = 1, 2.$$

But then the dimension of the affine hull of the incidence vectors of all such matchings M' is at most $\dim(X) - 2 = |E'| - 2$. If then Y denotes X's orthogonal complement, we have $Z \subset (H \cap X) \times Y$, whence $\dim(Z) \leq (|E'| - 2) + (|E| - |E'|) = |E| - 2$, and Z cannot be a facet.

So, G' is connected, and we can apply Lemma 2.3 to (G', E', a'), where a' is the restriction of a to E'. By our assumption, there is no vector $u \in V$ such that $|\delta(u) \cap M| = 1$ for all $M \in \mathcal{M}^*$. Since $M' \in \mathcal{M}^*(a')$ if and only if $M' \in \{M \cap E': M \in \mathcal{M}^*\}$ there is certainly no vector $u' \in V'$ such that $|\delta(u') \cap M'| = 1$ for all $M' \in \mathcal{M}^*(a')$. Hence, by Lemma 2.3, $|M'| = \lfloor |V'|/2 \rfloor$ for all $M' \in \mathcal{M}^*(a')$. If $|V'|$ were even then every node would be contained in every such matching M', but no node with this property exists. Thus $|V'|$ is odd, and, since G' contains at least one edge, $|V'| \geq 3$. Then

$$a_0 = \lfloor |V'|/2 \rfloor, \qquad a_i = \begin{cases} 1, & \text{if } e_i \in E', \\ 0, & \text{if } e_i \notin E \end{cases}$$

defines a solution of (2.15); so (2.14) is of the form (2.9). This completes the proof of Theorem 2.1. $\square$

Note that, as for bipartite matchings, we have now a complete (even though in general redundant) description of the matching polytope for general graphs. However, unlike for bipartite matchings (see (2.3)), the number of linear constraints is typically exponential in the number of nodes. It turns out, however, that this is not really important for devising a fast algorithm for MATCHING. Using the linear description of the matching polytope, Edmonds [6] was able to produce a primal-dual $O(|V|^4)$ algorithm (later reduced to $O(|V|^3)$ in [19]) for solving MATCHING. Since the focal point of this article is different, we will not give the algorithmic details here. This is a small pity since the proof of Theorem 2.1 is really the hard part. Readers familiar with the primal-dual framework may try to develop the algorithm on their own; others may consult [29] or [24].

3   Polyhedral Combinatorics and Branch-and-Cut

In contrast to the nicely tractable weighted matching problem, there is a large number of combinatorial optimization problems $(S, \mathcal{F}, c)$ for which only relatively few of the facets of the incidence polytope are known. Some of them can only be described in ways that are algorithmically intractable, and are hence useless for all practical purposes. However, even partial information can be utilized, and we will outline the general approach and the underlying rationale in this section.

The principal algorithmic approach building on polyhedral combinatorics is as follows: Start with an explicitly given subset of valid (preferably facet inducing) inequalities for $P(\mathcal{F})$. They define the feasible region P of a linear program. Since $P(\mathcal{F}) \subset [0, 1]^{|S|}$ we can always assume that $P \subset [0, 1]^{|S|}$, i.e., P is bounded. Solve the linear programming problem over P to find either that $P = P(\mathcal{F}) = \mathcal{F} = \emptyset$—a case in which you terminate immediately in disgust—or to obtain an optimal vertex x^* of P. If x^* is an integer solution, it is in fact the incidence vector of a feasible solution F of $\mathcal{F}$ of maximum weight—again you terminate, cheerfully this time. Otherwise, identify—if possible—a valid inequality for $P(\mathcal{F})$ that is violated by x^*. Add this inequality to the linear program, solve the new program and so forth. This process is continued until infeasibility of the problem has been shown, or until an optimum solution has been found, or until the quite disappointing ("dead end") situation is reached that no more violated inequalities for $P(\mathcal{F})$ can be found. We know that some such inequalities still exist—$P(\mathcal{F})$ is still a proper subset of the current polytope P—we are simply not able to *find* one. In this case we may resort to a *branch-and-bound* technique to split the problem into subproblems by prescribing or forbidding certain substructures. Note that, since $P \supset P(\mathcal{F})$, the number $c^T x^*$ is an upper bound for the weight of an optimal solution, a fact that can be utilized in the branch-and-bound framework. See [29] for an account of the general branch-and-bound approach, and see [27] and [30] for more about branch-and-cut. Figure 11 gives a more formalized description of the algorithm.

In general the number of facets of $P(\mathcal{F})$ is gigantic: The best current bounds on the maximum number $f(n)$ of facets of 0-1-polytopes in dimension n are $c_1^n \leq f(n) \leq c_2^{n \ln n}$ for some constants c_1, c_2 [31], [18]. Even the number of known (and tractable) facet defining inequalities is subject to the combinatorial explosion. Hence any practical algorithm

procedure *branch-and-cut*
    Select an initial set of (facet inducing) inequalities of $P(\mathcal{F})$
    that describe a polytope P.
    repeat
        begin
        (1) Solve the linear programming problem $\max\{c^T x : x \in P\}$
            to decide if $P = \emptyset$, or else,
            to obtain an optimal vertex x^* of P.
        (2) **if** $P = \emptyset$ **then stop** (the problem is infeasible).
            if x^* is the incidence vector of an element of $\mathcal{F}$,
            then stop (an optimal solution is found).
        (3) **if** a (facet inducing) inequality $a^T x \leq b$ for $P(\mathcal{F})$
            can be found that is violated by x^*,
            then add it to the current linear program.
        (4) **if** no such inequality can be found,
            exit repeat-loop
        end
    Use a branching step to split $(S, \mathcal{F}, c)$ into subproblems.
    Repeat the procedure for each subproblem.
end *branch-and-cut*

Figure 11: A branch-and-cut algorithm for solving $(S, \mathcal{F}, c)$.

can only utilize a small fraction of the (known) inequalities. Doesn't this mean that algorithms based on facet inducing inequalities are already in principle doomed to be impractical by the sheer number of such constraints? The remainder of this section will be used to show that there is hope: at least in theory, only very few inequalities are actually needed to construct an edge path from some vertex of $P(\mathcal{F})$ to an optimal vertex. (An *edge path* is just a path in the *edge graph* or *1-skeleton* of $P(\mathcal{F})$, the graph whose nodes are the vertices and whose edges are the edges of $P(\mathcal{F})$.) For two given vertices of P, there is always an edge path connecting the two, actually there are at least $\dim(P(\mathcal{F}))$ many disjoint (except, of course, for their endpoints) such paths [13]. As before, the cardinality of a shortest such path is the *distance* of the vertices. The maximum $\Delta(P)$ of the distances of any two vertices of P is called the (combinatorial) *diameter* of P.

Suppose v is an arbitrary vertex of a polytope P in $\mathbf{R}^d$, and v^* is a vertex that is optimal with respect to a given linear objective function $x \mapsto c^T x$. What then is (an upper bound for) the minimum number of incqualities that are needed to describe an edge path from v to v^*? The number of vertices

involved can be bounded above by $\Delta(P) + 1$, and—simply by linear algebra—each vertex is already determined by some d facet inducing inequalities of P.

This does not mean that, locally at a vertex, P is completely determined by d inequalities; in fact, many more constraints can be satisfied with equality (this is the phenomenon known as (primal) *degeneracy* in the realm of the simplex algorithm). Note that among the sets of d constraints that are satisfied with equality for v^* there is even a choice that has the property that no point of the cone that is determined by these d inequalities has a better objective function value than v^*. In geometric terms, optimality of v^* means that c is contained in the cone spanned by the outer normals of the halfspaces of those facet defining inequalities of P that are satisfied by v^* with equality (this is *dual feasibility* in linear programming terms). Hence the existence of the mentioned special choice of d of them is just a consequence of Caratheodory's Theorem. See [13] for Caratheodory's theorem and the geometry of polytopes and see [9] for the geometry of linear programming.

As a consequence of these considerations we see that, in theory, it always suffices to consider (an appropriate set of) at most $(\Delta(P(\mathcal{F})) + 1) \cdot |S|$ inequalities. But, how large is $\Delta(P(\mathcal{F}))$? Note that all polytopes $P(\mathcal{F})$ that are relevant in our context are 0-1-*polytopes*: all vertex coordinates are 0 or 1. Hence the following result of Naddef [23] is applicable.

Theorem 3.1 *For any $d \in \mathbf{N}$ and every 0-1-polytope P in $\mathbf{R}^d$ we have $\Delta(P) \leq \dim P$.*

Proof We use induction on the dimension k of P. If $k \leq 1$, there is nothing to show. So suppose $k \geq 2$, and that the assertion holds for all polytopes of dimension at most $k-1$. Let P be a 0-1-polytope of dimension k lying in (some) $\mathbf{R}^d$, let V denote its vertex set, and let $v_0, v_1 \in V$ with distance $\Delta(P)$. Clearly, v_0 and v_1 differ in at least one coordinate; so let us just assume that the last coordinate of v_0 is 0, while the last coordinate of v_1 is 1. (This assumption is no restriction of generality!) For $i = 0, 1$ let $H_i = \{(x_1, \ldots, x_d)^T \in \mathbf{R}^d : x_d = i\}$, and set $P_i = P \cap H_i$ and $V_i = V \cap H_i$. Geometrically, P_0 and P_1 are just the faces of P that lie in the "bottom" and "top" facet of the cube $[0,1]^d$, and V_0 and V_1 are their vertex sets, respectively. Note that $\dim(P_i) \leq k - 1$ so the induction hypothesis implies that $\Delta(P_i) \leq \dim(P_i)$ for $i = 0, 1$.

If we can now show that there is a vertex $w \in P_0$ which is connected to v_1 by an edge, then we are done: since $\Delta(P_0) \leq \dim(P_0) \leq k - 1$ there is a path from v_0 to w of length at most $k - 1$ so, after adding the edge wv_1, we have a path of length at most k that connects v_0 and v_1. By the choice of v_0 and v_1, their distance is $\Delta(P)$, hence $\Delta(P) \leq k$.

So, let us try to find such a vertex w. The most natural idea is certainly to try to "take a point in P_0 as close to v_1 as possible." And, in fact, this works! More specifically, for $v \in V_0$ define $D(v)$ to be the subset of $\{1, \ldots, d\}$ of coordinates in which v and v_1 differ. Then choose $w \in V_0$ so as to minimize $|D(v)|$ (see Figure 12).

Why is wv_1 then an edge of P? Suppose that this is not the case. Then there is no closed halfspace that contains P and whose bounding hyperplane H meets P exactly in wv_1. But this means (by a simple application of a standard separation theorem, see [13]) that $\mathrm{conv}(V \setminus \{w, v_1\})$ and wv_1 have a point p in common. Obviously $p \notin V$, so there are a subset $\hat{V}$ of $V \setminus \{w, v_1\}$ and positive coefficients

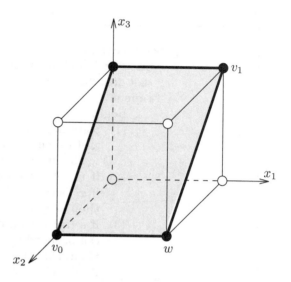

Figure 12: The shaded object depicts a 0-1-polytope with diameter 2; for v_0 and v_1 as indicated, a possible choice for w is marked; in this example, w is unique and $D(w) = \{2, 3\}$.

λ_w, λ_{v_1}, and λ_v for $v \in \hat{V}$ such that

$$\lambda_w + \lambda_{v_1} = 1,$$
$$\sum_{v \in \hat{V}} \lambda_v = 1, \qquad \text{and}$$
$$p = \lambda_w w + \lambda_{v_1} v_1 = \sum_{v \in \hat{V}} \lambda_v v.$$

Since the last coordinate of p is (strictly) less than 1, there must exist a vertex $\hat{v}$ in $\hat{V} \cap V_0$. Further, p has a coordinate 0 or 1 precisely when w and v_1 have the corresponding coordinate both 0 or 1, respectively, and this implies that all $v \in \hat{V}$ agree with w and v_1 on all coordinates $\{1, \ldots, n\} \setminus D(w)$. So, in particular $D(\hat{v}) \subset D(w)$. But w and $\hat{v}$ are two different 0-1-vectors, so $D(\hat{v})$ and $D(w)$ cannot be identical. Hence $|D(\hat{v})| < |D(w)|$, which is a contradiction to the choice of w, and we are done with the induction step. $\square$

The unit cube shows that the bound in Theorem 3.1 is best possible; in fact, equality holds exactly for the affine images of the unit cube [35, Theorem 3.11]. For particular classes of polytopes, this bound may, however, be significantly smaller; see, e.g., Theorem 4.1.

Returning to our previous estimate, Theorem 3.1 means that, in theory, of all the large number of facet inducing inequalities of an arbitrary incidence polytope $P(\mathcal{F})$, only at most $(|S|+1) \cdot |S|$ are needed to describe an edge path from an arbitrary starting vertex to an optimal vertex. Let us do a simple calculation to see what this bound becomes for the matching polytope $\mathbf{M}_G$ of a graph $G = (V, E)$ with $|V| \geq 2$.

$$(|E| + 1) \cdot |E| \leq \left[\binom{|V|}{2} + 1 \right] \cdot \binom{|V|}{2} \leq \frac{1}{4}|V|^4.$$

It may come as a surprise, but this "conservative bound" is precisely of the order of Edmond's original $O(|V|^4)$ algorithm for MATCHING.

The main practical problem is that, in general, nobody knows how to find small sets of inequalities for a given incidence polytope $P(\mathcal{F})$ that are "just right." Anyway, the observation that some $(|S| + 1) \cdot |S|$ inequalities will do is really at the heart of the hope for practical relevance of polyhedral combinatorics, a hope that has proven not to be elusive.

4   The Traveling Salesman Polytope

We turn now to one of the most famous problems in combinatorial optimization, the *traveling salesman problem*. Informally it is the following task:

> *Given a set of cities, in which order should a salesman visit them on a round trip, entering each exactly once, so as to minimize the total distance traveled?*

Again, problems of this kind come in many disguises. In VLSI, for instance, the "salesman" may be a machine arm and the "cities" may be points on a microchip; the machine performs certain actions at these (e.g., soldering, drilling), and the "traveled distance" may just be the total completion time. A comprehensive study of the traveling salesman problem, its history and a great variety of examples for its practical relevance can be found in the book [20].

Let us "strip off" the features of special interpretation and rephrase the traveling salesman problem in graph theoretic terms. But let us account for the fact, all too familiar from our urban traffic, that traveling from a site A to a site B may take much longer than traveling from B to A, since there may be one-way streets, different speed limits etc. So, we model our problem now by way of *directed* graphs.

A *directed graph* (or *digraph*) D is a pair (V, A) consisting of a set $V = \{v_1, \ldots, v_n\}$ of *nodes* and a set A of *ordered* pairs, called *arcs* of distinct elements of V. We write the arc issuing from a node

v and ending in a node w now as $\overrightarrow{vw}$ since $\overrightarrow{vw}$ has to be distinguished from $\overrightarrow{wv}$. A *weight* function is now a real-valued function on A. As in undirected graphs we can define paths and cycles, but these are directed now. A cycle that contains all nodes of D exactly once is called a *Hamiltonian cycle* or a *tour* in D. So, in particular, a tour T in D is a subset of A such that there exists a permutation σ of V with

$$A = \{\overrightarrow{v_{\sigma(i)} v_{\sigma(i+1)}} : i = 1, \ldots, n-1\} \cup \{\overrightarrow{v_{\sigma(n)} v_{\sigma(1)}}\}.$$

We then write $(\sigma(1), \ldots, \sigma(n))$ for T. If $D = (V, A, c)$ is a weighted digraph, the weight of a tour is called its *length*. We can now state the (asymmetric) traveling salesman problem as follows:

(ASYMMETRIC) TRAVELING SALESMAN PROBLEM.

Input: A weighted digraph $G = (V, A, c)$.

Task: Find a tour in G of minimum length or determine that G does not contain any tour.

Figure 13 shows a weighted digraph on four nodes (the reader may guess what the labels of the nodes stand for); in the copies in the middle and on the right tours have been highlighted, the one in the middle is the shortest.

We could now proceed as we did for MATCHING: define the incidence polytope, and try to determine its facial structure. It turns out, however, that the polytopes that we obtain are much more complicated now. For this reason we normalize the problem first in such a way that we do not have to consider different polytopes for each graph but can study a *universal incidence polytope* that depends only on the number of nodes. Observe first that we may assume that all weights are positive. (Otherwise, just add a sufficiently large number s to each weight. The length of each tour is then increased by the constant $s|V|$; so a tour that was shortest before persists in being shortest in the altered problem.) Then the total weight t_D of a given graph D, i.e., the sum of the weights of all arcs, is an upper bound on the length of any tour in D. How is the problem then changed if we add a new arc a to D with length t_D? Any tour in the augmented graph D' that contains the new arc a has length exceeding t_D. So any algorithm for solving

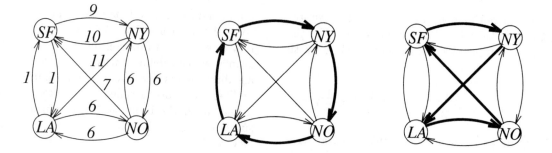

Figure 13: A weighted digraph with a "short" and a "long" tour.

the traveling salesman problem for D' can be used for solving the problem for D: if D' does not admit a tour, D does not either; if the length of a shortest tour in D' exceeds t_D then again D does not admit a tour; and if the length of a shortest tour in D' is at most t_D, then any shortest tour in D' is already a shortest tour in D. Hence we may add all missing edges and assume that the underlying graph is the *complete digraph* on the vertex set V, i.e., A is the set of all ordered pairs of distinct nodes of V. Since it is only the cardinality of V that is relevant, we may assume that $V = \{1, 2, \ldots, n\}$, and denote the complete digraph on V by D_n. This leads to the following problem.

"UNIVERSAL" (ASYMMETRIC) TRAVEL-ING SALESMAN PROBLEM.

Input: *A positive integer n and a positive weight function c on the arc set of the complete digraph D_n.*

Task: *Find a tour in D_n of minimum length.*

In terms of our triple $(S, \mathcal{F}, c)$, S is now the set of arcs of D_n (for the appropriate n) and $\mathcal{F}$ is the set of all tours in D_n. The corresponding incidence polytope $P(\mathcal{F})$ is then called the (asymmetric) *traveling salesman polytope*, and it is denoted by $\mathbf{T}_n$ in the following. Note that $\mathbf{T}_n$ is situated in $\mathbf{R}^{n(n-1)}$ since the complete digraph on n nodes has exactly $n(n-1)$ arcs, and each arc carries a component of the incidence vectors. For $n = 3$, there are only two possible tours; therefore, $\mathbf{T}_3$ is a rather trivial segment in $\mathbf{R}^6$. So, as an example let us consider the case $n = 4$, see Figure 14. Since we labeled the

nodes 1, 2, 3, 4, the ordered pairs

$$s_{12} = \overrightarrow{12}, \ s_{13} = \overrightarrow{13}, \ s_{14} = \overrightarrow{14},$$
$$s_{21} = \overrightarrow{21}, \ s_{23} = \overrightarrow{23}, \ s_{24} = \overrightarrow{24},$$
$$s_{31} = \overrightarrow{31}, \ s_{32} = \overrightarrow{32}, \ s_{34} = \overrightarrow{34},$$
$$s_{41} = \overrightarrow{41}, \ s_{42} = \overrightarrow{42}, \ s_{43} = \overrightarrow{43}$$

constitute the edge set E. The possible tours are

$$(1, 2, 3, 4), \quad (1, 2, 4, 3), \quad (1, 3, 2, 4),$$
$$(1, 3, 4, 2), \quad (1, 4, 3, 2), \quad (1, 4, 2, 3).$$

Thus the vertices of $\mathbf{T}_4$ are

$$
\begin{aligned}
t_1 &= (1, 0, 0, 0, 1, 0, 0, 0, 1, 1, 0, 0)^T, \\
t_2 &= (1, 0, 0, 0, 0, 1, 1, 0, 0, 0, 0, 1)^T, \\
t_3 &= (0, 1, 0, 0, 0, 1, 0, 1, 0, 1, 0, 0)^T, \\
t_4 &= (0, 1, 0, 1, 0, 0, 0, 1, 0, 1, 0)^T, \\
t_5 &= (0, 0, 1, 1, 0, 0, 0, 1, 0, 0, 0, 1)^T, \\
t_6 &= (0, 0, 1, 0, 1, 0, 1, 0, 0, 0, 1, 0)^T.
\end{aligned}
\tag{4.1}
$$

We mentioned before that we were aiming at a universal polytope for each number n of nodes. Why is $\mathbf{T}_n$ this polytope and how are the incidence polytopes of subgraphs of D_n related to $\mathbf{T}_n$? Suppose we delete in the graph D_4 of Figure 14 the arcs s_{12} and s_{42}, denote the new digraph by D, the

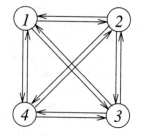

Figure 14: The complete digraph on four vertices, D_4.

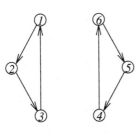

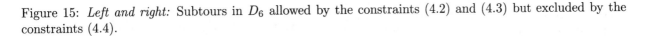

Figure 15: *Left and right:* Subtours in D_6 allowed by the constraints (4.2) and (4.3) but excluded by the constraints (4.4).

set of tours in D by $\mathcal{F}_D$ and the incidence polytope $P(\mathcal{F}_D)$ by P_D. How can we obtain the incidence vectors corresponding to $\mathcal{F}_D$ from $t_1, \ldots, t_6$? Since the edges s_{12} and s_{42} are missing, the corresponding entries (the first and the eleventh in our ordering of the arcs of D_4) have to be 0. So all we have to do is pick those vectors from $\{t_1, \ldots, t_6\}$ which do have these 0-entries—these are the vectors t_3 and t_5—and delete the two corresponding components (to obtain vectors in $\mathbf{R}^{10}$). This way we obtain all incidence vectors corresponding to $\mathcal{F}_D$, i.e., all vertices of P_D. But this means (since $\mathbf{T}_n \subset [0,1]^n$) that the incidence polytopes of all tours in arbitrary subgraphs of D_n are just faces of $\mathbf{T}_n$ obtained by intersecting $\mathbf{T}_n$ with appropriate coordinate subspaces (ignoring the fact that these faces are still embedded in $\mathbf{R}^{n(n-1)}$). In this sense, $\mathbf{T}_n$ is universal!

It is easy to check that while embedded in $\mathbf{R}^{12}$, the dimension of $\mathbf{T}_4$ is 5. (Since $\mathbf{T}_4$ has exactly the six vertices $t_1, \ldots, t_6$, it is actually a 5-dimensional simplex.) In general,

$$\dim(\mathbf{T}_n) = (n-1)^2 - n;$$

see [12], while the number of vertices of $\mathbf{T}_n$ is $(n-1)!$. Unlike for the matching polytope, there is no full linear description of the traveling salesman polytope known, and there are good reasons to believe that, for sufficiently large n, there never will be. Let us however give a few classes of known valid inequalities. First there are the obvious constraints

$$0 \leq x_a \leq 1 \qquad \text{for all } a \in A, \qquad (4.2)$$

and also the inequalities that require each node to be the start and endpoint of exactly one arc, respectively,

$$\sum_{\substack{j=1,\ldots,n \\ j \neq i}} x_{\overrightarrow{ij}} = 1 \quad \text{for all } i \in V \qquad (4.3)$$

and

$$\sum_{\substack{i=1,\ldots,n \\ i \neq j}} x_{\overrightarrow{ij}} = 1 \quad \text{for all } j \in V.$$

Constraints (4.2) and (4.3) do not exclude subtours, which are undesirable substructures as indicated in Figure 15.

The following *subtour elimination constraints* exclude these substructures: for each subset C of V with $2 \leq |C| \leq n-1$, we have the inequality

$$\sum_{a \in A(C)} x_a \leq |C| - 1, \qquad (4.4)$$

where (in analogy to our notation in the case of undirected graphs)

$$A(C) = \{\overrightarrow{ij} : i, j \in C, i \neq j\}.$$

Just as a matter of familiarizing the reader with this new class of constraints: which of them are violated by the incidence vector of the "solution" depicted in Figure 15 (right)? Obviously the two subtour elimination constraints $x_{\overrightarrow{12}} + x_{\overrightarrow{25}} + x_{\overrightarrow{51}} \leq 2$ and $x_{\overrightarrow{34}} + x_{\overrightarrow{46}} + x_{\overrightarrow{63}} \leq 2$!

Many more valid inequalities for $\mathbf{T}_n$ have been determined, among them those known as *comb inequalities*, *clique tree inequalities* or *hypohamiltonian inequalities*. As for the subtour elimination constraints, these names refer to certain substructures of D_n. We will not go into any details here; the reader is referred to [12] for definitions and properties of these objects. Let us just mention that, in terms of their algorithmic complexity, some of the substructures known to produce valid inequalities are no easier to deal with than the whole traveling salesman problem; hence the corresponding inequalities seem useless for all practical purposes.

The $2n$ constraints (4.3) are satisfied with equality but the codimension of $\mathbf{T}^n$ is $2n-1$. So at least one of them is redundant and, in fact, exactly one is: the affine hull of $\mathbf{T}_n$ is just the intersection of any $2n - 1$ of the hyperplanes induced by the constraints (4.3).

Which of the constraints (4.2) or (4.4) are facet inducing? Obviously, the constraints $x_a \leq 1$ are redundant since they are implied by the nonnegativity constraints, together with the 2-subtour elimination constraints $x_{\overrightarrow{ij}} + x_{\overrightarrow{ji}} \leq 1$. The nonnegativity constraints $x_{\overrightarrow{ij}} \geq 0$, on the other hand, do determine facets for $n \geq 5$, see [12]. We invite the reader to investigate the case $n = 4$ using the list of incidence vectors (4.1); this is actually a quickie. When $n \geq 5$ the subtour elimination constraints (4.4) define facets of $\mathbf{T}_n$ if $2 \leq |C| \leq n - 2$, although two different such inequalities $\sum_{a \in A(C)} x_a \leq |C| - 1$ and $\sum_{a \in A(C')} x_a \leq |C'| - 1$ are equivalent if and only if $C' = V \setminus C$, see [12]. (Check this equivalence for the two constraints induced by Figure 15; see the discussion following this figure.) This shows already that the number of facets of $\mathbf{T}_n$ is exponential in n.

Let us also give a warning here: the equivalence of a system of (pairwise different) constraints being facet inducing or being minimal (or as we called it before: irredundant) breaks down if the dimension of the polytope is smaller than the dimension of the space of incidence vectors. This is so since the affine hull of a facet Z of a polytope P is of dimension $\dim(P) - 1$, but it is a hyperplane of $\mathbf{R}^d$ only if $\dim(P) = d$. If $\dim(P) < d$, there are infinitely many hyperplanes that contain Z. So the proof that some inequality $a^T x \leq b$ is facet inducing by way of determining $\dim(P)$ many affinely independent vertices of P that lie in the hyperplane $\{x \in \mathbf{R}^d : a^T x = b\}$ is not an "irredundancy test" anymore. On the other hand, the second procedure that we mentioned in Section 2 (discarding the constraint in question and showing that the polytope changes) is still an irredundancy test.

As had to be expected, the number of facet inducing inequalities of $\mathbf{T}_n$ is (except for a few small cases: unknown, but definitely) tremendously high. However, as we have seen already, an optimal selection determining an edge path to an optimum is pretty small for general 0–1-polytopes. While the diameter of such a polytope may be as large as its dimension—in our case $\dim(\mathbf{T}_n) = (n-1)^2 - n$—the diameter of the traveling salesman polytope is actually *much* smaller. Clearly, $\Delta(\mathbf{T}_3) = \Delta(\mathbf{T}_4) = 1$, but the complete graphs on three and four nodes are not particularly rich in structure either. However, the diameter is not so much larger in the general case. We close this section with the statement of the corresponding result of Padberg and Rao [28] and with a few remarks.

Theorem 4.1 $\Delta(\mathbf{T}_3) = \Delta(\mathbf{T}_4) = \Delta(\mathbf{T}_5) = 1$ *and* $\Delta(\mathbf{T}_n) = 2$ *for all* $n \geq 6$.

Now, isn't that surprising! Theorem 4.1 implies that an optimal solution is always "just around the corner" whichever tour we begin with! In view of the lack of a full linear description of $\mathbf{T}_n$ it is quite remarkable that one can prove such a sharp structural result. Readers should however understand why they do not recall having seen a *justified* headline in their local paper celebrating that the traveling salesman problem can be solved efficiently: Theorem 4.1 is extremely nice, but there is one fundamental catch: nobody knows how to find the corner to peek around. Hard to believe, but it is actually worse: even deciding whether two given vertices of $\mathbf{T}_n$ are adjacent is (in a very precise sense) just as hard as the whole traveling salesman problem, see [26]. That means, it is unlikely that it will ever be possible to even check efficiently in general whether the incidence vectors of two given tours are joined by an edge of the traveling salesman polytope.

Despite all these difficulties many large problems (involving a few thousand cities) have been solved by branch-and-cut algorithms in practice [27]. The techniques based on the branch-and-cut algorithm are also useful to provide upper and lower bounds for problems of sizes so large that there is not much hope of ever solving them to optimality. These bounds allow to specify the "error margin," proving that a found feasible solution is, if not optimal, at least not too far off. Finding good solutions for very large problems involves other techniques as well. For a recent treatment of the *heuristic* side of the traveling salesman problem see [30].

5   The Order Polytope

Our last example of a "polyhedral approach" to combinatorial problems is different in nature. In the previous paragraphs, the combinatorial objects (like matchings or tours) were encoded as the vertices of the corresponding polytope. The following example will show that this approach can be extended, and other combinatorial objects may be associated with other structures of polytopes.

Searching and *sorting* routines belong to the fundamental tools used most frequently in computing. Suppose we have a list of distinct numbers in increasing order,

$$r_1 < r_2 < \cdots < r_{n-1} < r_n,$$

and a new number s (distinct from all the others) has to be added to the list. How can this be organized most efficiently when only comparisons are allowed, i.e., questions of the kind

$$\text{"Is } s < r_i\text{?"}$$

Obviously, *binary search* is a good strategy: We compare s first with an element closest to the "middle" of the list to exclude half of the possibilities, then we (approximately) halve the remaining possibilities again, etc. After just $O(\log n)$ questions the new element s can be inserted at its rightful place in the list. A similar kind of strategy can be used if we are just given an unsorted set $\{r_1, \ldots, r_n\}$ of distinct objects and we want to sort it. While comparison of each pair requires $\binom{n}{2}$ questions, a *divide-and-conquer* strategy to recursively partition the set into two sets of (approximately) equal size, sort these subsets and merge the sublists manages with $O(n \log n)$ pairwise comparisons. This is, of course, all well known and well studied and an impressive number of books in computer science are devoted just to searching and sorting; see, e.g., [17].

The examples in the previous paragraph are extreme in the following sense. In the first example, *complete* information about $\{r_1, \ldots, r_n\}$ in terms of the order relation $<$ is available, while in the second *no* such a priori information is given. In the latter case this means, on the positive side, that we are completely free in devising our comparison-based search algorithm. But what happens if none of these extreme cases occurs: we do not have full information but we are not completely ignorant either. This is actually quite typical in practice. But what would be a good strategy then? Applying the same principle, we would like to make a comparison that, no matter what the answer is, reduces the "remaining possibilities" to about one half. What are these "remaining possibilities"?

In the case of our first example (of full information), the possibilities are just the $n + 1$ possible positions of s, or equivalently the $n + 1$ different lists that can be compiled by placing s at each possible position of our ordered list $r_1 < r_2 < \cdots < r_{n-1} < r_n$. In the second example (where no a priori information is given) the possibilities are all $n!$ different completely ordered lists that exist for the n elements of $\{r_1, \ldots, r_n\}$. In general, the number of possibilities is the number of those totally ordered different lists that are compatible with the a priori information. It is this set that we would like to approximately halve by selecting a suitable pair of elements to undergo comparison. Before we formalize the problem, let us consider a simple example.

Let the set to sort consist of four distinct objects

r_1, r_2, r_3, r_4, and let us use the notation $r_i \prec r_j$ to indicate that according to a given criterion r_i is "smaller" than r_j. The objects could for instance be politicians, baseball teams, archeological findings, etc., and the relation "$\prec$" could mean "is less popular than," "loses to" or "is older than." The goal is again to bring all four objects into an increasing order. Now suppose we know already that $r_1 \prec r_3$ and $r_4 \prec r_2$. Then the correct ordering must be among the following six possibilities:

$$r_1 \prec r_3 \prec r_4 \prec r_2, \quad r_1 \prec r_4 \prec r_3 \prec r_2,$$
$$r_1 \prec r_4 \prec r_2 \prec r_3, \quad r_4 \prec r_1 \prec r_3 \prec r_2,$$
$$r_4 \prec r_1 \prec r_2 \prec r_3, \quad r_4 \prec r_2 \prec r_1 \prec r_3.$$

The following list shows all (reasonable) choices for pairs of objects that we can use for a next comparison:

$$\{r_1, r_2\}, \quad \{r_1, r_4\}, \quad \{r_2, r_3\}, \quad \{r_3, r_4\}.$$

Which one should we choose? According to our "halving the remaining possibilities" strategy, we should pick a pair $\{r_i, r_j\}$ with the property that the larger of the two numbers of remaining possibilities that we get, one for the outcome $r_i \prec r_j$, one for the outcome $r_j \prec r_i$, is minimal among all such queries. So, what happens if we use the pair $\{r_1, r_2\}$? Then we are left with the following compatible completely ordered lists:

if $r_1 \prec r_2$:	$r_1 \prec r_3 \prec r_4 \prec r_2,$
	$r_1 \prec r_4 \prec r_3 \prec r_2,$
	$r_1 \prec r_4 \prec r_2 \prec r_3,$
	$r_4 \prec r_1 \prec r_3 \prec r_2,$
	$r_4 \prec r_1 \prec r_2 \prec r_3;$
if $r_2 \prec r_1$:	
	$r_4 \prec r_2 \prec r_1 \prec r_3.$

This means that the pair $\{r_1, r_2\}$ divides the six possibilities in the ratio 5:1. But we can do better! When we use the pair $\{r_1, r_4\}$ we get the following possibilities:

if $r_1 \prec r_4$:	$r_1 \prec r_3 \prec r_4 \prec r_2,$
	$r_1 \prec r_4 \prec r_3 \prec r_2,$
	$r_1 \prec r_4 \prec r_2 \prec r_3;$
if $r_4 \prec r_1$:	$r_4 \prec r_1 \prec r_3 \prec r_2,$
	$r_4 \prec r_1 \prec r_2 \prec r_3,$
	$r_4 \prec r_2 \prec r_1 \prec r_3.$

This is a 3:3 ratio, and this is clearly optimal. The polytope that we are going to introduce will encode all possible extensions of the given partial information. But let us first introduce the problem more precisely.

Since the specific objects that are to be ordered are not relevant (our apologies to the politicians and baseball teams), we will simply assume that the set in question is $N = \{1, \ldots, n\}$ for some $n \in \mathbf{N}$. A *partial order* $\prec$ on N is a binary relation on N with the following properties:

(Non-reflexivity)   The relation $i \prec i$ never holds.

(Transitivity)   If $i \prec j$ and $j \prec k$ then $i \prec k$.

If $\prec$ is a partial order on N, then the pair $\mathcal{O} = (N, \prec)$ is called a *partially ordered set*, or *poset*. A *total* or *linear order* on N is a partial order which in addition satisfies

(Comparability)   If $i \neq j$, then either $i \prec j$ or $j \prec i$.

If $\mathcal{O} = (N, \prec)$ is a poset then any total order on N that is compatible with $\prec$ is called a *linear extension* of $\prec$ on N. We will also speak (in a slight abuse of language) of a *linear extension of $\mathcal{O}$*. Let $E(\mathcal{O})$ denote the set of linear extensions of $\mathcal{O}$.

With this notation, the problem of designing divide-and-conquer algorithms illustrated above is just the question of finding a pair for comparison that partitions the set $E(\mathcal{O})$ of linear extensions of a given partially ordered set $\mathcal{O}$ into two subsets E_1 and E_2 of similar size. What is the best (worst-case) ratio

$$\max_{\mathcal{O}} \max \left\{ \frac{|E_1|}{|E_2|}, \frac{|E_2|}{|E_1|} \right\}$$

that can always be achieved? The answer is: not known! It has been conjectured in [7] that this ratio is 2, and the best known upper bound is 8/3. This bound is still good enough for all practical purposes. So the main question that remains is: Is there an efficient procedure for finding a good pair? And very closely related: How hard is it to check that a given pair leads to a *balanced* partition of $E(\mathcal{O})$? Now, this is clearly the question of whether linear extensions can be counted efficiently. The problem of counting linear extensions has other applications (including some involving rankings of alternatives in social sciences) where the "balancing aspect" is not relevant but where it is desirable to know just how many linear extensions there are when the relation $i \prec j$ is added versus the number of linear extensions when the given preferences are augmented by $j \prec i$.

Let us now see where polytopes come into the picture. Let us take the example $N_3 = \{1, 2, 3\}$,

$\mathcal{O}_0 = (N_3, 2 \prec 1)$. We associate with every element i of N_3 a coordinate x_i in $\mathbf{R}^3$. Then we introduce the constraints

$$0 \leq x_1 \leq 1, \quad 0 \leq x_2 \leq 1, \quad 0 \leq x_3 \leq 1. \quad (5.1)$$

The inequalities (5.1) describe the cube $[0, 1]^3$ and represent the information contained in the set N_3. The order relation $2 \prec 1$ is encoded in the inequality

$$x_2 \leq x_1. \quad (5.2)$$

The set of feasible points of (5.1) and (5.2) is called the *order polytope* of $\mathcal{O}_0$; it is depicted in Figure 16.

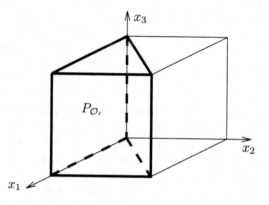

Figure 16: The order polytope for $\mathcal{O}_0 = (\{1, 2, 3\}, 2 \prec 1)$.

For an arbitrary poset $\mathcal{O} = (N, \prec)$, the *order polytope* $P_{\mathcal{O}}$ of $\mathcal{O}$ was introduced by Stanley [34] as follows:

$$P_{\mathcal{O}} = \{(x_1, \ldots, x_n)^T \in [0, 1]^n :$$
$$x_i \leq x_j \text{ for all } i, j \in N \text{ with } i \prec j\}. \quad (5.3)$$

Our goal is to identify the linear extensions of $\mathcal{O}$ as substructures of $P_{\mathcal{O}}$. Let us again consider our example $\mathcal{O}_0$. Here we have

$$E(\mathcal{O}_0) = \{3 \prec 2 \prec 1, \; 2 \prec 3 \prec 1, \; 2 \prec 1 \prec 3\}. \quad (5.4)$$

Let us take the first extension $3 \prec 2 \prec 1$ and add to the inequalities (5.1) and (5.2) the corresponding constraint

$$x_3 \leq x_2. \quad (5.5)$$

Then, as depicted in Figure 17, a simplex is cut off from $P_{\mathcal{O}_0}$. In fact, the plane given by the equation $x_3 = x_2$ splits $P_{\mathcal{O}_0}$ into two parts, the simplex corresponding to $3 \prec 2 \prec 1$ and a polytope P with five vertices that corresponds to the two remaining linear extensions $2 \prec 3 \prec 1$ and $2 \prec 1 \prec 3$. This becomes clearer if we apply the additional cut $x_1 = x_3$ that splits P according to the possible order relations $1 \prec 3$ and $3 \prec 1$. The full picture is depicted in Figure 17.

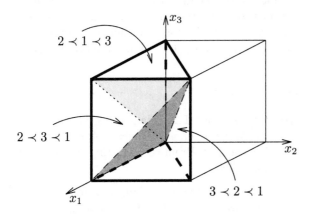

Figure 17: A dissection of the order polytope $P_{\mathcal{O}_0}$ into simplices.

Note that the three simplices partition the order polytope and are all of the same volume. Let us now show that this is true in general. For the formal description of the simplices it is convenient to identify a linear extension with a permutation π of N or, equivalently with a vector $(\pi(1), \pi(2), \ldots, \pi(n))$ which has the property that

$$i, j \in N \wedge i \prec j \quad \Longrightarrow \quad \pi^{-1}(i) < \pi^{-1}(j).$$

For instance, $(2, 3, 1)$ means $2 \prec 3 \prec 1$. Then consider for a given permutation π of N the polytope

$$T_\pi = \{(x_1, \ldots, x_n)^T \in [0, 1]^n :$$
$$0 \leq x_{\pi(1)} \leq x_{\pi(2)} \leq \cdots \leq x_{\pi(n)} \leq 1\}. \quad (5.6)$$

Observe that T_π is a simplex, and that all the constraints that define $P_{\mathcal{O}}$ are also constraints of T_π; hence $T_\pi \subset P_{\mathcal{O}}$. Further,

if π_1 and π_2 are different linear extensions of $\mathcal{O}$,
$$(5.7)$$

then $\operatorname{int}(T_{\pi_1}) \cap \operatorname{int}(T_{\pi_2}) = \emptyset, \quad (5.8)$

and also

$$\bigcup_{\pi \in E(\mathcal{O})} T_\pi = P_{\mathcal{O}}. \quad (5.9)$$

Property (5.7) follows simply from the fact that two different linear extensions differ at least in the order

of one pair $\{i, j\}$ of elements; hence the systems of inequalities defining the two simplices look like

$$\cdots \leq x_i \leq \cdots \leq x_j \leq \cdots \quad \text{and}$$
$$\cdots \leq x_j \leq \cdots \leq x_i \leq \cdots, \quad (5.10)$$

and the simplices can intersect only in points of the hyperplane given by the equation $x_i = x_j$. Property (5.9) is even simpler to verify, and we leave it to the reader to convince her/his spouse. Hence the simplices T_π, where $\pi \in E(\mathcal{O})$, form a dissection of the order polytope $P_{\mathcal{O}}$; see again Figure 17.

Finally note that each simplex T_π can be transformed into T_{id} by a permutation matrix, hence all simplices T_π are of equal volume, and a trivial computation yields

$$V(T_\pi) = \frac{1}{n!}. \quad (5.11)$$

Properties (5.7)–(5.11) show that each linear extension π of $\mathcal{O}$ contributes $1/(n!)$ to the volume of $P_{\mathcal{O}}$, and therefore

$$|E(\mathcal{O})| = n! \, V(P_{\mathcal{O}}).$$

Thus we have proved the following theorem.

Theorem 5.1 *The linear extensions $E(\mathcal{O})$ of a partially ordered set are in one-to-one correspondence with the simplices in a suitable dissection of the order polytope. All these simplices have the same volume.*

What is the use of this theorem for us? Certainly it reduces the problem of counting linear extensions to the problem of computing the volume of the corresponding order polytope. In a sense, one is reluctant to expect this reduction to be particularly useful for any practical purposes, since Brightwell and Winkler [2] showed that the problem of computing the number of linear extensions of a given partially ordered set $\mathcal{O} = (\{1, \ldots, n\}, \prec)$ is algorithmically intractable. And therefore the volume of polytopes that are given as the intersection of finitely many halfspaces must be hard to compute; see, e.g., the survey [10] on (mixed) volume computation for a precise form of this statement. Certainly, the fact that volume computation is at least as hard as counting linear extensions is a nice and quite remarkable result. But can't we get anything positive out of Theorem 5.1? Yes, we can! And the rest of this section will indicate how.

The above mentioned hardness result applies only for *deterministic* algorithms. If an algorithm is allowed, in the course of the computations, to "throw dice" and take certain actions randomly, then volume computations can be performed (in this randomized framework) quite efficiently. In fact, Dyer, Frieze and Kannan [4] constructed an algorithm which, for a given polytope P and given positive rationals β, ϵ, computes an approximation $\hat{V}$ of $\text{Vol}(P)$ such that

$$\text{Prob}\left(\left|\frac{\hat{V}}{\text{Vol}(P)} - 1\right| < \epsilon\right) \geq 1 - \beta.$$

The running time of the algorithm is polynomial in the three quantities $1/\epsilon$, $\log(1/\beta)$, and the binary size of the representation of P, i.e., the number of bits needed to encode all facet defining inequalities. Roughly speaking, the algorithm superimposes a fine chess board tiling to P and constructs a random walk on the cubical tiles in order to approximate a uniform distribution. Once we have a nearly uniform distribution, we can approximate the volume of P by the ratio of hits versus misses when points are thrown in at random. Certainly, this description is very coarse, and a lot of quite involved technical details have to be filled in, see [4], [3], [15], [21], [22], and [10] for more details, and further consequences of this result. What is certainly important for our context is that this randomized algorithm can, by virtue of Theorem 5.1, also be used as a randomized procedure for counting linear extensions of posets. And an efficient randomized algorithm can certainly just as well be utilized in the applications that we indicated at the beginning of this section as efficient deterministic approaches could (would they exist!). Since the simplices T_π of the dissection of $P_\mathcal{O}$ have all equal volume it is actually possible to use this dissection rather than a superimposed chess board tiling to construct the random walk. This approach of Karzanov and Khachiyan [16] leads to an improved randomized algorithm for counting the number of linear extensions. The approximately uniform distribution on the points of $P_\mathcal{O}$ yields an approximately uniform distribution on the linear extensions of $\mathcal{O}$. This allows one to select linear extensions randomly, which provides an even better approach to handling some of the practical problems mentioned before.

6   Epilogue

As we have seen, polyhedral methods are useful for solving hard problems in combinatorial optimization and also for studying certain problems from combinatorics. It turns out that there is an abundance of examples where the geometric aspect

(which may be well hidden to the naked eye) is really the key to the solution of a seemingly nongeometric problem. Examples of relevant classes of polytopes include the *Newton polytopes* in computer algebra, or the *zonotopes* which enjoy a wide spectrum of applications ranging from graph theory to production theory in economics. Certainly, polyhedral methods are not omnipotent, but if a problem has a chance of allowing a geometric interpretation one might as well try to find and then to exploit it.

References

[1] R.K. Ahuja, T.L. Magnanti, and J.B. Orlin, *Network Flows: Theory, Algorithms and Applications*, Prentice Hall, Englewood Cliffs, N.J., 1993.

[2] G. Brightwell and P. Winkler, *Counting linear extensions*, Order **8** (1991), no. 3, 225–242.

[3] M.E. Dyer and A.M. Frieze, *Computing the volume of convex bodies: A case where randomness provably helps*, Probabilistic Combinatorics and Its Applications (B. Bollobás, ed.), Proc. Symp. Appl. Math., vol. 44, Amer. Math. Soc., 1991, pp. 123–169.

[4] M.E. Dyer, A.M. Frieze, and R. Kannan, *A random polynomial-time algorithm for approximating the volume of convex bodies*, J. Assoc. Comput. Mach. **38** (1991), no. 1, 1–17.

[5] J. Edmonds, *Maximum matching and a polyhedron with 0,1-vertices*, J. Res. National Bureau of Standards B **69** (1965), 125–130.

[6] J. Edmonds, *Paths, trees and flowers*, Canad. J. Math. **17** (1965), 449–467.

[7] M. Fredman, *How good is the information theory bound in sorting?*, Th. Comput. Sci. **1** (1976), 335–361.

[8] M.R. Garey and D.S. Johnson, *Computers and Intractability: A Guide to the Theory of NP-Completeness*, Freeman, San Francisco, 1979.

[9] P. Gritzmann and V. Klee, *Mathematical programming and convex geometry*, Handbook of Convex Geometry (P.M. Gruber and J.M. Wills, eds.), vol. A, North-Holland, Amsterdam, 1993, pp. 627–674.

[10] P. Gritzmann and V. Klee, *On the complexity of some basic problems in computational convexity: II. Volume and mixed volumes*, Polytopes: Abstract, Convex and Computational

(T. Bisztriczky, P. McMullen, R. Schneider, and A. Ivić Weiss, eds.), NATO ASI Ser., Ser. C, Math. Phys. Sci., vol. 440, Kluwer, Dordrecht (Netherlands), 1994, pp. 373–466.

[11] M. Grötschel, L. Lovász, and A. Schrijver, *Geometric Algorithms and Combinatorial Optimization*, Algorithms and Combinatorics, vol. 2, Springer, Berlin, 1987.

[12] M. Grötschel and M. Padberg, *Polyhedral theory*, The Traveling Salesman Problem (E.L. Lawler, J.K. Lenstra, A.H.G. Rinnoy Kan, and D.B. Shmoys, eds.), Wiley-Interscience, Chichester, 1985, pp. 251–305.

[13] B. Grünbaum, *Convex Polytopes*, Wiley-Interscience, London, 1967.

[14] J. Kepler, *Prodomus dissertationum cosmographicum, continens Mysterium Cosmographicum*, Ph.D. thesis, Tübingen, 1596, Frankfurt 1621, reprinted in: Johannes Kepler Gesammelte Werke, Band 1, C.H. Beck'sche Verlagsbuchhandlung, München, 1938.

[15] L.G. Khachiyan, *Complexity of polytope volume computation*, New Trends in Discrete and Computational Geometry (J. Pach, ed.), Springer, Berlin, 1993, pp. 91–101.

[16] A. Karzanov and L.G. Khachiyan, *On the conductance of order Markov chains*, Order **8** (1991), no. 1, 7–15.

[17] D.E. Knuth, *The Art of Computer Programming III: Sorting and Searching*, Addison-Wesley, Reading, Mass., 1974.

[18] U.H. Kortenkamp, J. Richter-Gebert, A. Sarangarajan, and G.M. Ziegler, *Extremal properties of 0/1-polytopes*, Discrete Comput. Geom. **14** (1997).

[19] E.L. Lawler, *Combinatorial Optimization: Networks and Matroids*, Holt, Rinehart and Winston, New York, 1976.

[20] E.L. Lawler, J.K. Lenstra, A.H.G. Rinnoy Kan, and D.B. Shmoys (eds.), *The Traveling Salesman Problem*, Wiley-Interscience, Chichester, 1985.

[21] L. Lovász, *How to compute the volume?*, Jahresber. Deutsche Math. Verein. (1992), 138–151.

[22] L. Lovász, *Random walks on graphs: a survey*, Combinatorics: Paul Erdős is 80, Vol. II (D. Miklós, V.T. Sós, and T. Szőnyi, eds.), János Bolyai Mathematical Society, Budapest, 1995.

[23] D. Naddef, *The Hirsch conjecture is true for $(0,1)$-polytopes*, Math. Prog. **45** (1989), 109–110.

[24] G.L. Nemhauser and L.A. Wolsey, *Integer and Combinatorial Optimization*, Wiley-Interscience, Chichester, 1988, pp. 636–638.

[25] M. Padberg, *Linear Optimization and Extensions*, Springer, Berlin, 1995.

[26] C.H. Papadimitriou, *The adjacency relation on the traveling salesman polytope is NP-complete*, Math. Prog. **14** (1978), 312–324.

[27] M. Padberg and M. Grötschel, *Polyhedral computations*, The Traveling Salesman Problem (E.L. Lawler, J.K. Lenstra, A.H.G. Rinnoy Kan, and D.B. Shmoys, eds.), Wiley-Interscience, Chichester, 1985, pp. 307–360.

[28] M. Padberg and M.R. Rao, *The travelling salesman problem and a class of polyhedra of diameter two*, Math. Prog. **7** (1974), 32–45.

[29] C.H. Papadimitriou and K. Steiglitz, *Combinatorial Optimization: Algorithms and Complexity*, Prentice-Hall, Englewood Cliffs, NJ, 1982.

[30] G. Reinelt, *The Traveling Salesman: Computational Solutions for TSP Applications*, Springer, Berlin, 1994.

[31] G. Rote, *An upper bound on the number of facets of a 0-1-polytope*, Manuscript, 1996.

[32] A. Schrijver, *Theory of Linear and Integer Programming*, Wiley & Sons, Chichester, 1986.

[33] R. Schneider, *Convex Bodies: The Brunn-Minkowski Theory*, Encyclopedia of Mathematics and its Applications, vol. 44, Cambridge Univ. Press, 1993.

[34] R. Stanley, *Two poset polytopes*, Discrete Comput. Geom. **1** (1986), 9–23.

[35] G.M. Ziegler, *Lectures on Polytopes*, Graduate Texts in Mathematics, vol. 152, Springer, Berlin, 1995.

5331A